RATIONAL TUNNELLING, SUMMERSCHOOL, INNSBRUCK, 2003

ADVANCES IN GEOTECHNICAL ENGINEERING AND TUNNELLING

8

General editor:

Dimitrios Kolymbas

University of Innsbruck, Institute of Geotechnic and Tunnel Engineering

In the same series (A.A.BALKEMA):

1. D. Kolymbas, 2000, *Introduction to h poplasticit* , 104 pages, ISBN 90 5809 306 9
2. W. Fellin, 2000, *Rütteldruckverdichtung als plastod namisches Problem,* (*Deep vibration compaction as a plastod namic problem*), 344 pages, ISBN 90 5809 315 8
3. D. Kolymbas & W. Fellin, 2000, *Compaction of soils, granulates and powders - International workshop on compaction of soils, granulates, powders*, Innsbruck, 28-29 February 2000, 344 pages, ISBN 90 5809 318 2

In the same series (LOGOS):

4. Christoph Bliem, 2001, *3D Finite Element Berechnungen im Tunnelbau, (3D finite element calculations in tunnelling)*, 220 pages, ISBN 3-89722-750-9
5. D. Kolymbas, ed. (2001), *Tunnelling Mechanics, Eurosummerschool, Innsbruck*, 2001, 403 pages, ISBN 3-89722-873-4
6. M. Fiedler (2001), *Nichtlineare Berechnung von Plattenfundamenten (Nonlinear Anal sis of Mat Foundations),* 163 pages, ISBN 3-8325-0031-6
7. W. Fellin (2003), *Geotechnik - Lernen mit Beispielen*, 230 pages, ISBN 3-8325-0147-9

RATIONAL TUNNELLING SUMMERSCHOOL, INNSBRUCK, 2003

Edited by

Dimitrios Kolymbas

University of Innsbruck, Institute of Geotechnic and Tunnel Engineering

The first three volumes have been published by Balkema
and can be ordered from:

A.A. Balkema Publishers
P.O.Box 1675
NL-3000 BR Rotterdam
e-mail: orders@swets.nl
website: www.balkema.nl

Bibliographic information published by the Deutsche Nationalbibliothek

The Deutsche Nationalbibliothek lists this publication in the Deutsche Nationalbibliografie; detailed bibliographic data are available in the Internet at http://dnb.d-nb.de .

ISBN 978-3-8325-0350-5
ISSN 1566-6182

Logos Verlag Berlin
Comeniushof, Gubener Str. 47,
10243 Berlin
Tel.: +49 030 42 85 10 90
Fax: +49 030 42 85 10 92
INTERNET: http://www.logos-verlag.de

PREFACE

Tunnelling started as an empirical art and is now on the way to become a scientifically founded branch of engineering. 'Scientifically founded' means that the methods of tunnelling can not only be annunciated by some few authorities but also rationally explained, checked and—what counts in universities and continuing education institutions—learned.

In the Summerschool 'Rational Tunnelling', an international group of select specialists addressed an audience from 14 nations with lectures intending to illuminate various aspects of tunnelling. The present volume contains the notes of these lectures.

As the organizer of the event and editor of this volume, I wish to thank the authors, Mrs. Christine Neuwirt for the excellent organization and financial planning of the Summerschool, Mr. Josef Wopfner who prepared the final version of the present volume, Mrs. Myriam Berthold and Mr. Markus Mähr for valuable contributions to the organization.

Innsbruck, August 2003

Dimitrios Kolymbas

ACKNOWLEDGEMENT

The financial support of the Summerschool by the following organisation is gratefully acknowledged:

Federal Ministry for Education, Science and Culture
Minoritenplatz 5
A 1014 Vienna
Austria

TABLE OF CONTENTS

Constitutive models for numerical simulations 27

Ivo Herle

SCL Tunnel design in soft ground- insights from monitoring and numerical modelling 61

Chris Clayton, Alun Thomas, Pierre van der Berg

Part II Water problems 133

Problems of TBMs in water bearing ground 135

Lars Babendererde

Estimating groundwater inflow into hard rock tunnels – – the problem of permeability 155

Jack Raymer

Part III Measurements 185

Geotechnical instrumentation of tunnels 187

Helmut Bock

Part 1: Performance monitoring for tunnel design verification 187

Part IV Management aspects 235

What tends to go wrong in tunnelling 237

Sir Alan Muir Wood

Application of design-build contracts to tunnel construction 255

Robert A. Robinson

Cost and schedule management for major tunnel projects with reference to the Vereina tunnel and the Gotthard base tunnel 277

Felix Amberg, Bruno Röthlisberger

Part V Additional aspects

Hard rock TBM advance rates 355

Michael Alber

Tunnel refurbishment 387

Anton W. Ackermann , Christopher Hunt

Part I

Design of tunnels

Some principles for the design of lining

D. Kolymbas, M. Mähr, T. Pornpot

Institute of Geotechnical and Tunnel Engineering, University of Innsbruck, Techniker Str. 13, A-6020 Innsbruck, Phone: + 43 512 507 6670, Fax: + 43 512 507 2996
URL: http://geotechnik.uibk.ac.at,
e-mail: geotechnik@uibk.ac.at

1 Introduction

The lining of a tunnel never receives the stress which initially prevailed in the rock. Luckily, the initial (or primary) stress is reduced by deformation that occurs during the excavation but also after the installation of the lining (in this article 'lining' is understood as the shell of shotcrete, which is placed as soon as possible after excavation). Here we shall consider the important phenomenon that deformation of the ground (soil or rock) implies a reduction of the primary stress. This is a manifestation of *arching*. Since the deformation of the ground is connected with the deformation of the lining, it follows that the load acting upon the lining depends on its own deformation. This is always the case with soil-structure interaction and constitutes an inherent difficulty as the load is not an independent variable. Thus, the question is not 'which is the load acting upon the lining', but rather 'which is the dependence between load and deformation'.

The consideration of deformation in tunnelling is a merit of NATM. Of course, the principle 'load is reduced by deformation' is to be applied cautiously. Exaggerated deformation can become counterproductive (fig. 1 c) leading to a strong increase of load upon the bearing construction. To point on this was another merit of NATM. The collapse addressed by NATM is the softening (and the related loosening) of geomaterials. It should be emphasized, however, that it is not meant the gentle stress reduction subsequent to the peak, as this is obtained with soil mechanics laboratory tests on remoulded or reconstituted soil samples. Much more is meant the drastic strength reduction observed in poor rock due to loss of structural cohesion.

It is an (odd) tradition in civil engineering to distinguish between deformation and failure (collapse) of a structure. It is, however, impossible to find a genuine difference between these two notions. Virtually, failure is nothing but an overtly large deformation. At any rate, large deformations are to be avoided. How to achieve this in underpinning/tunnelling? There are two ways: Either early and rigid support (which is not economic) or by keeping the size of the excavated cavities small. The latter option is pursued in tunnelling. There are two ways to do this:

- partial excavation instead of full face excavation
- small advance steps.

Of course, too small excavation steps wouldn't be economic. So, the art of tunnelling consists in keeping the excavation steps as large as possible and exploiting the strength of the ground.

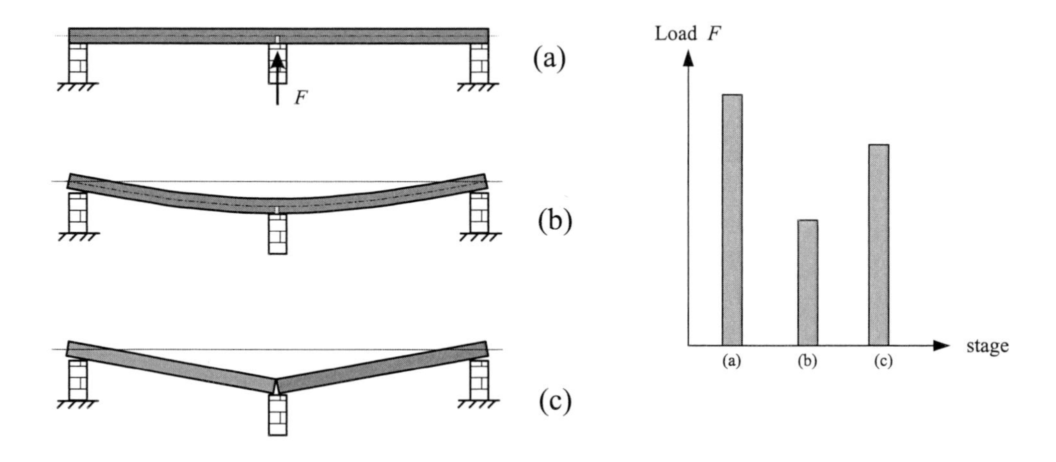

Figure 1: Stages of underpinning. The initial load F is reduced if the column yields (stage b). Finally the beam collapses (stage c), and the load icreases again.

2 Principles of arching

The equilibrium equation of continuum mechanics written in cylindrical coordinates reveals the mechanism of arching in terms of a differential equation. For axisymmetric problems, as they appear in tunnels with circular cross section, the use of cylindrical coordinates (fig. 2) is advantageous. In axisymmetric deformation, the displacement vector has no component in θ-direction: $u_\theta \equiv 0$. The non-vanishing components of the strain tensor

$$\varepsilon_{rr} = \frac{\partial u_r}{\partial r}, \ \varepsilon_{\theta\theta} = \frac{1}{r}\frac{\partial u_\theta}{\partial \theta} + \frac{u_r}{r}, \ \varepsilon_{zz} = \frac{\partial u_z}{\partial z}, \ \varepsilon_{r\theta} = \varepsilon_{\theta r} = \frac{1}{2}\left(\frac{1}{r}\frac{\partial u_r}{\partial \theta} - \frac{u_\theta}{r} + \frac{\partial u_\theta}{\partial r}\right)$$

$$\varepsilon_{rz} = \varepsilon_{zr} = \frac{1}{2}\left(\frac{\partial u_r}{\partial z} + \frac{\partial u_z}{\partial r}\right), \ \varepsilon_{\theta z} = \varepsilon_{z\theta} = \frac{1}{2}\left(\frac{1}{r}\frac{\partial u_z}{\partial \theta} + \frac{\partial u_\theta}{\partial z}\right)$$

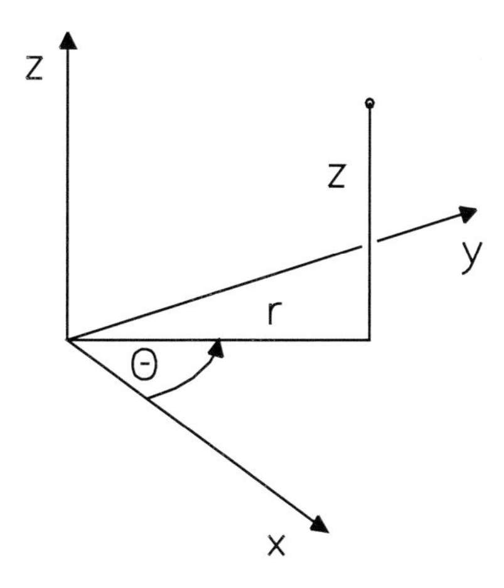

Figure 2: Cylindrical coordinates r, θ, z.

reduce, in this case, to

$$\varepsilon_r = \frac{\partial u_r}{\partial r}\,,\ \varepsilon_\theta = \frac{u_r}{r}\,,\ \varepsilon_z = \frac{\partial u_z}{\partial z}\quad ,$$

where u_r und u_z are the displacements in radial and axial directions, respectively. The stress components σ_r, σ_θ, σ_z are principal stresses (see fig. 3). The equation of equilibrium in r-direction reads:

$$\frac{\partial \sigma_r}{\partial r} + \frac{\sigma_r - \sigma_\theta}{r} + \varrho \boldsymbol{g} \cdot \boldsymbol{e}_r = 0 \tag{1}$$

and in z-direction:

$$\frac{\partial \sigma_z}{\partial z} + \varrho \boldsymbol{g} \cdot \boldsymbol{e}_z = 0 \quad . \tag{2}$$

Herein, ϱ is the density, ϱg is the unit weight, e_r and e_z are unit vectors in r- und z-directions. The second term in equation (1) describes arching. This can be seen as follows: If r points to the vertical direction z, then equ. (1) reads:

$$\frac{\mathrm{d}\sigma_z}{\mathrm{d}z} = \gamma - \frac{\sigma_x - \sigma_z}{r} \quad .$$

Herein, the term $(\sigma_x - \sigma_z)/r$ is responsible for that fact that σ_z does not increase linearly with depth (i.e. $\sigma_z = \gamma z$). In case of arching, i.e. for $(\sigma_x - \sigma_z)/r > 0$, σ_z increases underproportionally with z. Note that this term, and thus arching, exists only for $\sigma_r \neq \sigma_\theta$. This means that arching is due to the ability of a material to sustain deviatoric stress, i.e. shear stress. No arching is possible in fluids. This is why soil/rock 'forgives' shortages of support, whereas (ground)water is merciless.

The equilibrium equation in θ-direction, $\frac{1}{r} \cdot \frac{\partial \sigma_\theta}{\partial \theta} = 0$, is satisfied identically, as all derivatives in θ-direction vanish.

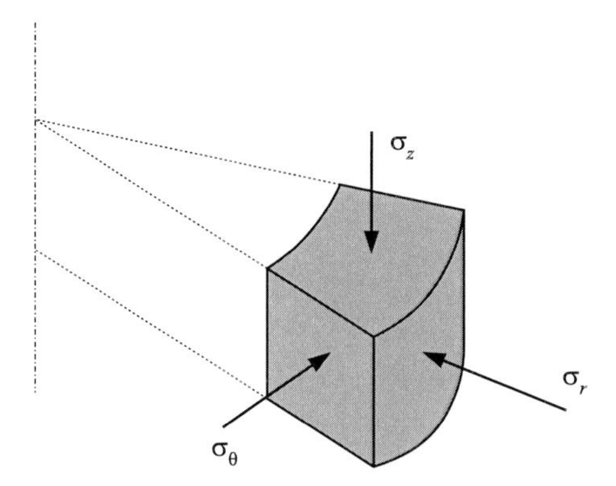

Figure 3: Components of the stress tensor in cylindrical coordinates

The arching term can be easily explained as follows: Consider the volume element shown in fig. 4. The resultant A of the radial stresses reads

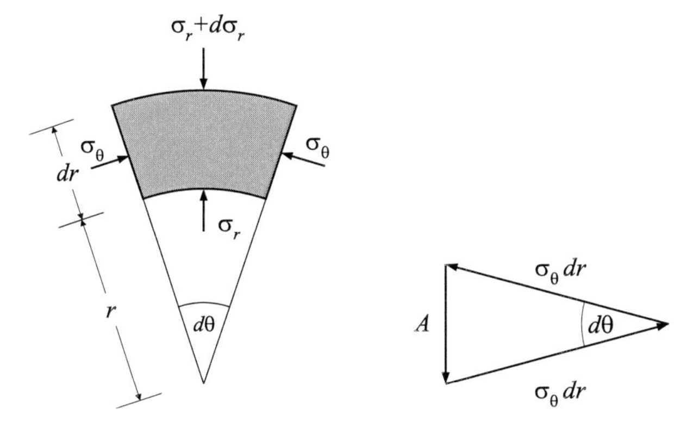

Figure 4: Equilibrium of the volume element in r-direction

$$A = (\sigma_r + d\sigma_r) \cdot (r + dr)d\theta - \sigma_r r d\theta \approx \sigma_r dr d\theta + d\sigma_r r d\theta$$

and should counterbalance the resultant of the tangential stresses σ_θ. The vectorial sum of forces shown in fig. 4 yields

$$\frac{A}{\sigma_\theta dr} = d\theta \quad .$$

It then follows (for $\mathbf{g} \cdot \mathbf{e}_r = 0$):

$$\frac{d\sigma_r}{dr} + \frac{\sigma_r - \sigma_\theta}{r} = 0 \quad .$$

For more general stress fields (i.e. for not axisymmetric case) that correspond to plane strain conditions, equ. (1) is still valid. In this case r is the curvature radius of the stress trajectory. We denote the corresponding stress σ_ϑ, σ_r is the principal stress perpendicular to σ_ϑ.

With reference to the arching term $\frac{\sigma_\theta - \sigma_r}{r}$ attention should be paid to r. At the tunnel crown, r is often set equal to the curvature radius of the crown. However, this is not always true. If we consider the gradual change of σ_z- and σ_x-stresses above the crown with decreasing support pressure p, we notice that for $K = \sigma_x/\sigma_z < 1$ for primary stress the horizontal stress trajectory has the opposite curvature than the tunnel crown (fig. 5)

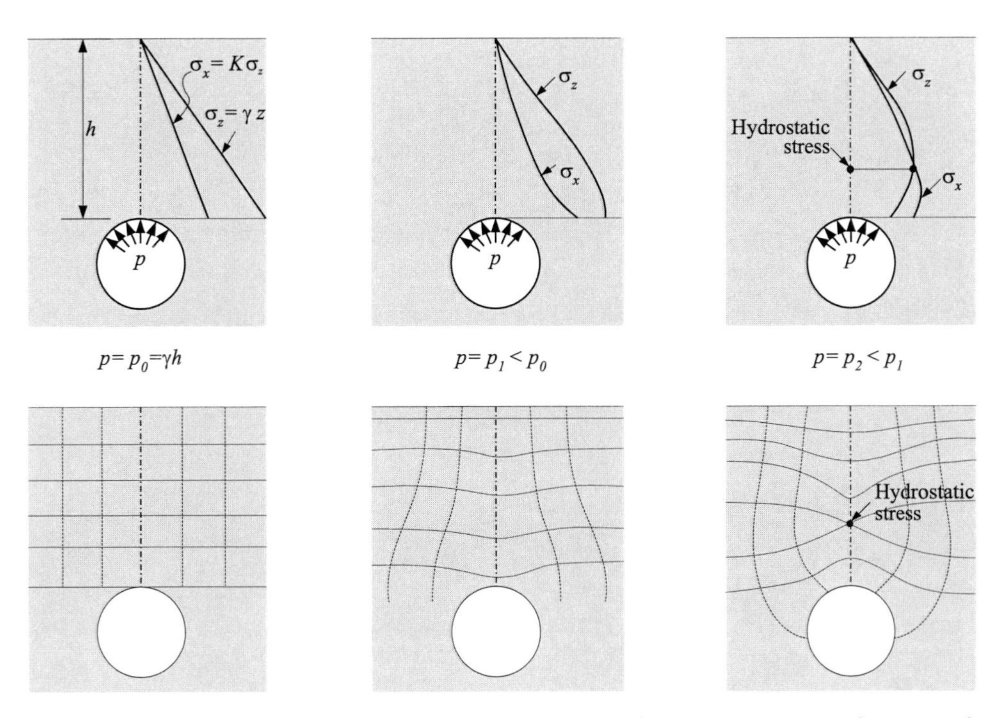

Figure 5: Stress distribution above the crown for various values of support pressure p, and corresponding stress trajectories. At the hydrostatic point (right), the curvature radius of the stress trajectory vanishes, $r = 0$.

2.1 Janssen's silo equation

In silos (i.e. vessels filled with granular material) the vertical stress does not increase linearly with depth. Silos are, therefore, an archetype for arching. The equation of JANSSEN (1895) is used for the design of silos. To derive it, let us look at a slim silo with a circular cross section (fig. 6).

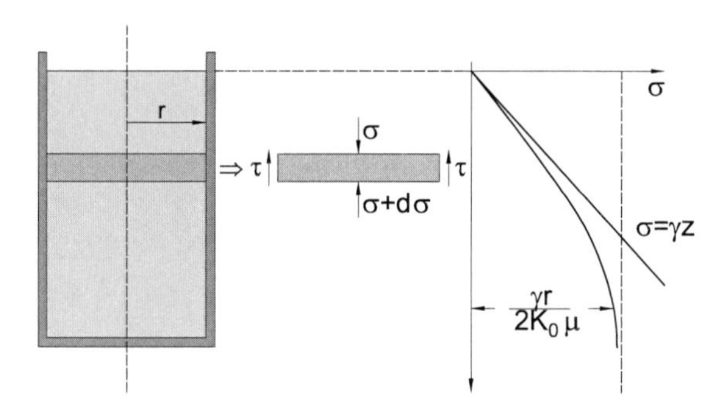

Figure 6: To the derivation of the equation of JANSSEN.

Upon a disk with the radius r and the thickness dz act the own weight $\pi r^2 \gamma dz$, the stress forces $\sigma \pi r^2$ and $(\sigma + d\sigma)\pi r^2$, as well as the shear force $\tau 2\pi r dz$ due to wall friction τ. The latter is proportional to the horizontal stress σ_H, $\tau = \mu \sigma_H$, and σ_H is assumed proportional to the vertical stress σ, i.e. $\sigma_H = K_0 \sigma$. K_0 is the earth-pressure-at-rest-coefficient[1], and μ is the wall friction coefficient. Equilibrium requires that the sum of these forces vanishes. In this way one obtains the differential equation

$$\frac{d\sigma}{dz} = \gamma - \frac{2K_0\mu}{r}\sigma \quad .$$

With the boundary condition $\sigma(z = 0) \stackrel{!}{=} 0$ it has the solution

$$\sigma(z) = \frac{\gamma r}{2K_0\mu}(1 - e^{-2K_0\mu z/r}) \quad . \tag{3}$$

Thus, the vertical stress cannot increase above the value $\gamma r/(2K_0\mu)$.

This derivation of equ. 3 also applies if the silo has no circular cross section. Then, r is the so-called hydraulic radius of the cross section:

$$\frac{A}{U} = \frac{r}{2} \quad ,$$

whereby A is the area and U the circumference of the cross section.

If the adhesion c_a acts between silo wall and granulate, then equ. 3 is to be modified as follows:

$$\sigma(z) = \frac{(\gamma - 2c_a/r)r}{2K_0\mu}(1 - e^{-2K_0\mu z/r}) \quad . \tag{4}$$

[1] according to JAKY is $K_0 \approx 1 - \sin\varphi$ for un-preloaded cohesionless materials

If the surface of the granulate is loaded with the load q per unit area, then the boundary condition at $z = 0$ reads $\sigma(z = 0) = q$. This leads to the equation

$$\sigma(z) = \frac{(\gamma - 2c_a/r)r}{2K_0\mu}\left(1 - e^{-2K_0\mu z/r}\right) + \frac{q\mu K_0}{r_0}e^{-2K_0\mu z/r} \quad . \tag{5}$$

The theory of JANSSEN points out that the granulate stored in silos 'hangs' partly at the silo walls by friction. This results in high vertical stresses in the silo walls, which often buckle. The mobilization of the shear stresses on the wall presupposes sufficiently large relative displacements between granulate and silo wall.

JANSSEN's equation is often used to assess arching above tunnels:

1. TERZAGHI regarded the range ABCD represented in fig. 7 as silo,[2] with the width b (for the plane deformation considered here the hydraulic radius is $r = b$), on the lower edge BC of which acts the pressure [1]. He thus obtained the following equation for the load p acting upon the roof of a tunnel with rectangular cross section:

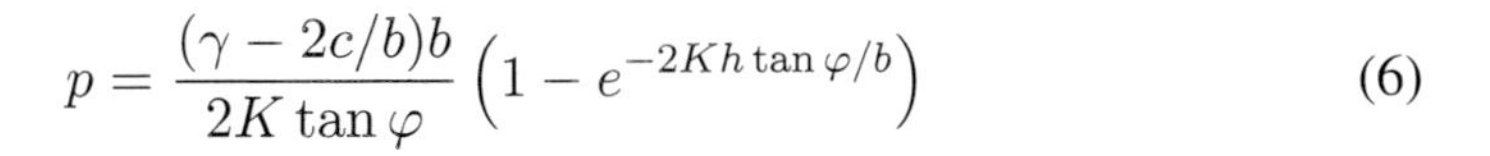

$$p = \frac{(\gamma - 2c/b)b}{2K\tan\varphi}\left(1 - e^{-2Kh\tan\varphi/b}\right) \tag{6}$$

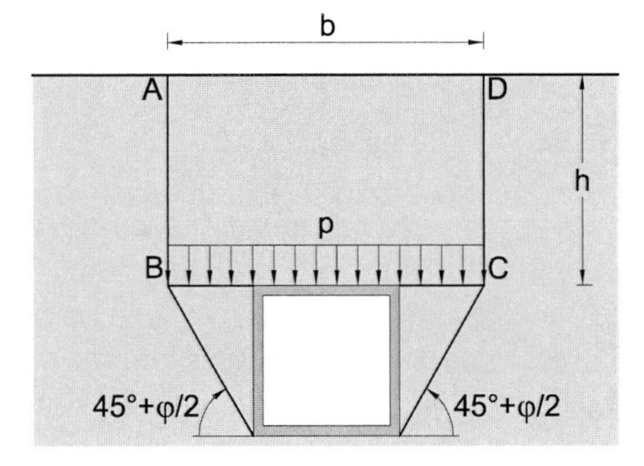

Figure 7: To the derivation of the equation of TERZAGHI.

2. To estimate the pressure (e.g. of a slurry shield) needed for the support of the tunnel face, JANSSEN's equation is used. The proof of face stability is often accomplished following a collapse mechanism originally proposed by HORN (fig. 8) [8]. To take into account the 3D-character of the collapse mechanism, the front ABCD of the sliding wedge is taken of equal area as the one of the

[2] Its delimitation by $45° + \varphi/2$ inclined lines is not clear.

tunnel cross section. On the sides BDI and ACJ is set cohesion and friction (in accordance with the geostatic stress distribution $\sigma_x = K\gamma z$). The vertical force V is computed according to the silo formula. The necessary support force S is determined by equilibrium consideration of the sliding wedge, whereby the inclination angle ϑ is varied until S becomes maximum. From the consideration of the relative displacements (fig. 8,c) it follows that at the sliding wedge acts also a horizontal force H, which is omitted by most authors [6]. One must consider that the applicability of the silo formula presupposes a full mobilization of the shear strength at the circumference of the prism slipping downward, a fact which would imply substantial settlements at the surface.

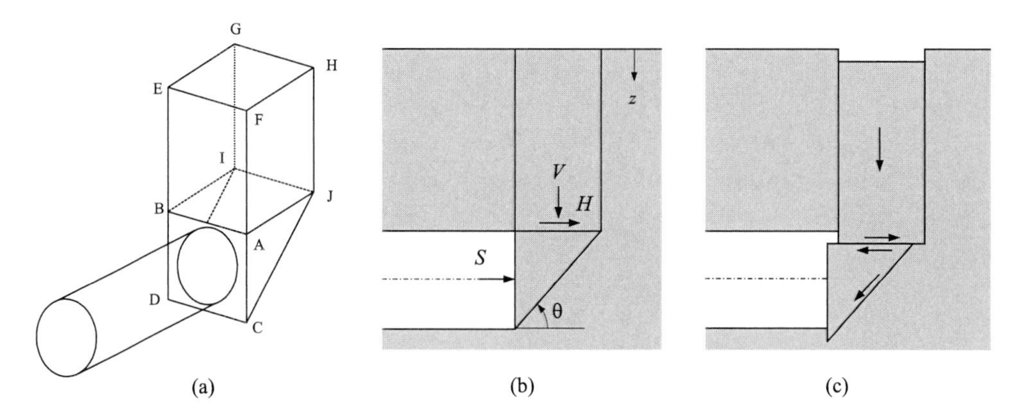

Figure 8: Mechanism of HORN to analyse face stability.

3. In top heading the upper part of the tunnel is excavated first and supported with shotcrete lining. This lining constitutes a sort of arch (or bridge) the footings of wich must be safely founded, i.e. the vertical force F exerted by the body ABCD (fig. 9) is to be introduced into the subsoil. To assess the safety against punching of the footings into the subsoil, ANAGNOSTOU [5][3] estimates F using JANSSEN's equation.

2.2 Trapdoor

The link between the equation of JANSSEN and tunnelling is established by the so-called trapdoor problem (fig. 10).

[3] see also [6], [8], [9] and the method of Broms and Bennemark as well as Tamez, cited in: M. Tanzini, Gallerie, Dario Flaccovio Editore, 2001

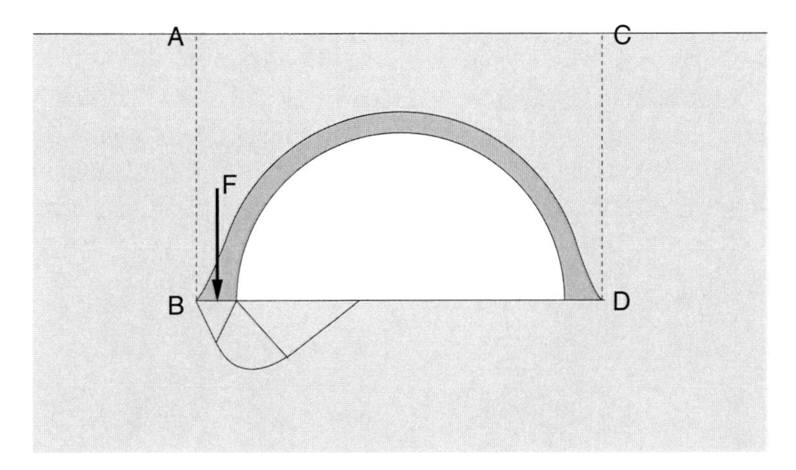

Figure 9: Foundation of top support.

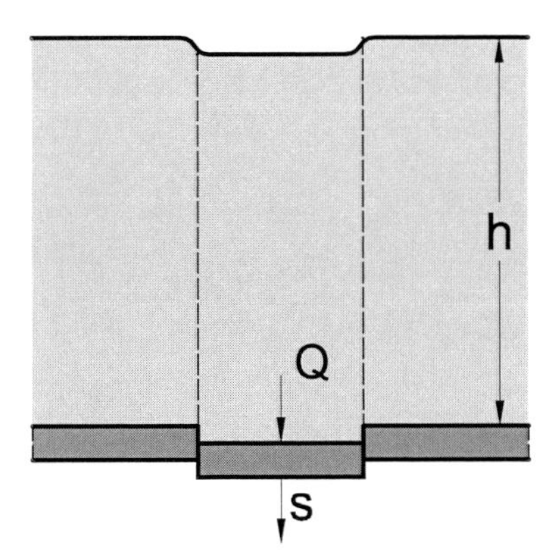

Figure 10: Trapdoor problem, downwards displacement of the trapdoor.

A trapdoor is moved downwards, whereas the force Q exerted by the overburden sand is being measured and plotted over the settlement s. The similarity between a trapdoor and a discharged silo becomes obvious if we consider the part of the soil that does not move downwards as the equivalent to the silo wall. However, the soil is compressible in horizontal direction, whereas the silo wall is considered as rigid. Thus, the analogy of the two problems is not complete. In fact, the stress distribution along the 'silo wall' of the trapdoor problem deviates from the one according to JANSSEN's equation. This is obtained with laboratory measurements [2] carried out with the steel-tape method (see fig. 11 and 12). The

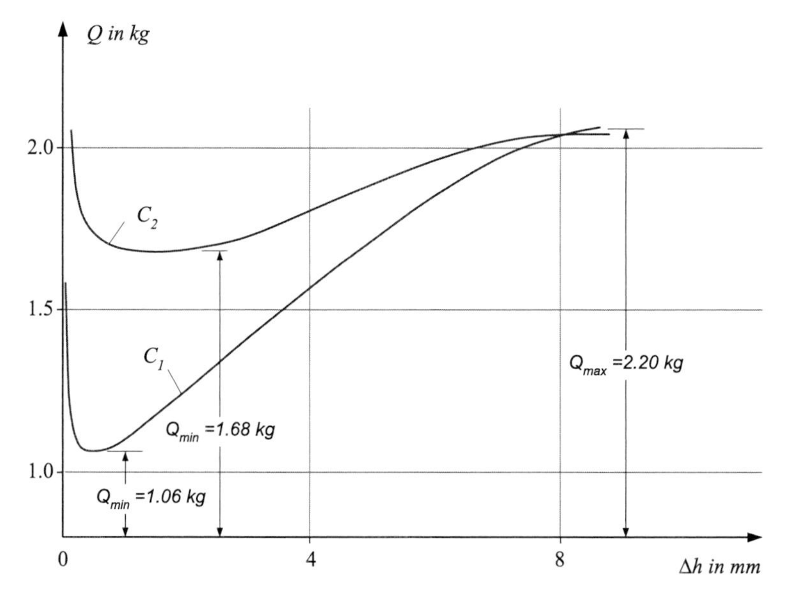

Figure 11: Experimentally obtained [2] relation between the vertical downward movement Δh of the trapdoor and the total vertical pressure Q. C_1: dense sand, C_2: loose sand.

measurements are confirmed by numerical results obtained with the code FLAC. Figures 14 to 22 show the vertical stress σ_z averaged over the trapdoor width b ($\overline{\sigma}_z := \frac{1}{b}\int_0^b \sigma_z dx = \frac{1}{b}Q$, see fig. 13) in dependence of the depth z (fig. 14 - 22). All curves have been obtained with $K = 1$. The deviations from JANSSEN's solution are obvious, especially for dilatant soil.

The observed Q-s-curve [2][4] is regarded as a model of the ground reaction line. The rising branch of this curve confirms the associated concept of the NATM. Considered as a proxy of NATM and disregarding the results shown in fig. 11, the rising branch has been attacked in a vehement way. It is interesting to note that the rising branch can also be obtained numerically using standard FEM schemes (e.g. FLAC with

[4] Similar results are reported in [4].

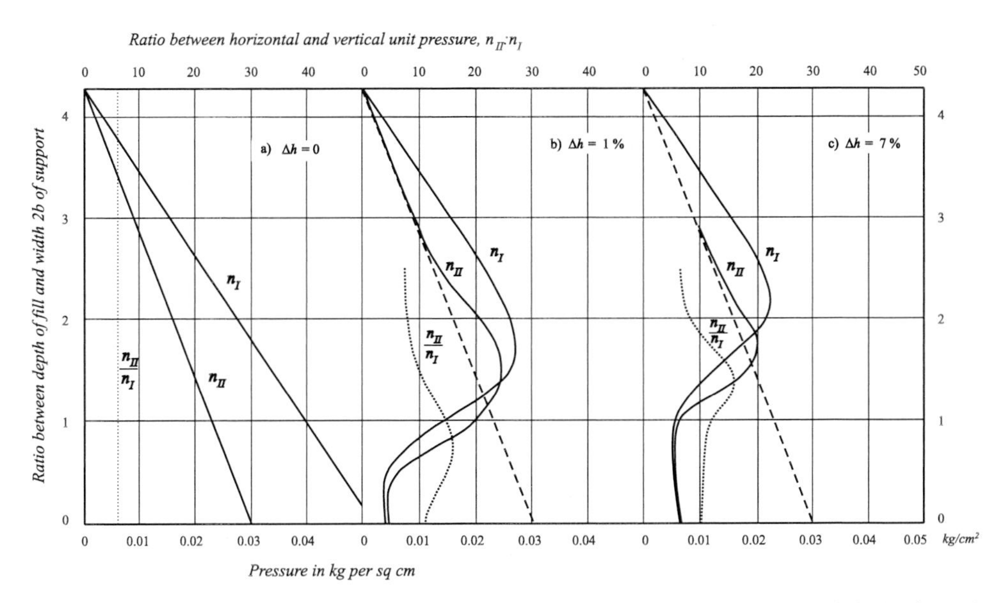

Figure 12: Experimentally obtained [2] distribution of the vertical pressures n_I and the horizontal pressures n_{II} over a plane, vertical section through the axis of the trapdoor (a) for the state preceding the downward movement of the trapdoor, (b) for the state corresponding to Q_{min}, and (c) for the state corresponding to Q_{max}.

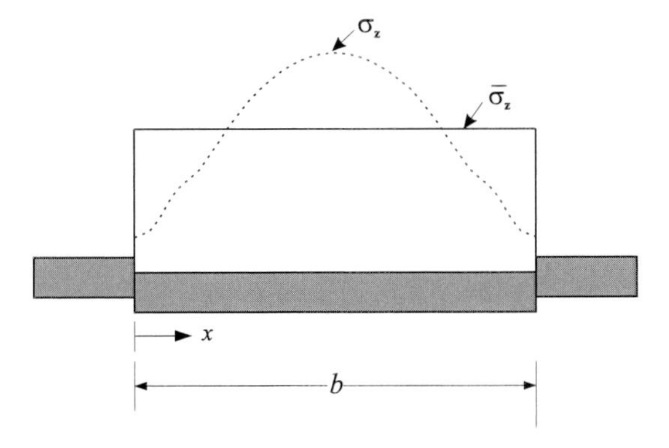

Figure 13: Vertical stress σ_z averaged over the trapdoor width b. Schematic representation.

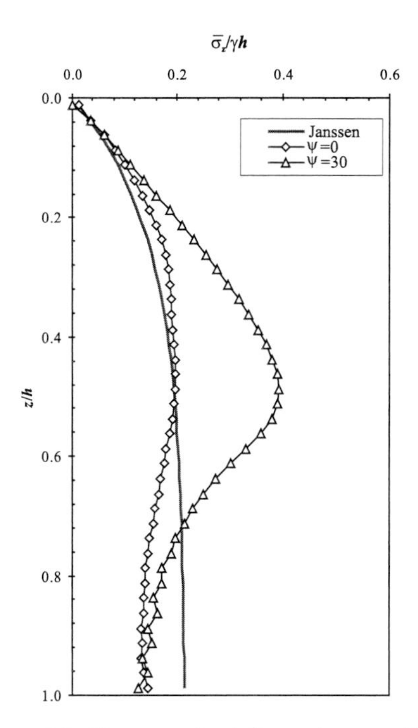

Figure 14: $\overline{\sigma}_z$ vs. z for $h/b = 4$, $\varphi = 30°$, $\psi = 0/30°$

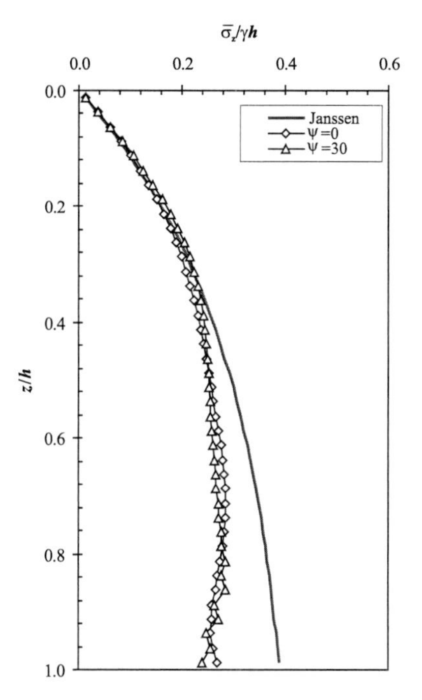

Figure 15: $\overline{\sigma}_z$ vs. z for $h/b = 2$, $\varphi = 30°$, $\psi = 0/30°$

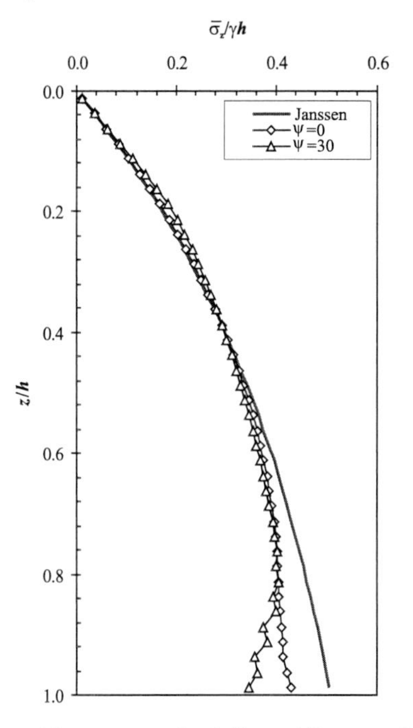

Figure 16: $\overline{\sigma}_z$ vs. z for $h/b = 4/3$, $\varphi = 30°$, $\psi = 0/30°$

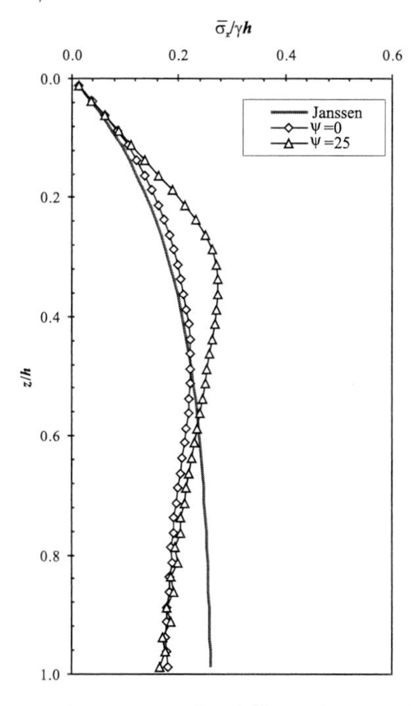

Figure 17: $\overline{\sigma}_z$ vs. z for $h/b = 4$, $\varphi = 25°$, $\psi = 0/25°$

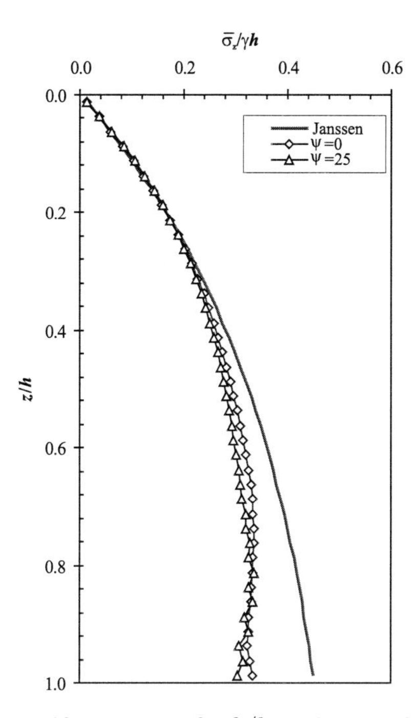

Figure 18: $\overline{\sigma}_z$ vs. z for $h/b = 2$, $\varphi = 25°$, $\psi = 0/25°$

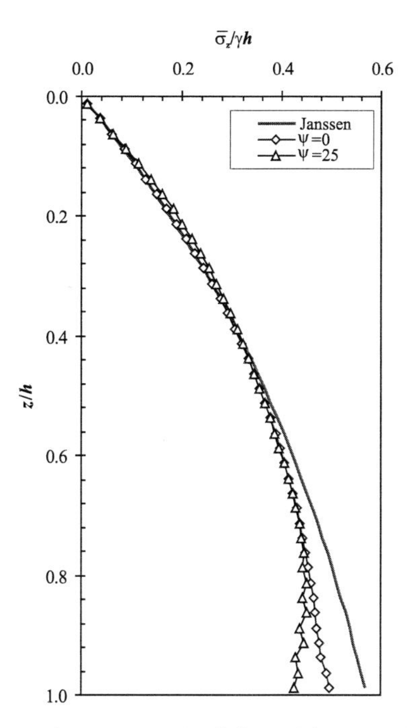

Figure 19: $\overline{\sigma}_z$ vs. z for $h/b = 4/3$, $\varphi = 25°$, $\psi = 0/25°$

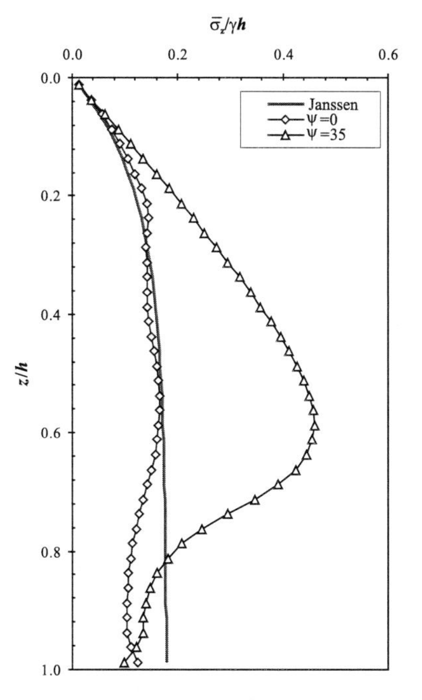

Figure 20: $\overline{\sigma}_z$ vs. z for $h/b = 4$, $\varphi = 35°$, $\psi = 0/35°$

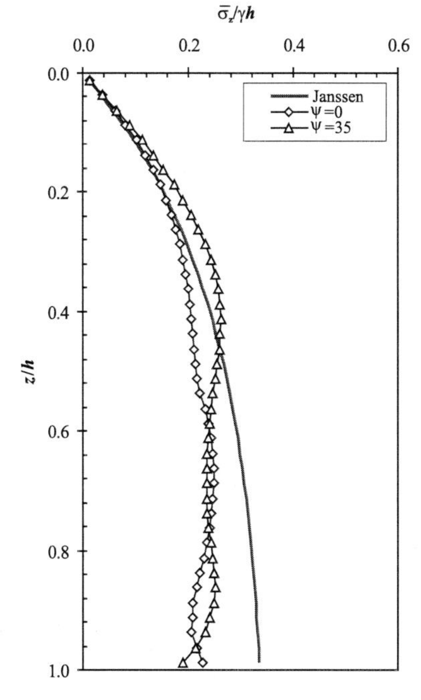

Figure 21: $\overline{\sigma}_z$ vs. z for $h/b = 4/2$, $\varphi = 35°$, $\psi = 0/35°$

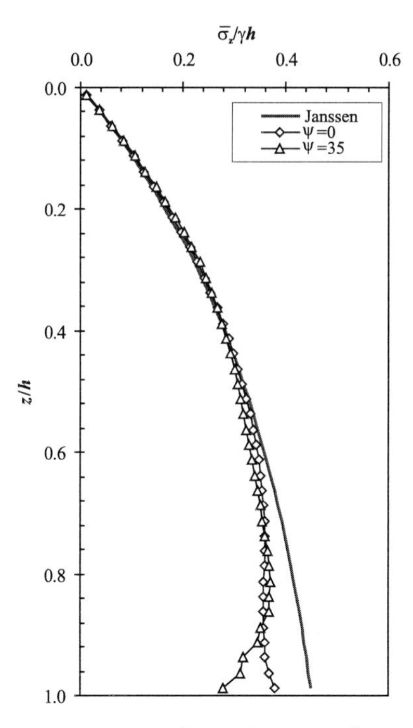

Figure 22: $\overline{\sigma}_z$ vs. z for $h/b = 4/3$, $\varphi = 35°$, $\psi = 0/35°$

MOHR-COULOMB constitutive equation). However, it can be easily seen that the results are mesh-dependent. The rising branch appears only with reducing the mesh size (fig. 23). This fact proves that the solution is mesh-dependent and, thus, the related numerical problem is ill-posed. Improved numerical approaches using so-called regularised approaches prove to be capable in describing the rising branch (fig. 24). Using a constitutive law with softening (i.e. decrease of strength beyond the peak) and a non local approach, the ground reaction lines shown in fig. 24 have been obtained [7].

3 Creep of shotcrete lining

In this section is shown that bending moments within the lining disappear due to creep. We consider the interaction of rock and the shotcrete lining of a cavity (2D problem). The shotcrete lining has the thickness d. The cross section of the cavity is circular or mouth profile. A mouth profile is composed of circular sections, the transition from a section to the adjacent one being smooth. The rock acts upon the lining with a displacement-dependent pressure $p = \Pi(u)$.

Aging (curing) of shotcrete (see [10]) implies a variable YOUNG's modulus:

$$E(t) = a\exp(c\,t^{-0.6})E_{28} \tag{7}$$

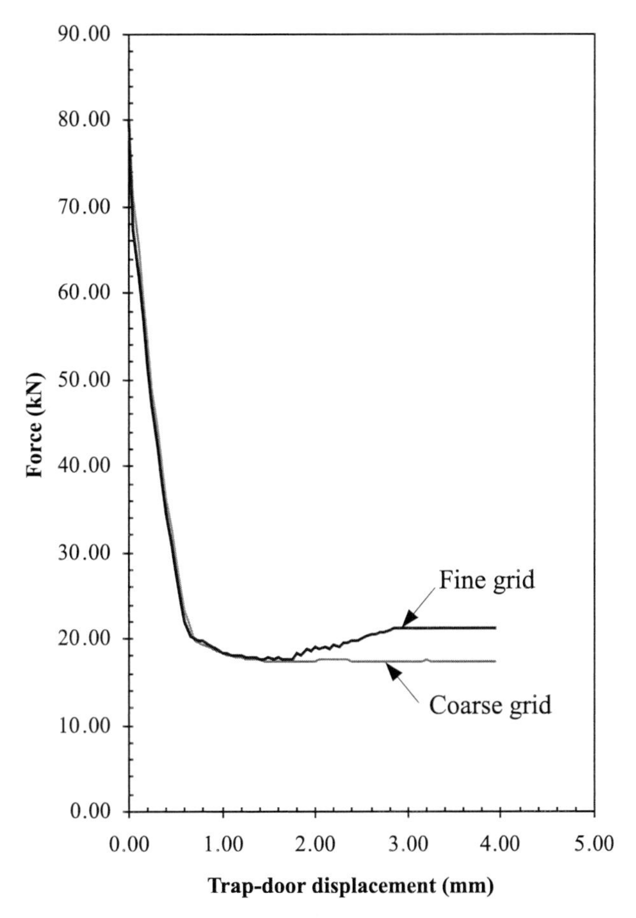

Figure 23: The rising branch disappears with reducing the mesh size. The shown curves were obtained numerically with FLAC and MOHR-COLOUMB elastoplastic constitutive law.

Creep of concrete (and, thus, also shotcrete), see [11] p. 24, may be assumed as:

$$\dot{\varepsilon} = \frac{1}{E(t)}\,\dot{\sigma} + \frac{\sigma}{E(0)}\,b\,e^{-\alpha t} \tag{8}$$

with
t: age of shotcrete
a, b, c, α: material parameters.

Regarding a beam and assuming a linear stress distribution across its section, we obtain the relation

$$\sigma = -\frac{N}{A} + \frac{M}{J}\,y \ , \tag{9}$$

hence

$$\dot{\sigma} = -\frac{\dot{N}}{A} + \frac{\dot{M}}{J}\,y \ . \tag{10}$$

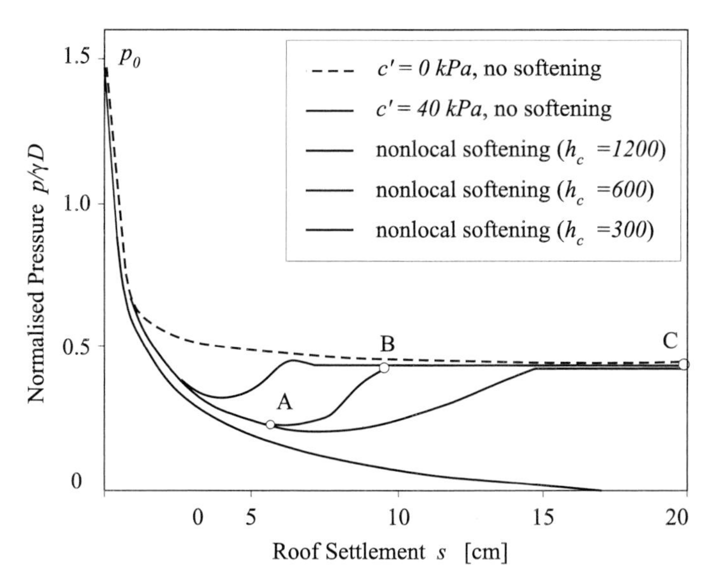

Figure 24: Ground reaction lines for softening rock. Numerically obtained with non-local approach [7].

Note that $\varepsilon < 0$ and $\sigma < 0$ for compression. Introducing (9) and (10) into (8) yields

$$\dot{\varepsilon} = \frac{1}{E(t)}\left(\frac{-\dot{N}}{A} + \frac{\dot{M}}{J}\,y\right) + \frac{1}{E(0)}\,b\,e^{-\alpha t}\left(\frac{-N}{A} + \frac{M}{J}\,y\right) \tag{11}$$

Evaluating (11) for $y = 0$ gives a relation for the strain ε of the central line of the lining:

$$\dot{\varepsilon} = \frac{-1}{E(t)}\,\frac{\dot{N}}{A} - \frac{1}{E(0)}\,b\,e^{-\alpha t}\,\frac{N}{A}\ . \tag{12}$$

Evaluating (11) at the inner and outer boundaries of the lining, i.e. for $y = y_1 = d/2$ and $y = y_2 = -d/2$, respectively, gives the boundary strains ε_1 and ε_2 and, thus, the curvature $\kappa = (\varepsilon_1 - \varepsilon_2)/d$ (cf. fig. 25).

For a beam with initial curvature (fig. 26) we obtain the following elongations of the central, upper and lower lines, respectively:

$$\begin{aligned} u_0 &= \varepsilon ds \\ u_{1,0} &= u_0 \cdot \frac{r + d/2}{r} = u_0\left(1 + \frac{d}{2r}\right) \\ u_{2,0} &= u_0\left(1 - \frac{d}{2r}\right) \end{aligned}$$

Thus we have:

$$
\begin{aligned}
d\vartheta &= \Big[(\varepsilon_1\, ds - u_{1,0}) - (\varepsilon_2 ds - u_{2,0})\Big]\,\frac{1}{d} \\
d\vartheta &= \left\{\left[\varepsilon_1\, ds - \varepsilon ds\Big(1+\frac{d}{2r}\Big)\right] - \left[\varepsilon_2 ds - \varepsilon ds\Big(1-\frac{d}{2r}\Big)\right]\right\}\frac{1}{d} \\
\kappa &= \frac{d\vartheta}{ds} = \frac{\varepsilon_1-\varepsilon_2}{d} - \frac{\varepsilon}{r}
\end{aligned}
$$

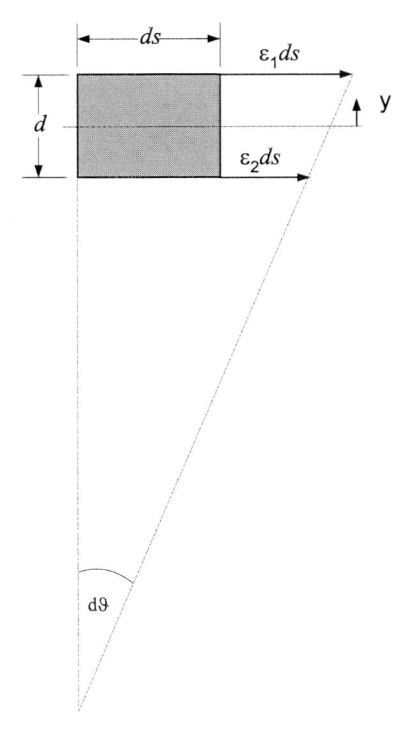

Figure 25: Relation between strains ε_1 and ε_2 and curvature $\kappa = 1/r$: $d\vartheta = (\varepsilon_1 - \varepsilon_2)ds/d \leadsto d\vartheta/ds = \kappa = (\varepsilon_1 - \varepsilon_2)/d$. This relation holds for a beam without initial curvature.

Hence

$$
\dot{\kappa} = \frac{\dot{\varepsilon}_1 - \dot{\varepsilon}_2}{d} - \frac{\dot{\varepsilon}}{r} \tag{13}
$$

From equ. (11) we obtain the expressions for $\dot{\varepsilon}_1 = \dot{\varepsilon}(y = d/2)$ and $\dot{\varepsilon}_2 = \dot{\varepsilon}(y = -d/2)$. Thus we finally have

$$
\dot{\kappa} = \frac{\dot{M}}{E(t)J} + \frac{M}{E(0)J} b e^{-\alpha t} + \frac{1}{rA}\left(\frac{\dot{N}}{E(t)} + \frac{N}{E(0)}\, b\, e^{-\alpha t}\right) . \tag{14}
$$

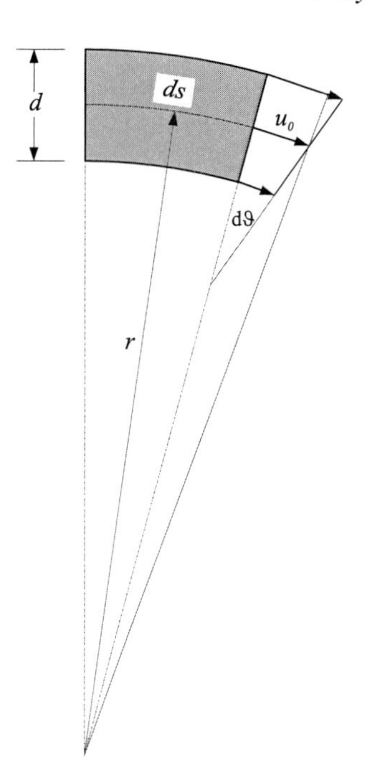

Figure 26: Beam with initial curvature.

The equilibrium equations for a curved beam read [3]:

$$
\begin{aligned}
\grave{Q} - N &= -pr \\
\grave{N} + Q &= -qr \\
\grave{M} &= r\,Q
\end{aligned}
$$

where ` denotes the derivative with respect to ϑ and r is the local radius of curvature. Using the arclength s as independent variable, we obtain

$$
Q' - N/r = -p \tag{15}
$$

$$
N' + Q/r = -q \tag{16}
$$

$$
M' = Q \tag{17}
$$

The prime denotes derivative with respect to arclength s. Eliminating Q yields:

$$
M'' - N/r = -p \tag{18}
$$

$$
N' + M'/r = -q \tag{19}
$$

or

$$
\begin{aligned}
r^2 \grave{\grave{M}} - N/r &= -p \\
r\grave{N} + \grave{M} &= -q
\end{aligned}
$$

p and q are the normal and tangential loads acting upon the lining, Q and N are the transversal and longitudinal forces within the lining, respectively.

Now we need a geometrical relation between the displacement, the strain and the curvature of the lining. We consider the center line of the lining as a closed plane curve with the polar representation $\mathbf{x}(\vartheta)$ and the arclength s.[5] In the course of a deformation, a material point of the lining undergoes the displacement $\mathbf{u}$.

We now assume that the initial cross section is circular with radius r_0 . We furthermore assume that the displacements u are oriented in radial direction towards the centre of the circle $(r = r_0 - u)$. If we now express the curvature of the deformed lining and neglect terms of higher order, we obtain

$$\kappa = \frac{1}{r_0} + \frac{u}{r_0^2} + u'' \ . \tag{20}$$

Obviously, $1/r_0$ is the initial curvature and u/r_0^2 is the change of the curvature due to the reduction of r_0 by the amount u. This reduction can be achieved with a uniform pressure Δp and is not related to bending moments. Only the last part (u'') is due to bending moments. To obtain the strain ε of the center of the lining we consider fig. 27.

$$\begin{aligned}
u_1 &= u \\
u_2 &\approx u + \frac{du}{ds}\,ds \\
AB &= ds = r d\vartheta \\
CD &= (r-u)\,d\vartheta \\
CE &= \sqrt{(ds)^2 + \left(\frac{du}{ds}\,ds\right)^2} \\
&= ds\sqrt{1 + \left(\frac{du}{ds}\right)^2} \\
&\approx ds\left(1 + \frac{1}{2}\left(\frac{du}{ds}\right)^2\right) \\
\varepsilon &= \frac{CE - AB}{AB} \approx -\frac{u}{r}\left(1 + \frac{1}{2}u'^2\right) \approx -\frac{u}{r}
\end{aligned}$$

[5]Note, that, in general, $x := |\mathbf{x}|$ needs not to coincide with the radius of curvature r. $r = x$ can be set for parts of mouth profiles, which are composed of circular segments. If, however, the lining has local deflections, then x does no more coincide with the radius of local curvature. The general relation is

$$\kappa = \frac{1}{r} = \frac{x^2 + 2\dot{x}^2 - \ddot{x}}{(x^2 + \dot{x}^2)^{3/2}}$$

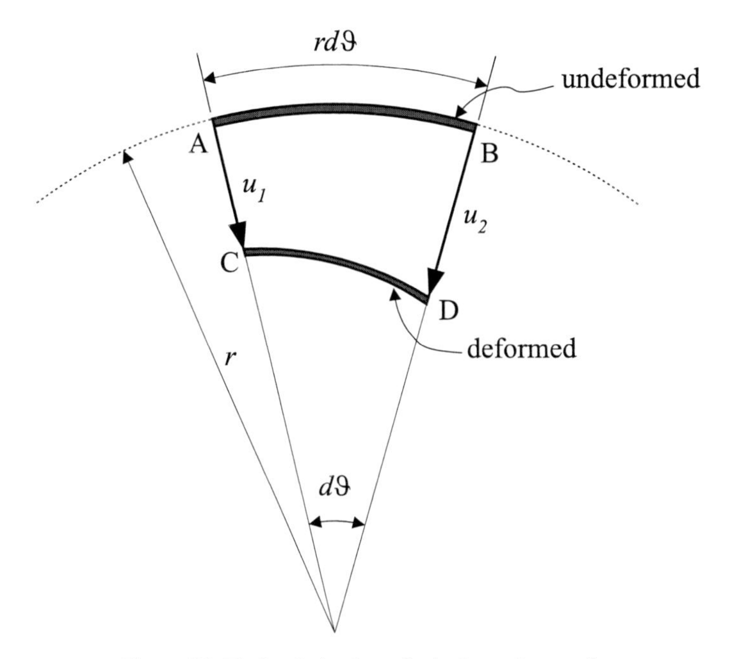

Figure 27: To the derivation of ε in dependence of u

Now we separate equ. (13) and attribute the bending moment M to u'' and the normal force N to u/r_0^2 . With $E := E(t)$, $E_0 := E(0)$ we obtain

$$\dot{u}'' = \frac{\dot{M}}{EJ} + \frac{M}{E_0 J}\, b\, e^{-\alpha t} \tag{21}$$

$$\dot{u}/r_0 = -\frac{\dot{N}}{EA} - \frac{N}{E_0 A}\, b\, e^{-\alpha t} \tag{22}$$

Assuming a WINKLER-interaction between rock and lining we may write $p = p_0 - ku$, hence $\dot{u} = -\dot{p}/k$.

With $E_1 := \frac{E_0}{b\, e^{-\alpha t}}$ we have:

$$-\frac{\dot{p}''}{k} = \frac{\dot{M}}{EJ} + \frac{M}{E_1 J} \tag{23}$$

$$\frac{\dot{p}}{kr_0} = \frac{\dot{N}}{EA} + \frac{N}{E_1 A} \tag{24}$$

Equations (23) and (24) are coupled by means of equ.

$$N = r_0\,(p + M'') \tag{25}$$

which results from (18) .

According to equ. (25), a vanishing spatial variation of M (i.e. $M'' \equiv 0$) implies that N and p are related as $N = r_0 p$. This is the known equation holding for cylindrical vessels that are loaded by pressurized liquids.

We now develop the functions N, p, M according to FOURIER:

$$\begin{aligned} N &= \sum_{i=0}^{n} N_i \cos(is/L) \\ p &= \sum_{i=0}^{n} p_i \cos(is/L) \\ M &= \sum_{i=0}^{n} M_i \cos(is/L) \ , \end{aligned}$$

where L is the circumferencial length of the lining. The coefficients N_i, p_i, M_i are functions of time t . We introduce the FOURIER series into equ. (23), (24), (25). Thus, for each order i we obtain three equations. For simplicity, we consider only $n = 1$, i.e. $i = 0$ and $i = 1$. From (25) it follows:

$$N_0 = r_0 \, p_0 \tag{26}$$

$$N_1 = r_0 \, (p_1 - M_1/L^2) \tag{27}$$

A constant bending moment M_0 does not interact with any external force and is, therefore, self-equilibrated. Since $M_0 \equiv 0$ at the time of shotcrete projection, we infer that M_0 vanishes permanently.

From (27) it follows:

$$M_1 = L^2 \, (p_1 - N_1/r_0)$$

Introducing this result into (23) yields:

$$\frac{1}{EJ} (\dot{p}_1 - \dot{N}_1/r_0) + \frac{1}{E_1 J}(p_1 - N_1/r_0) = -\frac{\dot{p}_1}{k} \ ,$$

hence

$$\frac{\dot{N}_1}{EJr_0} + \frac{N_1}{E_1 J r_0} = \dot{p}_1 \left(\frac{1}{k} + \frac{1}{EJ}\right) + p_1 \frac{1}{E_1 J} \tag{28}$$

From (24) we obtain

$$\dot{p}_1 = k r_0 \left(\frac{\dot{N}_1}{EA} + \frac{N_1}{E_1 A}\right) \ .$$

Introducing this result into (28) yields

$$\frac{\dot{N}_1}{EJr_0}+\frac{N_1}{E_1Jr_0}=kr_0\left(\frac{1}{k}+\frac{1}{E_0}\right)\left(\frac{\dot{N}_1}{EA}+\frac{N_1}{E_1A}\right)+p_1\frac{1}{E_1J}$$

$$\frac{\dot{N}_1}{E}+\frac{N_1}{E_1}=\frac{Jkr_0^2}{A}\left(\frac{1}{k}+\frac{1}{EJ}\right)\left(\frac{\dot{N}_1}{E}+\frac{N_1}{E_1}\right)+p_1\frac{r_0}{E_1}$$

$$\left(\frac{\dot{N}_1}{E}+\frac{N_1}{E_1}\right)\underbrace{\left(\frac{Jr_0^2}{A}+\frac{kr_0^2}{AE}-1\right)}_{B}=-p_1\frac{r_0}{E_1}$$

For the stationary case $\dot{N}_1=0$ we have

$$N_1B=-p_1\frac{r_0}{E_1}\quad\rightsquigarrow\quad p_1=-N_1BE_1/r_0$$

$$\rightsquigarrow\quad M_1=-N_1L^2/r_0\cdot(BE_1-1)$$

For $t\to\infty$ we have $\frac{1}{E_1}\to 0$, hence $N_1\to 0\,, p_1\to 0\,, M_1\to 0$. Thus, in the long range the bending moment dissappears and the load p becomes uniform.

Constant normal pressures and vanishing bending moments along lining sections of constant curvature imply a jump of normal force N at changes of curvature. Note that such changes may appear as edges (Fig. 28), they can, however, be smooth (Fig. 29). To enable jumps of N, one should provide elephant foots and/or micropiles at locations of curvature changes (Fig. 30)

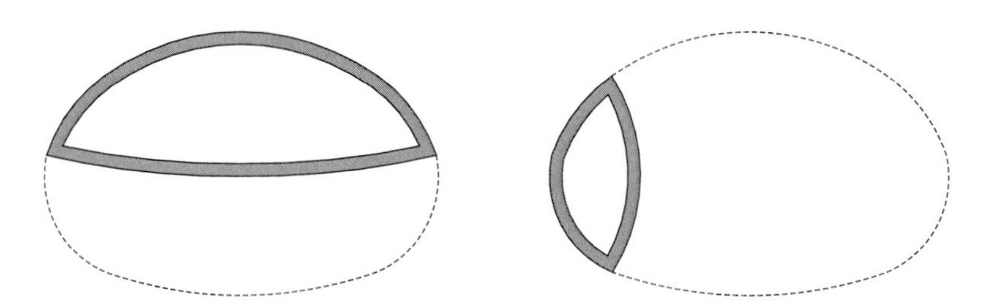

Figure 28: Linings with edges: (b) top excavation, (b) sidewall drift

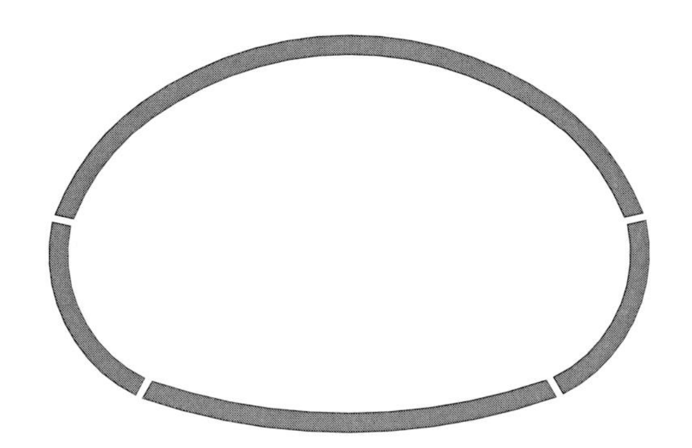

Figure 29: Curvature jumps in smooth lining (mouth profile)

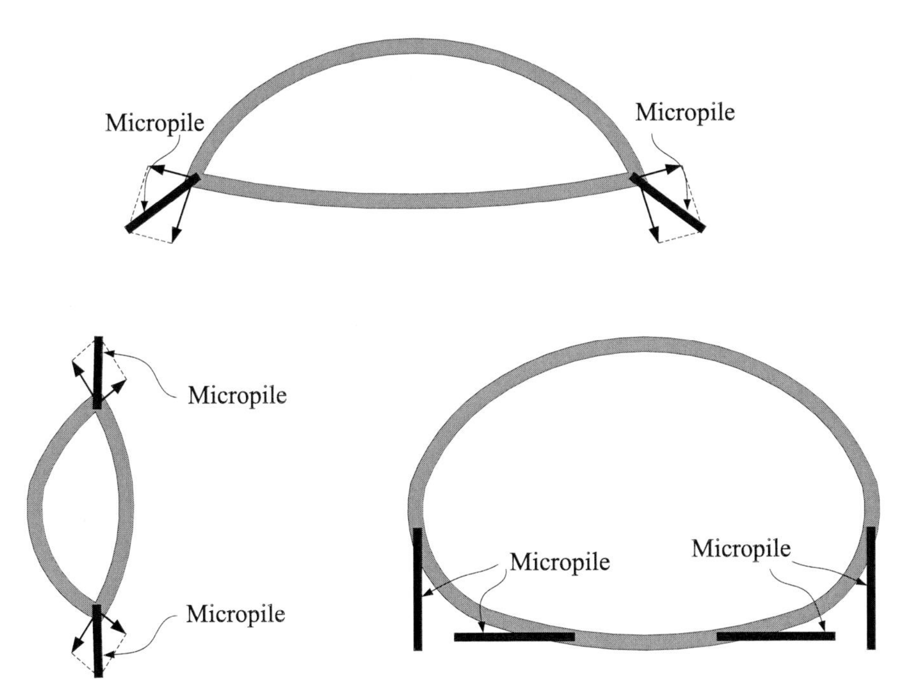

Figure 30: Micropiles to accomodate jumps of N at jumps of curvature

Bibliography

[1] K. Széchy, Tunnelbau, Springer-Verlag, Wien, 1969.

[2] K. Terzaghi: Stress distribution in dry and in saturated sand above a yielding trapdoor. Proceed. Int. Conf. Soil Mechanics, Cambridge Mass., 1936, Vol. 1, p. 307-311.

[3] F. Ziegler, Technische Mechanik der festen und flüssigen Körper. Springer, Wien, 1992.

[4] E. Papamichos, I. Vardoulakis and L. K. Heil, Overburden Modeling Above a Compacting Reservoir Using a Trap Door Apparatus. *Phys. Chem. Earth (A)* Vol. 26, No. 1-2, pp. 69-74, 2001

[5] G. Anagnostou, Standsicherheit der Ortsbrust beim Vortrieb von oberflächennahen Tunneln. Städtischer Tunnelbau: Bautechnik und funktionale Ausschreibung, Intern. Symposium Zürich, März 1999, 85-95

[6] P. A. Vermeer et al., Ortsbruststabilität von Tunnelbauwerken am Beispiel des Rennsteig Tunnels, 2. Kolloquium 'Bauen in Boden und Fels', TA Esslingen, Januar 2000

[7] P. Vermeer, Th. Marcher, N. Ruse, On the Ground Response Curve, *Felsbau*, 20 (2002), No. 6, 19-24

[8] J. Holzhäuser, Problematik der Standsicherheit der Ortsbrust beim TBM-Vortrieb im Betriebszustand Druckluftstützung, Mitteilungen des Institutes und der Versuchsanstalt für Geotechnik der TU Darmstadt, Heft **52**, 2000,49-62

[9] S. Jancsecz u.a., Minimierung von Senkungen beim Schildvortrieb ..., *Tunnelbau* 2001, 165-214, Verlag Glückauf.

[10] A. Pucher, Lehrbuch des Stahlbetonbaues, 3. Auflage, Springer Wien 1961, p.29

[11] Weber, J. W., Empirische Formeln zur Beschreibung der Festigkeitsentwicklung und Entwicklung des E-Moduls von Beton, *Beton und Festigkeitstechnik*, Heft 12, 1979

Constitutive models for numerical simulations

Ivo Herle

University of Innsbruck, Inst. of Geotechnical and Tunnel Engineering, Techniker Str. 13,
A-6020 Innsbruck, Austria
e-mail: ivo.herle@uibk.ac.at, URL: http://geotechnik.uibk.ac.at/staff/ivo.html

Abstract

A critical overview of some widespread constitutive models for soils and rocks is presented. Advantages and limitations of particular models with respect to the observed laboratory behaviour and applications in tunnelling are discussed. The presentation is accompanied with examples of the results from numerical simulations of boundary value problems.

1 Introduction

Numerical solutions in rock mechanics and tunnelling are popular since a long time. Powerful computers and impressive software enable to analyze complex problems fast and in detail. Beside already classical numerical methods like

- Finite Difference Method (FDM),
- Finite Element Method (FEM),
- Boundary Element Method (BEM),

a number of their derivatives, hybrids and novel approaches have appeared in recent years [45]. The most promising seem to be the Discrete Element Methods (DEM) [34] and meshless methods [8, 87].

It is a widespread opinion that modern numerical methods can help us to obtain better and more reliable results. Nevertheless, this is only partially true. Numerical methods provide just a tool for solving sets of equations (linear, algebraic, differential etc.) in a finite number of points (nodes) in space and time. They involve methods for interpolating values of variables between the nodes, direct and iterative schemes for searching roots of equations and integration methods. However, they do not directly consider material behaviour and therefore need constitutive models as an inevitable counterpart.

There is a large number of constitutive models (also called material models) at disposal in geomechanics. Their development still attracts an unremitting interest in research which is not surprising if we take into account many peculiarities of the soil and rock mechanical behaviour. Although there is certainly a necessity to look for better constitutive models, one misses comprehensive presentations of advantages and limitations of the existing models in an accessible way. This paper aims at filling that gap and offers a critical overview of the widespread continuum models suitable for soil and rock mechanics with emphasis on the behaviour of rocks.

The geological strata can be considered either as continuum or as discrete blocks (particles). The decision of the model framework can be difficult, especially in rock mechanics. Large rock blocks and systems of discontinuities seem to be predestinated for discontinuous (DEM) approach [55]. Calculation algorithms of DEM are simple: rigid or deformable blocks can move along slip planes and must satisfy the second law of Newton (force acting on a body is proportional to the acceleration of this body). Explicit numerical integration is able to produce results in most cases, if time steps are small enough and a damping is included. The method can be even coupled with hydromechanical and thermal effects. However, using DEM does not mean to get rid of the constitutive modelling. One must describe the behaviour of contacts between interacting blocks and one still needs constitutive models for deformable blocks.

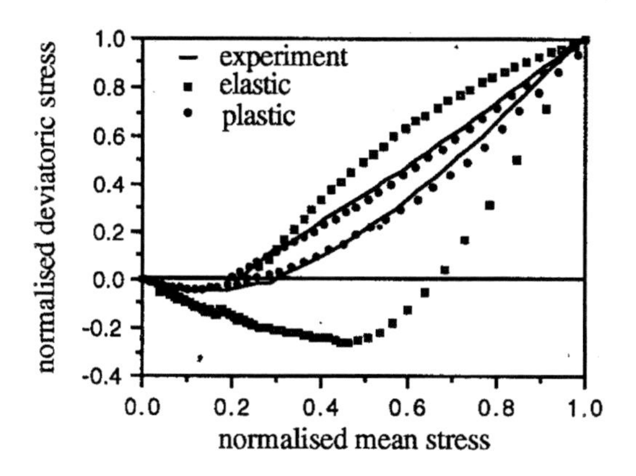

Figure 1: Comparison of stress paths from 3D-DEM simulations and from oedometer test (loading and unloading) with glass ballotini [82]. Influence of plastic deformation at contacts.

Constitutive models for discontinuities are much more difficult than for continua. Mostly very simple models are assumed which have not been validated against any experiments. Rare attempts to check the influence of discontinuity models on the overall behaviour show significant differences between results obtained with different models (Figs. 1 and 2). The determination of model parameters represents an additional difficulty which is usually bypassed by trial and error fitting of the overall response [17] (and thus returning to a quasi-continuum approach). Recognizing the

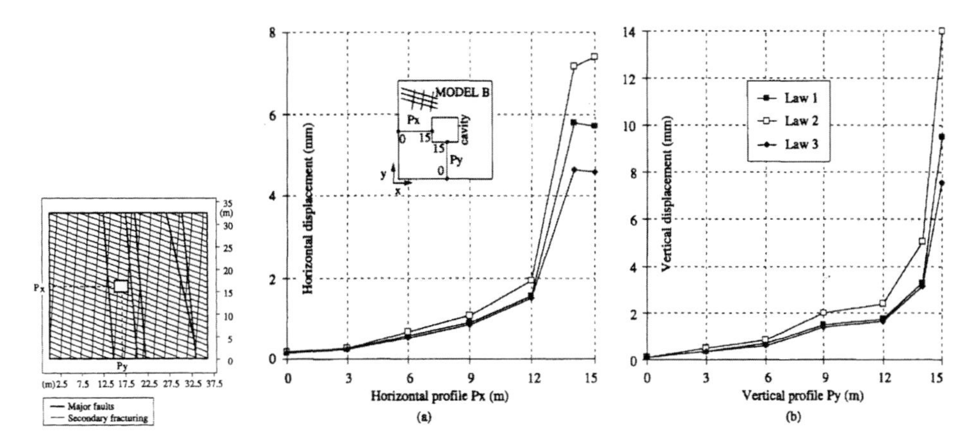

Figure 2: Influence of discontinuities models on displacements due to excavation [78].

difficulties in the discontinuum description, continuum models are being developed also for rock masses with discontinuities, see e.g. review in [77].

2 Observed behaviour

The mechanical behaviour of soils and rocks is usually studied in element tests on laboratory specimens. The specimens are assumed to represent infinitesimally small elements of the ground. Whereas they can provide sufficient information for soils and soft rocks, in case of jointed hard rocks additionally the behaviour of discontinuities must be investigated. Nevertheless, an understanding of the element behaviour remains an inevitable task for the description of the rock mass of any kind.

The observed behaviour enables us to create a mosaic of important features which should be taken into account in numerical calculations in soil and rock engineering. Although the experimental outcomes are often complex, we can still distinguish several basic patterns:

Nonlinearity and irreversibility of stress-strain curves

The relationship between stress and strain components in a particular direction is mostly nonlinear. This means that the stiffness does not remain constant and depends on some state variable(s). Moreover, unloading-reloading cycles often reveal a hysteresis which is a sign of energy dissipation, accumulation of irreversible strains and path-dependence. Even apparently linear part of the stress-strain curve does not automatically imply reversible (elastic) behaviour and may be but an inflexion part of an S-shaped curve resulting from the bedding errors and closing of microcracks in the first branch and of evolution of shear bands in the third branch [47, 81] (Fig. 3).

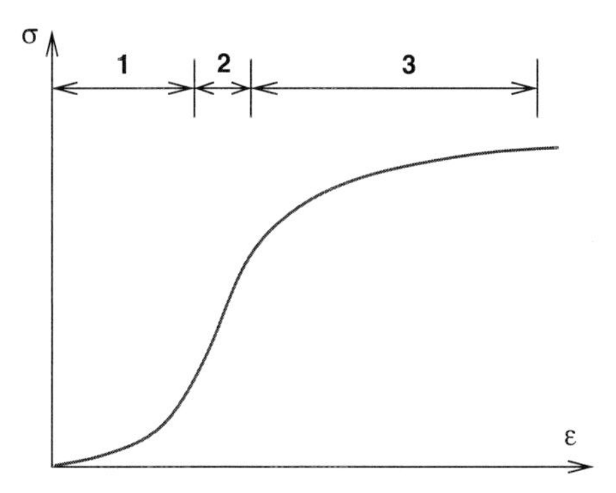

Figure 3: A typical S-shaped stress-strain curve with three distinct zones: 1 – bedding errors, closure of microcracks; 2 – apparently linear behaviour (overlapping of zones 1 and 3); 3 – development of microcracks, evolution of shear zones.

Experimental results of isotropic compression of Mushan sandstone show the stiffness growth with increasing pressure (Fig. 4). Plastic strains were determined from the unloading branch of the stress-strain curve. They reveal irreversible behaviour from the very beginning of the test. The stiffness at unloading (pressure decrease) and reloading (pressure increase) is higher than at primary loading and it abruptly increases if a change in the deformation sense takes place.

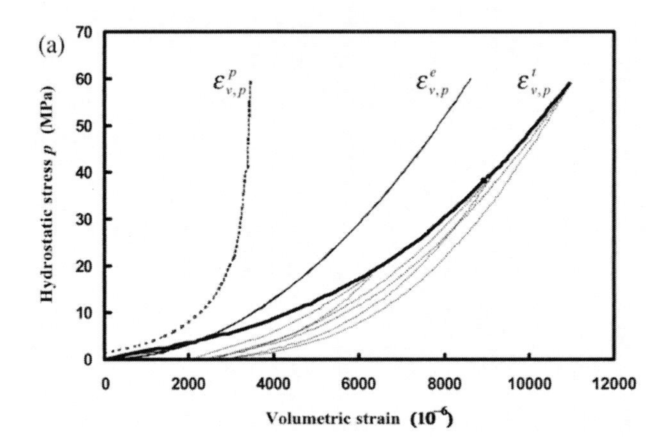

Figure 4: Nonlinear and irreversible behaviour of Mushan sandstone in isotropic compression [44]. Splitting into elastic and plastic strains is based on unloading-reloading cycles.

Also in triaxial tests with constant radial stress the stiffness is not constant. It decreases continuously with increasing axial stress (Fig. 5). There seems to be a linear part of the stress-strain curve at low deviator stresses but the magnitude and inclination of this linear part depends on the confining stress. For higher confining stresses this part becomes shorter and the response softer.

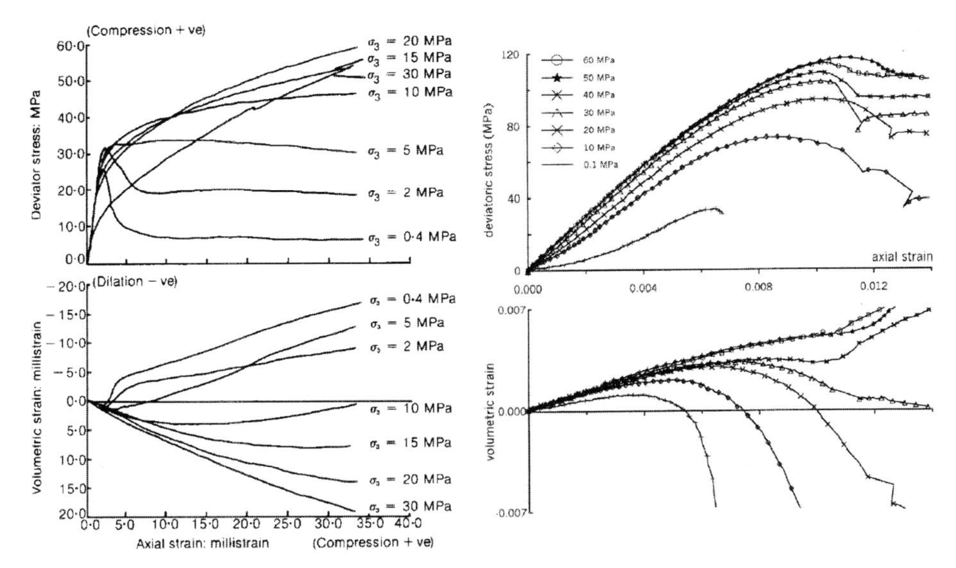

Figure 5: Behaviour of Bath limestone (left) [24] and Vosges sandstone (right) [9] in triaxial compression. Note the different sign convention of volumetric strains in both figures.

In order to obtain reliable measurements of the stress-strain response, modern laboratory instrumentation is inevitable. Local strain gauges mounted directly on the specimen yield a very different picture compared to the classical technique utilizing external displacement transducers (Fig. 6 left). Sensitive apparatuses confirm the nonlinearity of the stress-strain curve even in the range of small initial strains (Fig. 6 right). This phenomenon has been thoroughly investigated especially for soils [5].

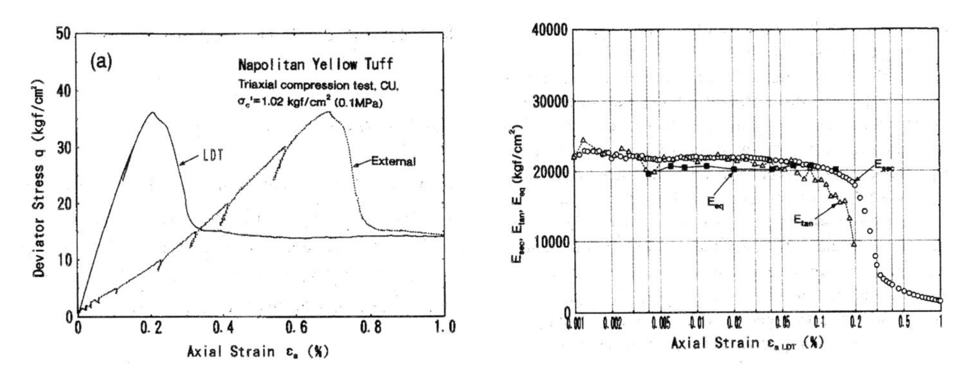

Figure 6: Importance of local measurement of deformation (left) and nonlinearity of stiffness (right) in triaxial tests of Neapolitan yellow tuff [81].

Shear-volumetric coupling

The stress-strain curve delivers only a part of information. Additional details on volumetric behaviour are also needed for the constitutive modelling. Fig. 5 suggests that there is a cross link between shear stresses and volumetric strains which is strongly dependent on the confining pressure. Whereas initial small volume decrease is followed by a volume increase (dilatancy) at lower confining stresses, a permanent volumetric compression (contractancy) is characteristic for higher confining stresses. However, one should take care in the evaluation of experimental results after the peak of a stress-strain curve. Nonhomogeneous (localized) deformation does not allow for standard interpretation in terms of stresses and strains! Moreover, the stiffness of the testing machine has also a remarkable influence on the results in the softening stage.

Stress-dependence

The mentioned examples of experimental results and other publications (e.g. [14, 44]) demonstrate a pronounced influence of the mean stress on the behaviour of rocks. The stiffness in isotropic compression increases with rising stresses (Fig. 4) and also the initial stiffness in triaxial compression is larger for higher confining stresses (Fig. 5). The same pattern can be also distinguished in triaxial extension tests (Fig. 7). Although the stress-dependence of stiffness is more pronounced for soft rocks and soils, it applies to hard rocks as well, especially if they are located in great depths. The range of stress variations in the rock mass during a tunnel construction can be enormous in the latter case .

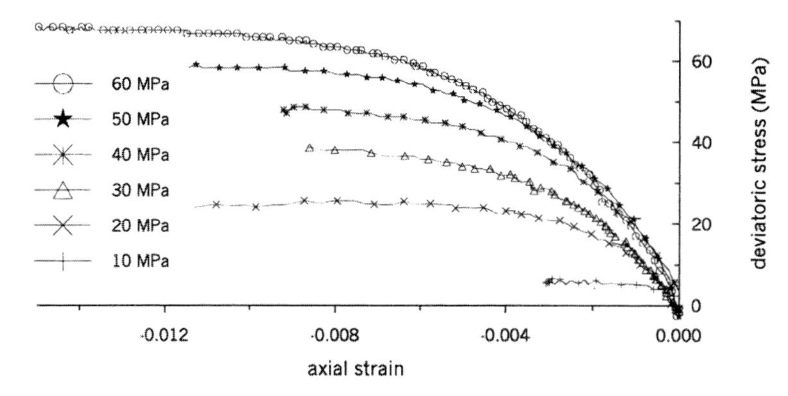

Figure 7: Behaviour of Vosges sandstone in triaxial extension [9].

Shear strength (maximal deviatoric stress) depends on the confining stress in a similar manner like stiffness: it becomes higher for higher confining stress. This stress-dependence is also nonlinear and results in a curved strength envelope. Many different types of soils [52, 3] and rocks follow this pattern (Fig. 8). The shear strength is also sensitive to the value of the intermediate principal stress [33].

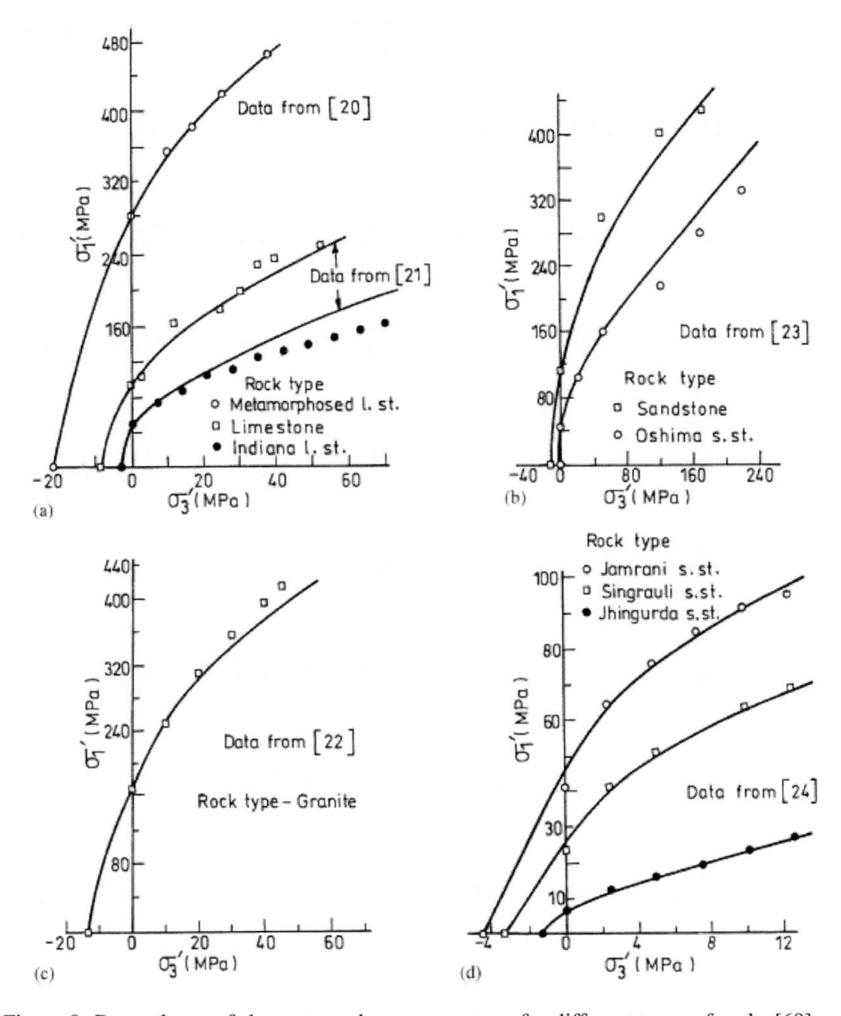

Figure 8: Dependence of shear strength on mean stress for different types of rocks [69].

Anisotropy

The mechanical response of geomaterials in the laboratory depends very often on the orientation of the specimen within a testing device. This is the consequence of anisotropic structure with planes of weakness due to bedding, sedimentation, metamorphosis or shear deformation of the rock.

The anisotropic structure has a pronounced effect on both, stiffness and strength. Fig. 9 depicts a variation of the peak deviatoric stress with the angle α between the major principal stress and the plane of weakness for various rocks.

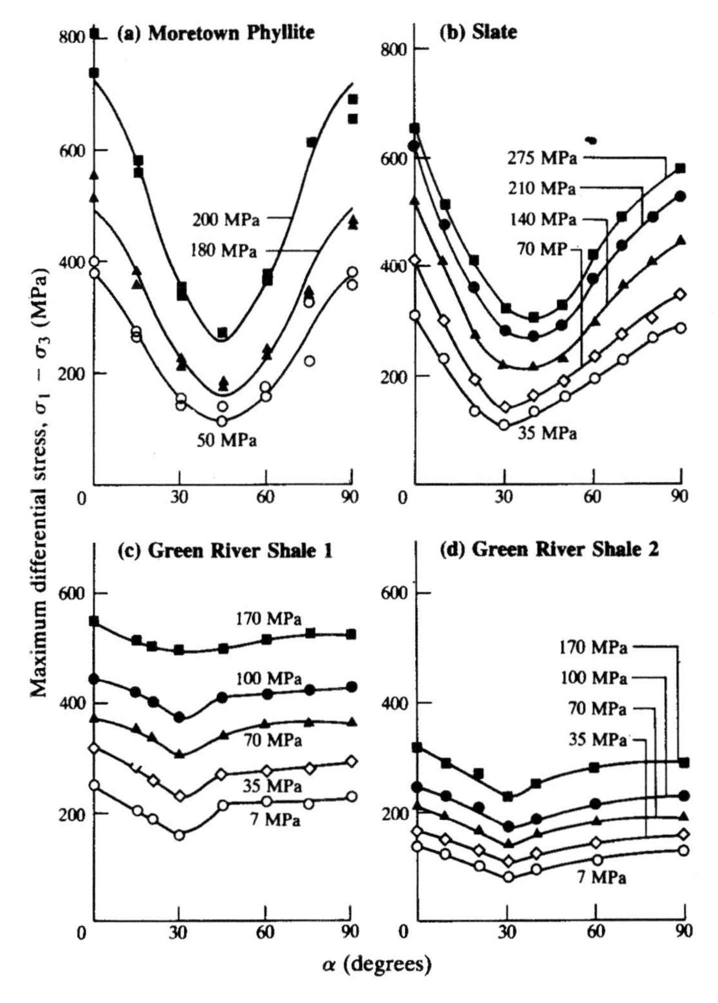

Figure 9: Strength anisotropy of various rocks in triaxial tests (α is the angle between major principal stress and plane of weakness) [14].

Other effects

Although the already outlined phenomena are complicated enough, there is a number of other effects which can have a pronounced impact on the ground behaviour on the element level: scale effects, time and rate effects, coupling effects (interaction between water, gas, temperature), weathering, geochemistry .... Further problems arise due to rock mass properties, e.g. distribution of discontinuities and initial stresses. Nevertheless, these topics are out of the scope of this paper and can be studied elsewhere (e.g. [26, 21, 31]).

In the following sections we will examine the applicability of the frequently used models for the reproduction of the experimentally observed behaviour and the influence of these models on numerical results of boundary value problems in tunnelling.

3 Elasticity

3.1 Linear isotropic elasticity

"The model of linear elasticity based on the generalized Hooke's law is still by far the most widely adopted assumption for the mechanical behaviour of rocks" [45]. Is thus the linear isotropic elasticity a sufficiently good model for rocks?

The model features are well-known and have been thoroughly investigated. There exist several closed-form and approximate solutions of tunnelling problems, starting from the classical Kirsch's solutions up to recent result, e.g. [70, 84, 25][1]. The model is attractive due to its simplicity and the low number of only two material parameters which can be expressed in different ways (Lamé constants, Young modulus, Poisson ratio, shear modulus or bulk modulus). The resulting symmetric stiffness matrix is welcome in numerical implementations.

However, the consequence of the isotropic formulation is the same model stiffness in all directions. Neither planes of weakness nor stress-induced anisotropy at higher deviatoric stresses are included.

The assumed linearity cannot be observed in experimental results (see Section 2). It implies that stiffness is independent of

- depth (stress level),
- direction of deformation, and
- deformation history

Stiffness in linear elasticity is mostly underpredicted at low deviatoric stresses and overpredicted at higher deviatoric stresses. Moreover, the deviatoric stresses are not bounded by any limit stress condition and can increase infinitely. Considering volumetric changes, only compaction can be modelled with Poisson ratios $0 \leq \nu \leq 0.5$. Dilatancy, which strongly influences limit stresses and formation of shear bands, cannot be taken into account. Shear stresses do not influence volumetric changes and variation of mean stresses does not produce any shear strains.

[1] Further closed-form solutions consider linear elasticity together with perfect plasticity, see e.g. [16, 49, 75] and the references therein.

Calibration, i.e. determination of material parameters (i.e. constants in the constitutive equation), is a crucial procedure for the application of any constitutive model. Model constants are uniquely related to the particular model and loose their meaning outside the model framework (thus, questions on physical meaning of model parameters are superfluous in this sense [48]).

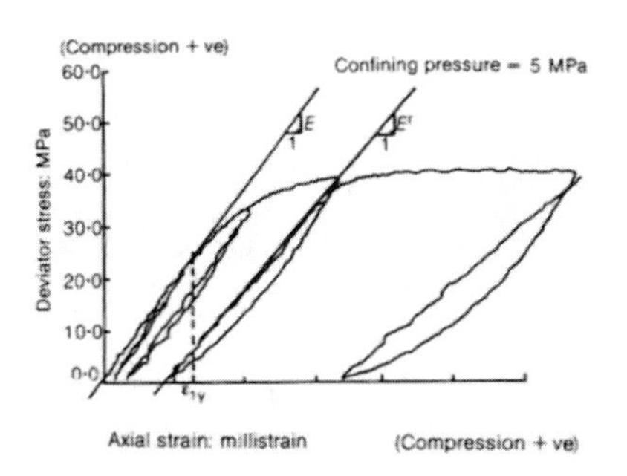

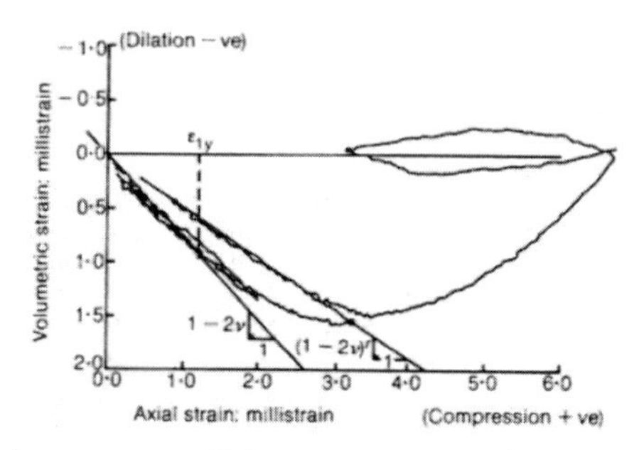

Figure 10: Usual way of determination of the elasticity parameters [24]: Young modulus follows from the quasi-linear part of the stress-strain curve and Poisson ratio from the volumetric curve.

Within linear elastic models, linear interpolation of the nonlinear behaviour is inevitably related to subjective decisions (leaving aside sophisticated optimization procedures which are also problematic in more-dimensional spaces). The usual way of calibration on one test curve seems to work (Fig. 10) but a calibration using outputs from several element tests (e.g. one compression and one shear test, or triaxial tests at different stress levels) is practically impossible (Figs. 17 and 18).

The usual treatment of this problem — back calculation of the model parameters from the measured in situ behaviour — points out to the unsuitability of the model for numerical modelling in geomechanics: the model parameters are not problem-independent constants characteristic for a certain ground but must be fitted to particular known results. Inability of extrapolations hinders predictions for various scenarios and limits the applications to a few checked cases and benchmarking.

3.2 Linear anisotropic elasticity

Assumption of isotropic soil or rock is rarely justified. Geological processes result in layered structures and preferred orientations of discontinuities and weak zones. Moreover, in situ stresses are mostly anisotropic ($K = \sigma_2/\sigma_1 \neq 1$) and at the same time the stiffness is pronouncedly stress-dependent which results in further loss of isotropy.

In a general case the elasticity theory establishes a relationship between six independent stress components and six independent strain components, i.e. 6×6 parameters are needed for the material description. Since the stiffness matrix should be symmetric (due to requirements on the elastic potential [29]), the number of parameters reduces to 21.

A further simplification can be obtained if the behaviour does not depend on rotations about the axis of symmetry. This case is called transverse isotropy or orthotropy or cross-anisotropy. Such a material is characterized by 5 parameters.

Considering normal stresses σ_{11}, σ_{22}, σ_{33}, shear stresses σ_{12}, σ_{13}, σ_{23}, normal strains ε_{11}, ε_{22}, ε_{33}, and shear strains ε_{12}, ε_{13}, ε_{23}, the following representation of the cross-anisotropic elastic model is possible (the vertical direction 2 corresponds to the symmetry axis) [27]:

$$\left\{\begin{array}{c}\varepsilon_{11}\\ \varepsilon_{22}\\ \varepsilon_{33}\\ \varepsilon_{12}\\ \varepsilon_{13}\\ \varepsilon_{23}\end{array}\right\} = \left[\begin{array}{cccccc} \frac{1}{E_h} & -\frac{\nu_{hv}}{E_h} & -\frac{\nu_{hh}}{E_h} & & & \\ -\frac{\nu_{hv}}{E_h} & \frac{1}{E_v} & -\frac{\nu_{hv}}{E_h} & & & \\ -\frac{\nu_{hh}}{E_h} & -\frac{\nu_{hv}}{E_h} & \frac{1}{E_h} & & & \\ & & & \frac{1}{G_{vh}} & & \\ & & & & \frac{1}{G_{vh}} & \\ & & & & & \frac{2(1+\nu_{hh})}{E_h} \end{array}\right] \left\{\begin{array}{c}\sigma_{11}\\ \sigma_{22}\\ \sigma_{33}\\ \sigma_{12}\\ \sigma_{13}\\ \sigma_{23}\end{array}\right\}$$

E_v is the elastic modulus in the vertical direction, E_h is the elastic modulus in the horizontal direction, G_{vh} is the shear modulus. The Poisson ratio ν_{hv} takes into account effects of the horizontal stress on the vertical strains, and ν_{hh} effects of the horizontal stress on the perpendicular horizontal strain.

However, only three model parameters can be determined in a standard triaxial test with vertical deformation along the sample symmetry axis [29]. Additional stress paths are needed for the remaining parameters. Often, additional restrictive assumptions are postulated to facilitate the calibration (e.g. the volumetric strains have the same sign as the applied stress [6] or equations are derived from the isotropic material considering different stiffness in vertical and horizontal direction [29]). In any case, the parameters are not totally independent and several conditions must be fulfilled in order not to violate thermodynamic criteria [63].

Two distinct values of the stiffness in perpendicular directions are the most important feature of the anisotropic elasticity in rock mechanics. Nevertheless, this model involves also shear-volumetric coupling, i.e. a link between volumetric strains and shear stress (or shear strains and mean stress, respectively). Consequently, an isotropic stress path can produce anisotropic strain response and vice versa.

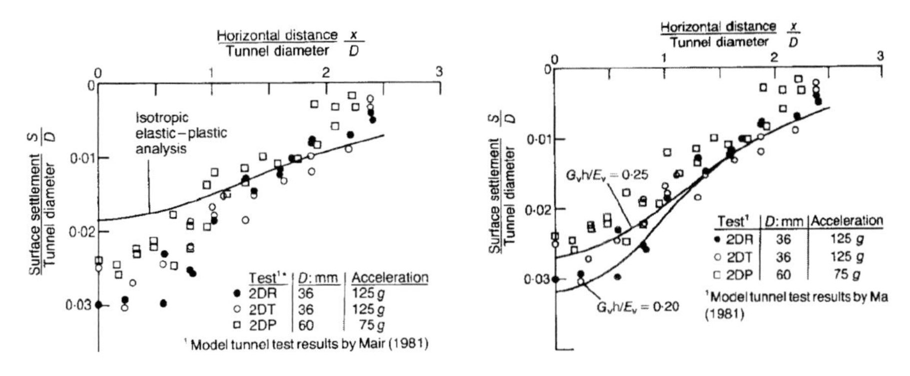

Figure 11: Influence of the anisotropy on the calculated settlement trough [58].

FE calculations of tunnel excavation in anisotropic soil show a significant effect of the G_{vh}/E_v ratio on the shape of the settlement trough [58]. Whereas the isotropic material (Fig. 11 left) yields a wide and only slightly curved settlement trough, the model with adjusted anisotropic parameters can well reproduce the settlement pattern from model tests (Fig. 11 right). However, this conclusion is not shared by other works which doubt such a strong influence of elastic anisotropy if realistic soil parameters are considered [32, 2].

3.3 Nonlinear elasticity

Nonlinearity of the stress-strain response is perhaps the most important feature of the behaviour of geomaterials. Still, it is not straightforward how to handle it. We can namely observe two opposite tendencies: compression of a sample is connected either with increasing or decreasing stiffness, depending on the amount of lateral confinement. (Fig. 12).

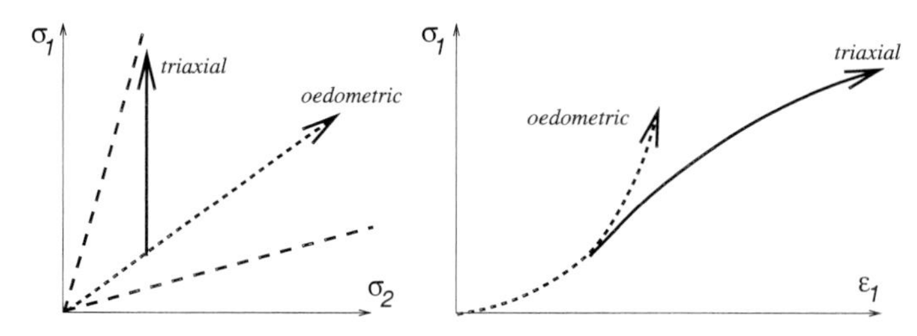

Figure 12: Different nonlinear responses for oedometric and triaxial compression.

This pattern can be modelled by different equations for each type of the behaviour. Terzaghi and later Ohde [61] proposed a pressure-dependent stiffness for oedometric compression

$$E = E_0 \left(\frac{\sigma}{\sigma_0}\right)^{\alpha}$$

(E_0 corresponds to E at stress σ_0, and α is a parameter) and Kondner [54] used a hyperbolic relationship between shear stress τ and shear strain γ

$$\tau = \frac{\gamma}{\alpha\gamma + \beta}$$

(α and β are material parameters). However, the fast development of numerical methods needed a general treatment applicable to all paths. This was accomplished by Duncan and Chang [23] who unified both equations with respect to requirements of a FE implementation. Their approach became popular and can still be found in many software packages. Although single curves can be modelled rather realistically, we must realize that there is not a thorough theory behind this approach. Perhaps the most remarkable deficiency is the lack of a consistency condition which results in a sharp jump in stiffness between *loading* and *unloading* for paths along the neutral direction. Corresponding numerical problems have been reported [79]. Moreover, in this model the volumetric response is not related to the nonlinearity of the stress-strain curve (assuming a single constant value of the Poisson ratio) and cannot take into account coupling between shear stress and volume changes and vice versa.

In spite of these well-known deficiencies one can find further proposals which are mere extensions of this concept. Especially, the recent trend in the description of the soil behaviour at small strains has brought several contributions in this direction, see e.g. [43, 13, 68]. They are based on fitting of experimentally measured stiffness by a single curve. A scalar relationship between a strain invariant and Young (or shear) modulus is then used in FE calculations. However, stiffness is in general not a scalar and it depends on the direction of deformation and other state variables. This means that experimental paths used for the derivation of such a scalar relationship may be very different from real conditions and the calculated stiffness does not need to be representative.

The nonlinear stress-strain behaviour is coupled with the volumetric behaviour and both effects cannot be separated from each other. Quasi-elastic models usually cannot deal with it (although there are exceptions, e.g. [44]). Especially close to the limit stress states, the direction of elastic strain increments becomes completely unrealistic. An insufficient modelling of the volumetric effects may be less important e.g. for excavation or footing problems [88], but it is critical in problems with a high degree of confinement like tunnels and anchors [60, 42, 40]. In case of undrained conditions, the volumetric constraint is maximal and calculations of the effective stresses with quasi-elastic models fail due to unrealistic pore pressure build-up.

In spite of these critical remarks, taking into account the stress-strain nonlinearity in any way usually represents an important improvement of the ground modelling. There is a number of publications suggesting a better coincidence between measured

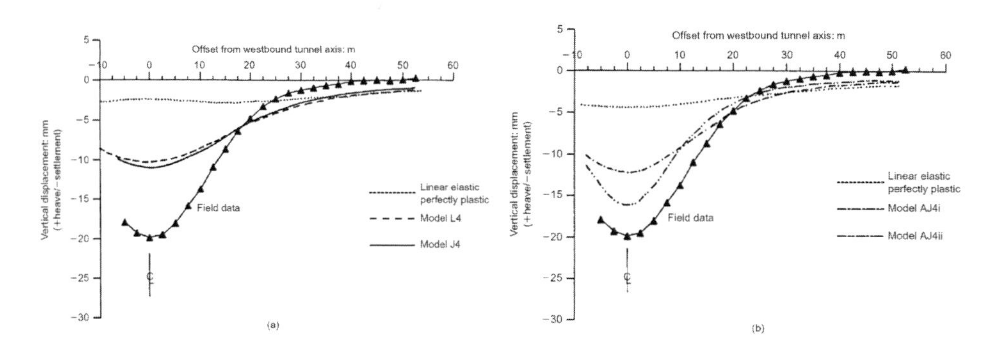

Figure 13: Influence of nonlinearity (left) and anisotropy (right) on the settlement trough [2]. The models L4, J4 are nonlinear but isotropic, the models AJ4i and AJ4ii are nonlinear and anisotropic.

in situ data and numerical simulations when using nonlinear models, although true predictions are generally missing. Considering the settlement profiles above tunnel excavations, nonlinear models produce more realistic, narrower and deeper profiles[2] compared to the linear elasticity [32, 13, 12]. Combining nonlinearity and anisotropy may yield even better results [2, 76], see Fig. 13.

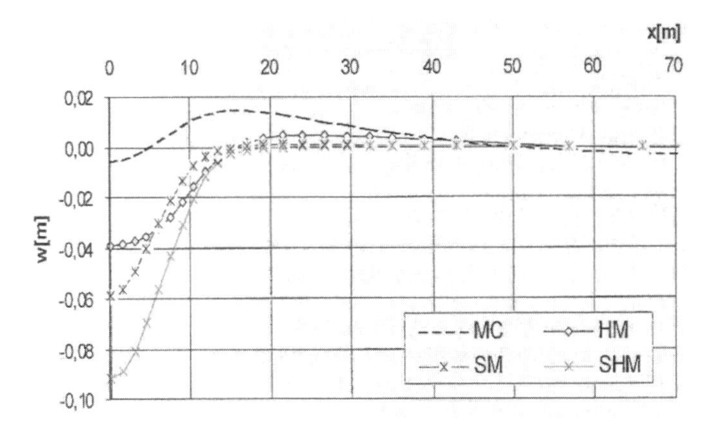

Figure 14: Influence of the model nonlinearity on the settlement profile [74]: MC...Mohr-Coulomb, HM...hardening plasticity, SM...MC & small strain model, SHM...HM & small strain model.

A strong influence of the degree of nonlinearity on the calculated shapes of the settlement profile above a tunnel excavation is shown in Fig. 14[3] [74]. A small-strain model coupled with hardening plasticity produces maximal settlements with the narrowest settlement trough.

[2]Narrow and deep profiles are characteristic for greenfield settlements. In case of overlying buildings, the settlement profile can become much wider and shallower [67]. Also the calculation of internal forces in the overlying structures is very conservative when the greenfield settlement is considered [59].

[3]The results were obtained with constitutive models involving plasticity. However, due to the monotonic deformation process the role of plasticity is limited mainly to the nonlinear stress-strain and volumetric response. Thus, one can also think of a hypothetical nonlinear elastic model.

Fig. 15 (left) documents the inability of the linear elastic-perfectly plastic model to take into account the amount of overburden z/d [74]. Increasing the overburden results in unrealistic large and widespread settlements. On the contrary, a highly nonlinear model (Fig. 15 right) can capture the role of overburden correctly. Deeper tunnels cause lower but wider settlement profiles than shallow tunnels.

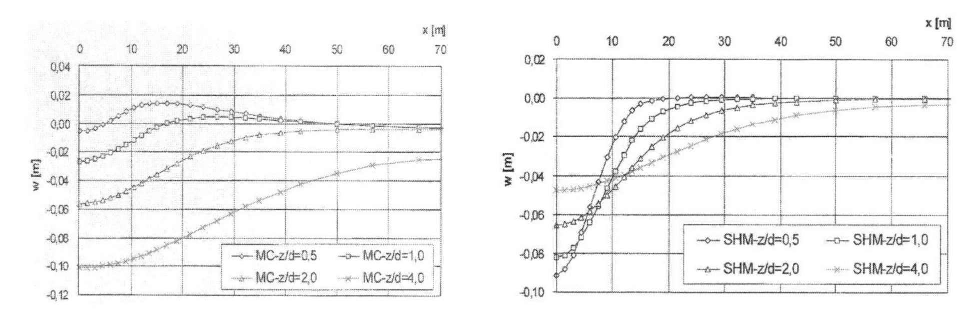

Figure 15: Influence of the overburden z/d on the settlement trough (left: Mohr-Coulomb model, right: small-strain hardening plasticity model) [74].

4 Plasticity

4.1 Perfect plasticity

The experimentally observed irreversibility means that the constitutive models for geomaterials should be formulated in rates (infinitesimally small increments). Finite relationships between stresses and strains cannot take into account the hysteretic and path-depend behaviour.

The theory of incremental plasticity brings a higher quality into the constitutive modelling compared to elasticity. The most important task — a bound for possible stress states — is accompanied by further effects, e.g. irreversibility, path-dependence or dilatancy.

Elastoplastic models still need elasticity as their basic ingredient. The crucial *assumption* splits a strain increment into elastic and plastic parts:

$$\dot{\varepsilon} = \dot{\varepsilon}^e + \dot{\varepsilon}^p \tag{1}$$

The stress increment $\dot{\boldsymbol{\sigma}}$ is then calculated from

$$\dot{\boldsymbol{\sigma}} = \mathbf{M}^e \dot{\varepsilon}^e = \mathbf{M}^e(\dot{\varepsilon} - \dot{\varepsilon}^p) \tag{2}$$

where $\mathbf{M}^e$ is the elastic stiffness matrix. In order to insert the elastic strain increment $\dot{\varepsilon}^e$ into the previous equation, the plastic strain increment $\dot{\varepsilon}^p$ must be found.

A necessary condition for the development of plastic strains is a stress state satisfying the yield function $f(\boldsymbol{\sigma}) = 0$. If the yield function coincides with a limit stress (failure) condition, one speaks of perfect plasticity. The Mohr-Coulomb or Drucker-Prager criteria are examples of yield functions in perfect plasticity. Whereas the Mohr-Coulomb condition is usually presented for a slip plane with the normal stress σ and the shear stress τ

$$f = \tau + \sigma \tan\varphi - c = 0\,,$$

the Drucker-Prager equation

$$f = q + p\alpha - k = 0$$

uses the stress invariants

$$p = -I_1/3$$

(I_1 is the first invariant of the stress tensor) and

$$q = \sqrt{3J_2}$$

(J_2 is the second invariant of the deviatoric stress tensor $\boldsymbol{\sigma}_{dev} = \boldsymbol{\sigma} - \frac{1}{3}I_1\mathbf{1}$).

The zero value of the yield function $f(\boldsymbol{\sigma}) = 0$ must be accompanied by a second condition if plastic strains should arise: Stress increments due to elastic strain increments must point outwards the surface $f(\boldsymbol{\sigma}) = 0$, i.e.

$$\frac{\partial f}{\partial \boldsymbol{\sigma}} \dot{\boldsymbol{\sigma}}^e > 0\,, \tag{3}$$

see also Fig. 16a. In all other cases only elastic strains develop (Fig. 16b and c).

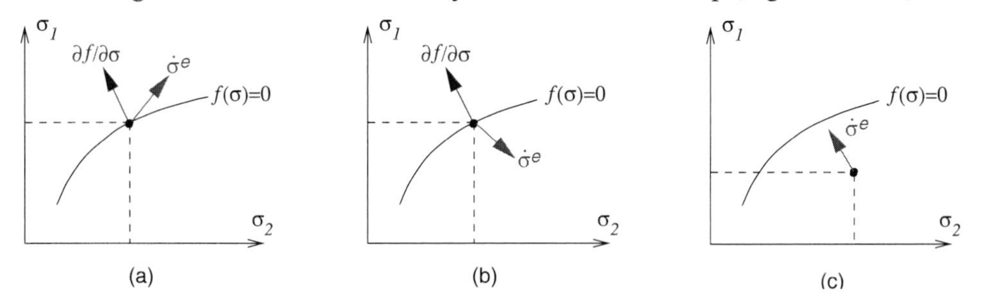

Figure 16: "Loading" case (a) produces plastic strains. In other cases (b and c) only elastic strains develop.

When plastic strains appear during deformation, their direction and magnitude must be determined. The theory of plasticity *assumes* that the direction of the plastic strain increment $\dot{\varepsilon}^p$ depends only on the particular stress state $\boldsymbol{\sigma}$ (flow rule) and is independent of the stress increment $\dot{\boldsymbol{\sigma}}$. Thus, defining the plastic potential $g(\boldsymbol{\sigma})$, the flow rule is given by the gradient of g:

$$\vec{\dot{\varepsilon}^p} = \frac{\dot{\varepsilon}^p}{|\dot{\varepsilon}^p|} = \frac{\partial g}{\partial \boldsymbol{\sigma}} \tag{4}$$

Many models consider $f = g$ (associated plasticity). Nevertheless, $f \neq g$ (non-associated plasticity) is more realistic for frictional materials. In the latter case, it is common to use the same equation for g like for f but with different material parameters. E.g. the Mohr-Coulomb model involves friction angle φ in the yield function f and dilatancy angle ψ in the plastic potential g. Generally $\varphi \neq \psi$.

At this stage, the magnitude λ of the plastic strain increment

$$\dot{\varepsilon}^p = \lambda \frac{\partial g}{\partial \boldsymbol{\sigma}} \tag{5}$$

is still unknown. λ can be determined from the consistency condition

$$\frac{\partial f}{\partial \boldsymbol{\sigma}} \dot{\boldsymbol{\sigma}} = 0 \tag{6}$$

which means that the stress increment cannot surpass the yield function. Hence, for perfect plasticity in case of plastic deformation the stress state may move only along the yield surface. Considering Eqs. (2) and (5), the multiplier λ can be calculated:

$$\lambda = \frac{\dfrac{\partial f}{\partial \boldsymbol{\sigma}} \mathbf{M}^e \, \varepsilon}{\dfrac{\partial f}{\partial \boldsymbol{\sigma}} \mathbf{M}^e \dfrac{\partial g}{\partial \boldsymbol{\sigma}}} \tag{7}$$

Thus, with f and g one can find for a given strain increment $\dot{\varepsilon}$ the corresponding stress increment $\dot{\boldsymbol{\sigma}}$. The outlined general framework is widespread not only in soil and rock mechanics but also in other material sciences.

4.2 Limitations of elastic perfectly plastic models

In spite of being extremely popular in soil and rock mechanics, elastic perfectly plastic models suffer from severe shortcomings. Many of them are related to the large elastic domain because plastic behaviour appears only at the limit stress state. Consequently, stress paths with constrained deformation (compression paths) remain in the elastic range. Other defects stem from the model formulation, e.g. infinite dilatancy during shearing or lack of anisotropic effects. Several further comments follow:

Calibration

The unrealistic description of the soil and rock behaviour with elastic perfectly plastic models is reflected in an ambiguous calibration. The calibration procedure underlies subjective decisions and can hardly lead to material parameters suitable for genuine numerical predictions. Such parameters depend, in reality, on the state of soil, i.e. they are stress- and density-dependent. This is a paradox situation where material *constants* are *state-dependent* and, thus, variable. For one and the same soil they must be adjusted according to the actual conditions. However, these conditions change in the course of deformation!

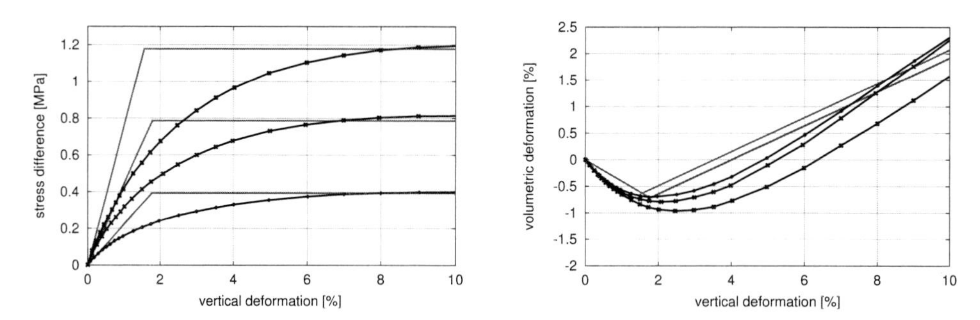

Figure 17: Calibration of the Mohr-Coulomb model (straight lines) on standard triaxial experimental curves of Hochstetten sand at different cell pressures [37].

Calibrating the Young modulus from the initial stiffness (Fig. 17 left) results in large deviations between experimental and model stress-strain curves although the volumetric behaviour, see Fig. 17 right, is reproduced slightly better. Note that different Young moduli were fitted to each experimental curve.

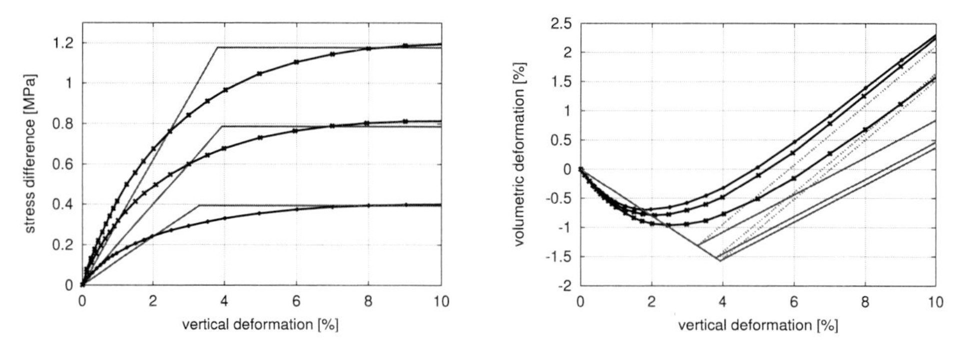

Figure 18: Another way how to calibrate the Mohr-Coulomb model (straight lines) on standard triaxial experimental curves of Hochstetten sand at different cell pressures [37].

When trying to improve the calibration by choosing lower Young moduli, see Fig. 18 left, the secant approximations of the experimental curves appear to be a better solution than the fitting of the initial stiffness. However, lower Young moduli mean larger elastic strains which result in an exaggerated range of compression, see Fig. 18 right. Thus, an improvement of one aspect is counteracted by extra discrepancy.

Nonlinearity

Many limit stress conditions (Mohr-Coulomb, Drucker-Prager, Tresca etc.) express a linear relationship between the shear strength and normal stresses, although experimental results reveal a clearly nonlinear behaviour. The nonlinearity of limit stresses has a decisive influence e.g. on slope stability calculations [3]. A linear extrapolation of the limit stress envelope can produce a serious overestimation of the safety factor F. For a simple slope geometry, a Mohr-Coulomb envelope may yield $F = 1.5$ whereas a nonlinear envelope $F = 0.97$ [4].

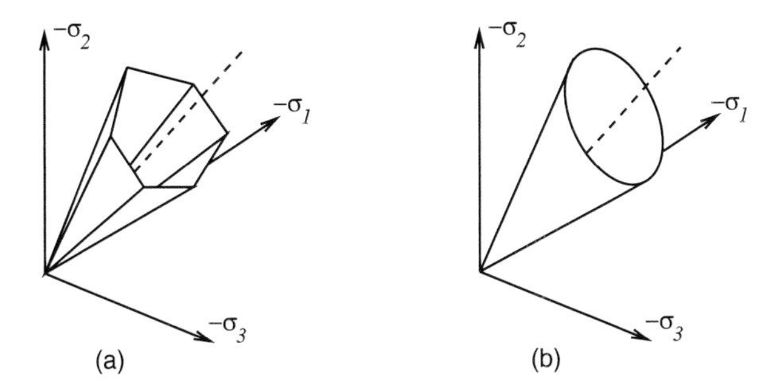

Figure 19: Yield functions in the principal stress space: (a) Mohr-Coulomb, (b) Drucker-Prager.

Further trouble can arise from the formulation of the limit stress condition in stress invariants. If the third stress invariant is not taken into account, the trace of the resulting yield surface in the deviatoric plane is of a Drucker-Prager type (Fig. 19). E.g. the family of Cam Clay models is formulated in p' and q invariants with a fixed critical stress ratio $q/p' = M$. This is unrealistic not only in 3D calculations. Even in axisymmetric or plane-strain problems, the obtained large difference between the shear strength in compression and extension can become dangerous. Using $M = 1.2$, we get

$$\sin\varphi_c = \frac{3M}{6+M} \Rightarrow \varphi_c = 30^\circ \text{ in compression, and}$$

$$\sin\varphi_e = \frac{3M}{6-M} \Rightarrow \varphi_e = 48.6^\circ \text{ in extension.}$$

In rock mechanics, a nonlinear limit stress locus proposed by Hoek and Brown [41] is popular. The maximal and minimal principal stresses σ_1 and σ_2 are scaled by the uniaxial compressive strength σ_c of the intact rock:

$$\frac{\sigma_1}{\sigma_c} = \frac{\sigma_3}{\sigma_c} + \sqrt{m\frac{\sigma_3}{\sigma_c} + s} \qquad (8)$$

(m and s are material parameters). Other criteria derived from this equation have been published as well, e.g. [69]. However, calibration procedures remain the weak point of these models. Until now, no convincing method has been established for the validation of the empirically estimated model parameters in the field circumstances.

Irreversibility (unloading)

Many boundary value problems are not characterized only by monotonous paths. Sharp bends or path reversals ("*unloading*") must be often taken into account. Although the theory of plasticity enables to distinguish between "loading" and "unloading", simple models are mostly unsuitable for realistic simulations of path reversals.

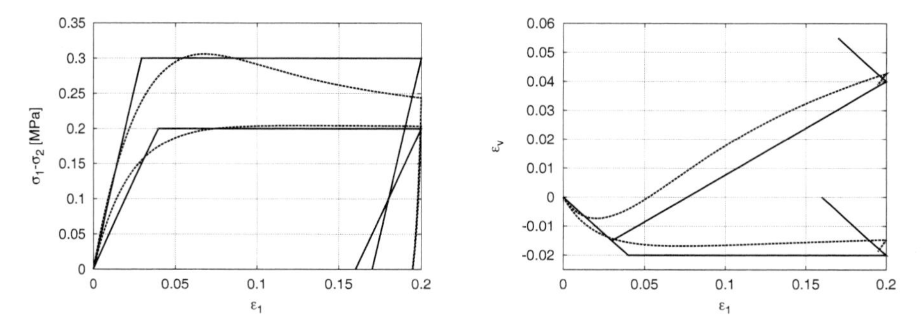

Figure 20: Full line — Mohr-Coulomb model: unrealistic stiffness and volume increase at unloading in standard triaxial test (dotted line — correct behaviour from hypoplastic calculation)

The stiffness produced by elastic perfectly plastic models is the same for primary loading (stress increase) and unloading (stress decrease) in the elastic range. Consequently, the calculated behaviour during unloading is too soft (Fig. 20 left). In the practice, "dirty tricks" are then applied to overcome this deficiency (assigning different values of Young moduli to regions with stress increase and decrease, respectively [86]) otherwise e.g. an exaggerated heave of the tunnel bottom due to excavation is obtained from simulations.

Another aspect concerns the volumetric behaviour after the path reversal. It was shown experimentally for granular soils that a pronounced contractancy sets on [28, 57]. However, elasticity predicts the opposite if the mean pressure decreases, see Fig. 20 right.

Consider further "unloading" into the elastic range. It is often recommended to use ν between 0.15 and 0.25 [56, 15]. However, $\nu = 0.2$ corresponds to the ratio of stress increments

$$\frac{\dot{\sigma}_3}{\dot{\sigma}_1} = \frac{\nu}{1-\nu} = 0.25$$

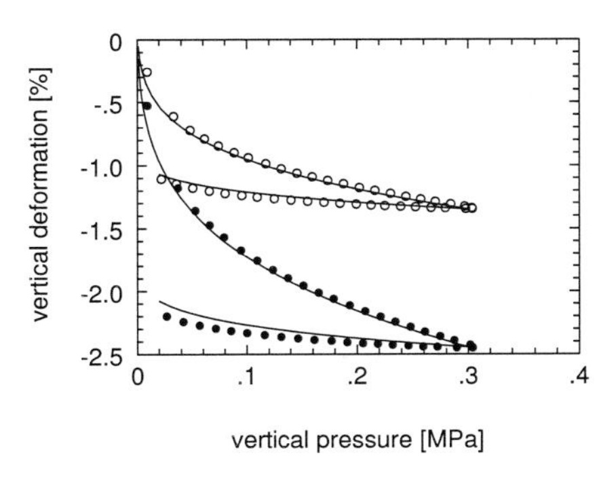

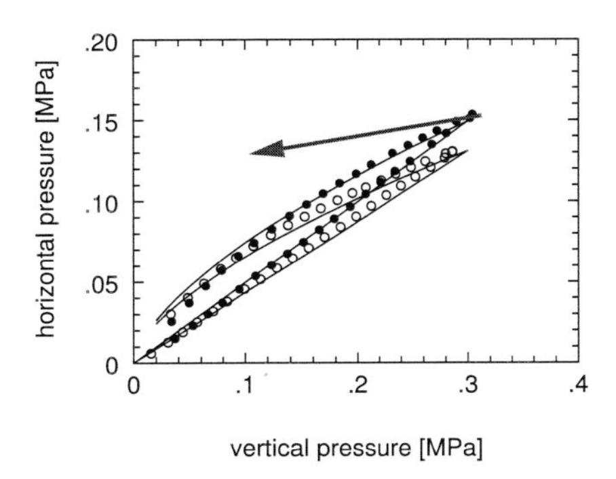

Figure 21: Elastic unloading (arrow) in oedometric test for $\nu = 0.2$. Points: experiments with Karlsruhe sand at two different densities [7, 50], full lines: hypoplastic model [39].

This ratio seems too low when compared with experimental results [7, 50, 19]: the calculated elastic stress path (arrow in Fig. 21 right) approaches fast the limit state. Many advanced elastoplastic models include this property [38].

4.3 Anisotropy

The *Multilaminate model* [72] can be considered as a concept which can account for induced stiffness anisotropy using an arbitrary constitutive model. Each material point is assumed to be intersected by a number of slip planes. The behaviour on these planes is described with a particular (e.g. Mohr-Coulomb) model. The global response is obtained by integration of the contributions from each contact plane. Anisotropic behaviour results from various stress paths along these planes.

Strength anisotropy is of the same importance like stiffness anisotropy. However, many criteria for the limit stress conditions are formulated in stress invariants and/or principal stresses which switches off the effects of anisotropy. General models formulated in stress tensors are needed [11, 85] but those are rarely advanced for practical applications. Still, there are exceptions based on standard approaches, like the *Jointed Rock* model [15]. It includes the cross-anisotropic elasticity linked with perfect plasticity (Mohr-Coulomb) model. A few planes characterized by dip and strike may be defined with different shear strengths (friction and dilatancy angle, cohesion and tensile strength). In this way, some preferential weak directions can be taken into account.

4.4 Advanced models

Some enhancements of the elastic perfectly plastic models are required in order to reach a better coincidence between the measured and calculated behaviour. Implementation of an additional yield surface — a cap — is perhaps the next useful step

[71]. Consequently, irreversible (plastic) strains develop also for compression paths remote from the limit stresses. However, a cap-type yield function cannot be fixed in the stress space like a failure-type yield function does. Thus, the cap yield surface must be changed during the deformation since the stress state cannot surpass the yield surface (Fig. 22). Additional state variables storing the actual shape and position of the yield surface are needed.

This approach is called hardening plasticity. In general, a non-fixed yield surface can change its shape and position either in an isotropic or in a kinematic way (or in a mixed mode). An additional scalar state variable is sufficient in case of isotropic hardening, where the yield surface keeps its shape and expands or shrinks around a fixed point. At least one tensorial variable is necessary for the kinematic hardening if the yield surface moves in the stress space without changing its size.

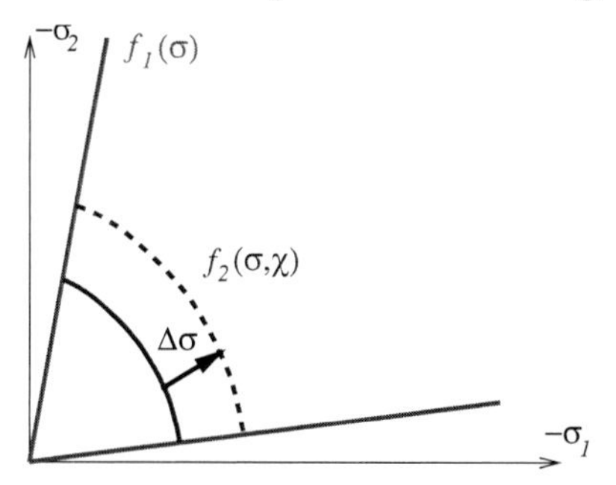

Figure 22: Isotropic hardening of a cap-type yield surface f_2 (χ is an additional state variable characterizing the actual position of the yield surface). The failure type yield surface f_1 is fixed in the stress space.

A large amount of constitutive models based on hardening plasticity has been proposed. Different shapes and numbers of the yield surfaces have been tried together with various hardening rules. Nevertheless, every model reveals some deficiencies and the research in this field remains active. Several critical reviews have been published [35, 22, 18, 36]. Another approaches, like fracture or damage mechanics, have been also tested. Their range of applicability concerns mainly hard and brittle rocks.

The complexity of advanced models is outweighted by a higher quality of the simulation results. Studies of different excavation methods can only be performed if the applied constitutive model takes into account nonlinearity and irreversibility prior to the limit states. E.g. a nonlinear elastic model coupled with the Cam Clay hardening plasticity model was used for the numerical simulations of the Heathrow trial tunnel [46]. It was shown in agreement with in situ measurements that the sequential excavation (SEM) yields the largest settlements, see Fig. 23.

With help of advanced models one can investigate phenomena in more detail than with simple models. A numerical analysis of a tunnel excavation with a nonlinear elastic-hardening plastic model ("weak rock") revealed a significant dilation zone

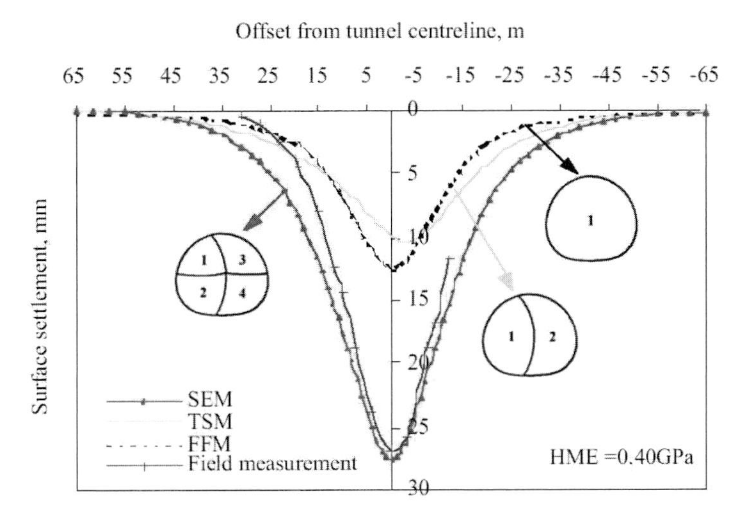

Figure 23: Study of the influence of the excavation method on surface settlements [46].

around the tunnel (Fig. 24) [44]. The magnitude and range of the dilation zone is reflected in realistically large inward displacements. The standard elastic perfectly plastic model is also able to produce dilation zone but of a much smaller extent due to the elastic range with a purely compressive volumetric behaviour. On the contrary, the weak rock model considers elastic shear-volumetric coupling and thus allows for dilatancy even in the elastic range.

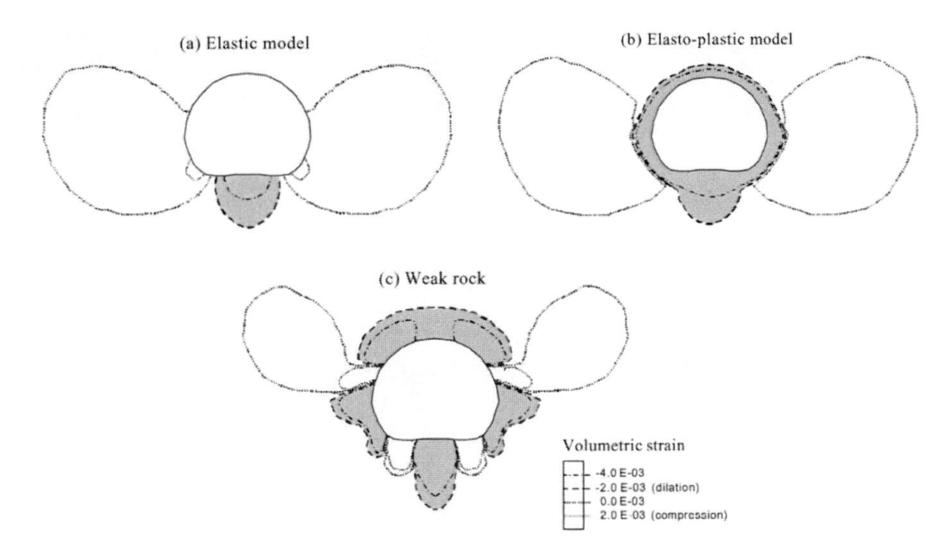

Figure 24: Volumetric deformation during tunnel excavation calculated with different constitutive models (dilation – shaded area) [44].

There are other advanced constitutive frameworks outside of the theory of plasticity as well. Hypoplasticity seems to be the most promising one [53, 20, 30, 80]. It was used e.g. for the numerical analysis of tail gap grouting, which can be simplified as a problem of convergence and expansion of a circular cavity, see Fig. 25 [51]. Hypoplasticity was compared with the Mohr-Coulomb model with respect to the surface settlement trough.

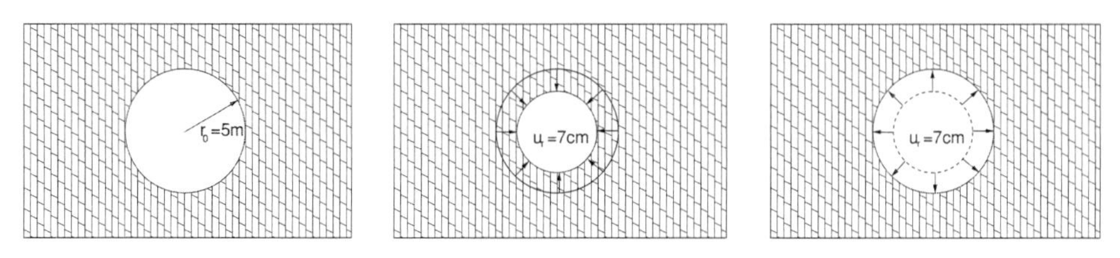

Figure 25: Simulation of tail gap grouting [51].

The calculation results in Fig. 26 suggest that the results of the convergence stage do not depend considerably on the used constitutive law. However, large qualitative differences can be noticed for expansion. The Mohr-Coulomb model predicts a heave of the surface whereas the hypoplastic calculation does not result in any substantial deformations in this stage. The latter output agrees with in situ observations. The reason for the differences between the models is in the different volumetric response for unloading (see Fig. 20).

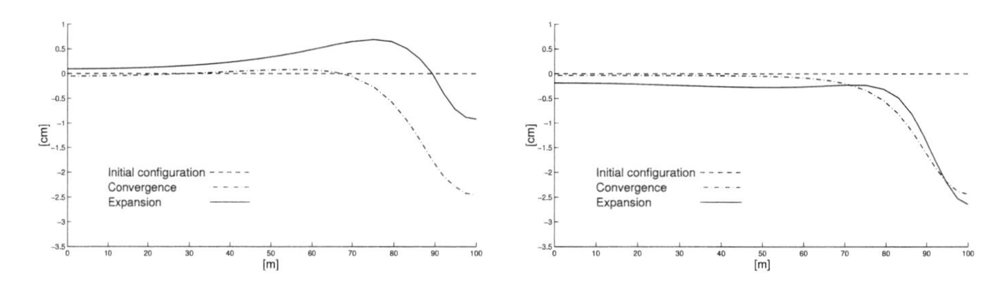

Figure 26: Surface deformation due to tail gap grouting [51]. Results with the Mohr-Coulomb (left) and the hypoplastic (right) model.

The utilization of advanced models requires a good understanding of the model performance and the range of its applicability. There is no correlation between the quality of the constitutive model and the quality of the numerical simulation under all circumstances. The calculated time-settlement curve in Fig. 27 represents an output from a 3D simulation of a tunnel excavation in a cemented sand [10]. Whereas the hypoplastic model predicts unrealistic large deformations, the elastic and elastoplastic calculations come close to the observed values.

The constitutive models were calibrated from triaxial and oedometer laboratory tests. The comparison between the measured and calculated triaxial curves is rather surprising, see Fig. 28. Hypoplasticity seems to match the test results much better than

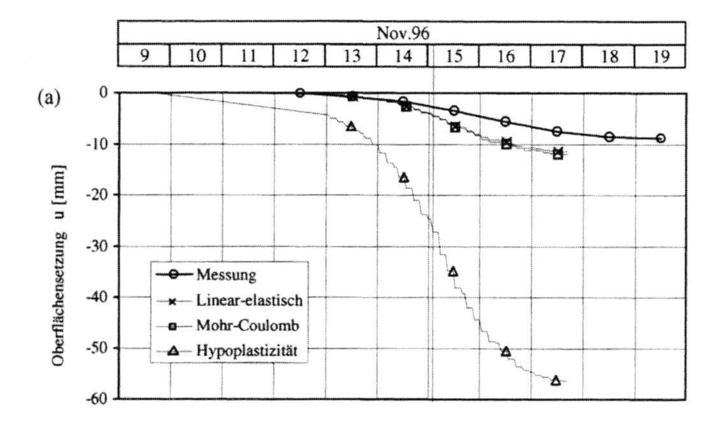

Figure 27: Calculated time-settlement curve from a 3D FE simulation of a tunnel excavation using different constitutive models [10].

the elasto-plastic model[4]! So why are the corresponding results of the FE analysis so disappointing?

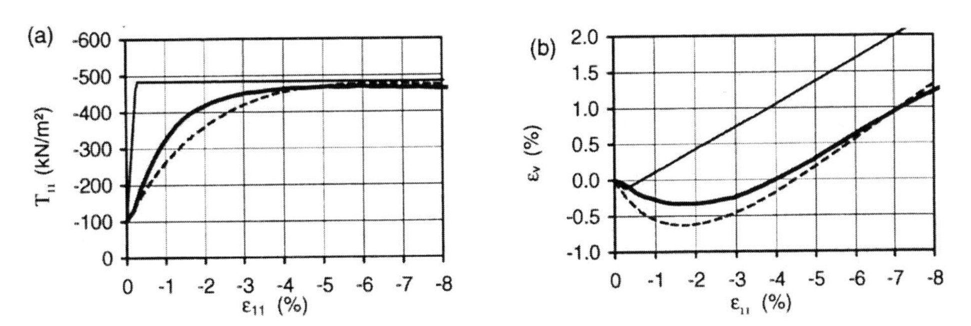

Figure 28: Recalculation of a triaxial test (thick full line) with Mohr-Coulomb (thin full line) and hypoplastic (dashed line) models [10].

The answer can be found if combining Fig. 27 and Fig. 28. The FE results suggest that the behaviour was almost elastic because the elastic and the Mohr-Coulomb solutions practically coincide. Thus, the strains in calculations remain very small — according to Fig. 28 elastic strains in the triaxial compression do not surpass 0.25%, otherwise plastic strains in the Mohr-Coulomb model would be initiated. However, the experimental curve in this small strain range shows an S-shape which is typical for bedding errors. The hypoplastic curve in Fig. 28 follows the unrealistic initial inclination of the experimental curve. Consequently, the chosen parameters of the hypoplastic model do not reflect the material behaviour in the strain range characteristic for the simulated problem (improper calibration).

[4]Young modulus of the Mohr-Coulomb model was determined from a single unloading branch in oedometer test.

5 Calculation benchmarks

Benchmarks are usually designed as numerical calculations of typical boundary value problems with relatively strict input specifications without need to calibrate constitutive models on experimental outputs. Considering FE simulations, the freedom is mostly limited to numerical aspects like the choice of mesh, time step or loading procedure. In most cases it is not purposed to compare calculation results with the real (measured) behaviour. Rather it is aimed to compare the calculation results to each other, in rare cases to analytical solutions.

Benchmarking is becoming increasingly popular in recent years. Compared to prediction competitions (where the calibration has to be done on the basis of experimental data and the obtained results are compared with in situ measurements), benchmarking is faster, simpler to evaluate and cheaper. However, it can never replace prediction competition's since it covers only a limited part of the simulation process leading to true predictions. Moreover, a kind of benchmarking, namely parametric studies or sensitivity analyses, should always accompany numerical simulations as they help detecting many sensitive factors.

5.1 INTERCLAY II project

The project was sponsored by the European Commission during 1991-1994 as a benchmark exercise dealing with numerical predictions of the mechanical behaviour of clay, especially concerning the long term features. Several major geotechnical companies and institutions were involved (for details, see the database of EU projects at [1]).

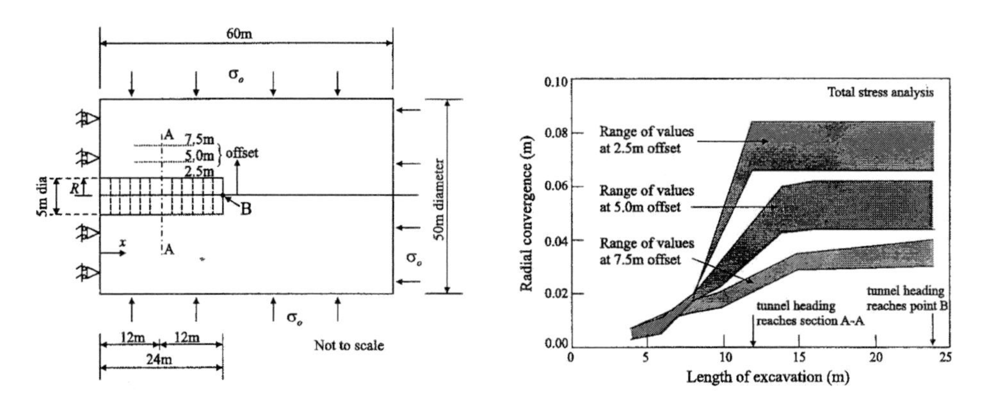

Figure 29: Advancing tunnel heading: results from the INTERCLAY II project [66]

"A comparison of the theoretical capabilities of the popularly used models and computer codes and their performance on simple, somewhat hypothetical problems" [1] belonged to the main tasks of the project. One of those hypothetical problems was the advancing tunnel heading. Purely cohesive incompressible soil characterized by Tresca soil model was assumed (E=240 MPa, μ=0.499 and s_u=1.0 MPa) [66]. Isotropic initial stress conditions (2500 kPa) including pore pressures, axisymmetric analysis and rigid permeable lining were prescribed (Fig. 29 left).

The results at Fig. 29 (right) reveal a large scatter in the predicted radial convergence although the constitutive model and stress conditions are rather simple.

5.2 Tunnel excavation (DGGT)

The Working Group 1.6 of the German Society for Geotechnical Engineering (DGGT) organized in the nineties a series of benchmarks with emphasis on plane-strain simulations of tunnel and deep excavations [73]. In case of the tunnel, the parameters of the linear elastic-perfectly plastic model (Mohr-Coulomb) were given, the initial state (without water) and the construction sequence were prescribed.

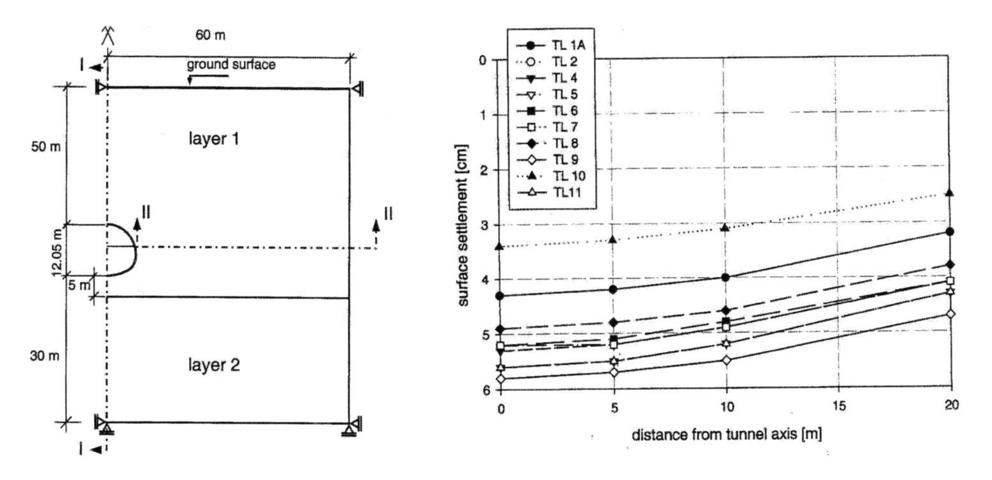

Figure 30: Surface settlements for the full face excavation [73]

The calculated surface settlement above the tunnel axis varies between 3.3 and 5.8 cm, see Fig. 30. When leaving aside two predictions with the lowest settlement, still there is about 20% difference between the predicted values. Even larger scatter was obtained for the calculated bending moments in the lining [73].

5.3 Tunnel excavation (COST)

As a part of the work done by the Working Group A of the COST C7 project, a relatively simple benchmark of tunnel analysis was performed [65]. Elastic and elastic-perfectly plastic solutions of a circular tunnel in the initially isotropic stress field were to calculate (Fig. 31). Elastic parameters (G=12 MPa and ν=0.495), undrained cohesion (s_u=130 kPa or 60 kPa, respectively) and full face excavation without and with lining, respectively, were prescribed. In case of the lining, a volume loss of 2% was given.

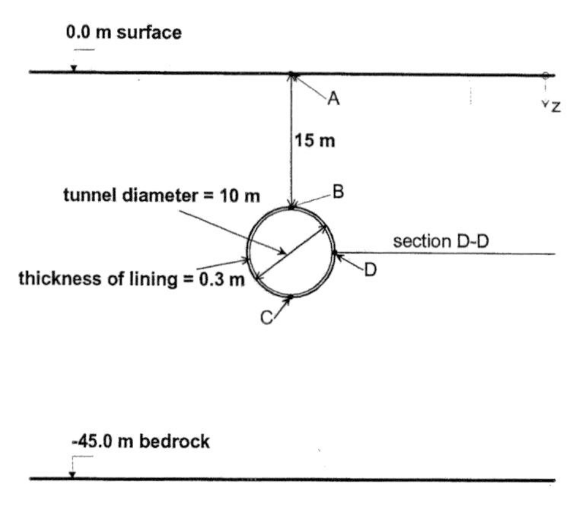

Figure 31: Excavation of a circular tunnel [65]

The vertical settlements obtained with elastic and elastic-perfectly plastic analyses are shown in Fig. 32. Considering first the elastic-perfectly plastic calculations (Fig. 32 left) and without taking into account two results (ST4 and ST9) lying far away from the others, there is about 15% difference in the value of the maximum settlement. This means that there was no substantial improvement in the calculated results compared to the DGGT tunnel benchmark (Fig. 30). Moreover, the linear elastic calculation (Fig. 32 right) revealed a similar scatter of the results which is very disappointing since one can find even an analytical solution for this problem [84]! Unfortunately, the differences cannot be attributed only to different sizes of the used FE meshes.

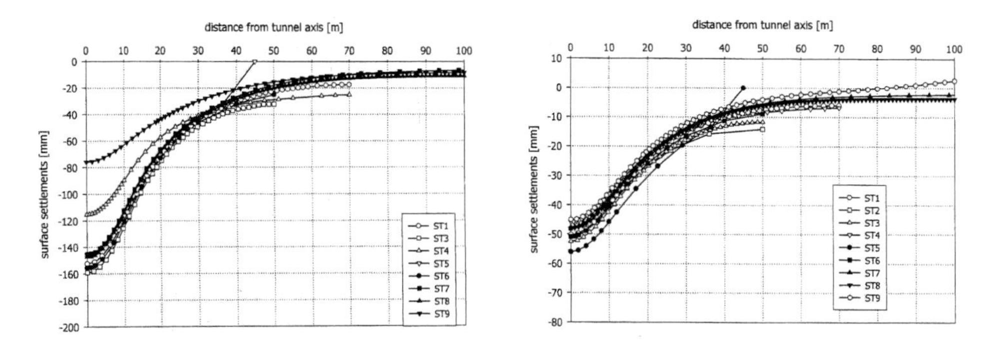

Figure 32: Surface settlements for the tunnel excavation: elastic-perfectly plastic (left) and linear elastic (right) model [65]

6 Conclusions

Contrary to analyses based on the practical experience only, numerical simulations enable predictions and extrapolations in new and complicated situations. They help above all understanding of processes in the ground. This task is more important than expecting exact values of stresses and strains from the simulation results. Numerical analyses are also closely linked to the observational method [62] widely used in geotechnical and tunnel engineering. The interaction between measurements, calculations and design modifications should make a closed feedback loop [67].

Constitutive models for continuum represent perhaps the most important part of the numerical simulations in soil and rock mechanics and tunnelling. There exists a large number of different approaches from simple (linear elasticity) to complex ones (hardening plasticity, hypoplasticity). Simple models are the most popular although they miss many important features of the ground behaviour. Consequently, such models can be only used for superficial analyses, parametric studies and benchmarking, but predictions or detailed studies are, in fact, impossible.

It is worth to use advanced models. Nevertheless, a good understanding of the constitutive modelling is needed for their successful application. Complex models do not automatically guarantee better results! Moreover, any model is valueless without a proper calibration. For routine engineering applications experts can develop design charts and "empirical" relationships based on the results from advanced models [76]. Analyses of the influence of surface structures on the settlement trough [64] or of the face stability in shallow tunnels [83] may serve as examples.

Benchmarks on tunnelling problems show that even for fixed constitutive models and parameters the numerical results can significantly scatter. Therefore, one should not be overconfident in any numerical results and a sound judgement based on knowledge and experience remains inevitable.

Acknowledgement

This work was supported by a *Marie Curie Fellowship* of the European Community programme *Improving Human Research Potential* under contract number HPMF-CT-2000-01108. Valuable comments by Prof. D. Kolymbas are highly appreciated.

Bibliography

[1] `http://www.cordis.lu`.

[2] T. Addenbrooke, D. Potts, and A. Puzrin. The influence of pre-failure soil stiffness on the numerical analysis of tunnel construction. *Géotechnique*, 47(3):693–712, 1997.

[3] J. Atkinson and D. Farrar. Stress path tests to measure soils strength parameters for shallow landslips. In *Proc. XI ICSMFE, San Francisco*, volume 2, pages 983–986, Rotterdam, 1985. A.A.Balkema.

[4] R. Baker. Inter-relations between experimental and computational aspects of slope stability analysis. *International Journal for Numerical and Analytical Methods in Geomechanics*, 27:379–401, 2003.

[5] G. Baldi, D. W. Hight, and G. E. Thomas. A reevaluation of conventional triaxial test methods. In R. T. Donaghe, R. C. Chaney, and L. S. Marshall, editors, *Advanced triaxial testing of soil and rock, ASTM STP 977*, pages 219–263. American Society for Testing and Materials, 1988.

[6] L. Barden. Stresses and displacements in a cross-anisotropic soil. *Géotechnique*, 11:198–210, 1963.

[7] E. Bauer. Zum mechanischen Verhalten granularer Stoffe unter vorwiegend ödometrischer Beanspruchung. Veröffentlichungen des Institutes für Bodenmechanik und Felsmechanik der Universität Fridericiana in Karlsruhe, 1992. Heft 130.

[8] T. Belytschko, Y. Krongauz, D. Organ, M. Fleming, and P. Krysl. Meshless methods: An overview and recent developments. *Journal of Physical Chemistry, American Chemical Society*, 139:3–47, 1996.

[9] P. Bésuelle, J. Desrues, and S. Raynaud. Experimental characterisation of the localisation phenomenon inside aVosges sandstone in a triaxial cell. *Int. J. of Rock Mechanics and Mining Sciences*, 37:1223–1237, 2000.

[10] C. Bliem. *3D Finite Element Berechnungen im Tunnelbau.*, volume 4 of *Advances in Geotechnical Engineering and Tunnelling*. Logos Verlag, Berlin, 2001.

[11] J. P. Boehler and A. Sawczuk. On yielding of oriented solids. *Acta Mechanica*, 27:185–206, 1977.

[12] J. Boháč, I. Herle, and D. Mašín. Stress and strain dependent stiffness in a numerical model of a tunnel. In *2nd Int. Conf. on Soil Structure Interaction in Urban Civil Engineering*, volume 2, pages 357–364, Zurich, 2002.

[13] M. Bolton, G. Dasari, and A. Britto. Putting small strain non-linearity into Modified Cam Clay model. In Siriwardane and Zaman, editors, *Computer Methods and Advances in Geomechanics*, pages 537–542. Balkema, 1994.

[14] B. Brady and E. Brown. *Rock Mechanics for Underground Mining*. Chapman & Hall, London, 2nd edition, 1994.

[15] R. B. J. Brinkgreve, editor. *Plaxis 2D — Version 8*. A.A.Balkema Publishers, 2002.

[16] E. Brown, J. Bray, B. Ladanyi, and E. Hoek. Ground response curves for rock tunnels. *Journal of Geotechnical Engineering Division ASCE*, 109(1):15–39, 1983.

[17] F. Calvetti, C. Tamagnini, and G. Viggiani. On the incremental behaviour of granular soils. In *NUMOG VIII*, pages 3–9. Swets & Zeitlinger, 2002.

[18] B. Cambou and C. Di Prisco, editors. *Constitutive modelling of geomaterials*. Hermes, 2000.

[19] R. Castellanza. Modelling weathering effects on the mechanical behaviour of bonded geomaterials. In *Constitutive Modelling and Analysis of Boundary Value Problems in Geotechnical Engineering*. Universitá degli Studi di Napoli Federico II, 2003. in print.

[20] R. Chambon, J. Desrues, W. Hammad, and R. Charlier. CLoE, a new rate-type constitutive model for geomaterials — Theoretical basis and implementation. *International Journal for Numerical and Analytical Methods in Geomechanics*, 18(4):253–278, 1994.

[21] N. D. Cristescu and U. Hunsche. *Time effects in rock mechanics.* Materials, Modelling and Computation. Wiley, 1998.

[22] F. Darve, editor. *Geomaterials: Constitutive equations and modelling.* Elsevier Applied Science, 1990.

[23] J. M. Duncan and C.-Y. Chang. Nonlinear analysis of stress and strain in soils. *Journal of the Soil Mechanics and Foundations Division ASCE*, 96(SM5):1629–1653, 1970.

[24] G. Elliott and E. Brown. Yield of a soft, high porosity rock. *Géotechnique*, 35(4):413–423, 1985.

[25] G. Exadaktylos and M. Stavropoulou. A closed-form elastic solution for stresses and displacements around tunnels. *Int. J. of Rock Mechanics and Mining Sciences*, 39:905–916, 2002.

[26] J. A. Franklin and M. B. Dusseault. *Rock Engineering.* McGraw-Hill, New York, 1989.

[27] R. Gibson. The analytical method in soil mechanics. *Géotechnique*, 24(2):115–140, 1974.

[28] M. Goldscheider. Dilatanzverhalten von Sand bei geknickten Vervormungswegen. *Mech. Res. Comm.*, 2:143–148, 1975.

[29] J. Graham and G. Houlsby. Anisotropic elasticity of a natural clay. *Géotechnique*, 33(2):165–180, 1983.

[30] G. Gudehus. A comprehensive constitutive equation for granular materials. *Soils and Foundations*, 36(1):1–12, 1996.

[31] Y. Guégen and V. Palciauskas. *Introduction to the physics of rocks.* Princeton University Press, Princeton, New Jersey, 1994.

[32] M. Gunn. The prediction of surface settlement profiles due to tunnelling. In *Predictive soil mechanics – Proc. Wroth memorial symposium*, pages 304–316. T.Telford, 1993.

[33] B. Haimson and C. Chang. A new true triaxial cell for testing mechanical properties of rock, and its use to determine rock strength and deformability of westerly granite. *Int. J. of Rock Mechanics and Mining Sciences*, 37:285–296, 2000.

[34] R. D. Hart. An introduction to distinct element modeling for rock engineering. In J. A. Hudson, editor, *Comprehensive rock engineering*, volume 2, pages 245–261. Pergamon Press, 1993.

[35] K. Hashiguchi. Macrometric approaches -static- intrinsically time-independent. In *Constitutive laws of soils, Proc. XI ICSMFE, Special report*, pages 25–65, San Francisco, 1985.

[36] I. Herle. Essentials of constitutive models for soils. In *Constitutive Modelling and Analysis of Boundary Value Problems in Geotechnical Engineering.* Universitá degli Studi di Napoli Federico II, 2003. in print.

[37] I. Herle. Numerical predictions and reality. In D. Kolymbas, editor, *GeoMath — Geomechanics and Mathematics.* Springer, 2003. in print.

[38] I. Herle, T. Doanh, and W. Wu. Comparison of hypoplastic and elastoplastic modelling of undrained triaxial tests on loose sand. In D. Kolymbas, editor, *Constitutive Modelling of Granular Materials*, pages 333–351, Horton, 2000. Springer.

[39] I. Herle and G. Gudehus. Determination of parameters of a hypoplastic constitutive model from properties of grain assemblies. *Mechanics of Cohesive-Frictional Materials*, 4(5):461–486, 1999.

[40] I. Herle and K. Nübel. Hypoplastic description of the interface behaviour. In G. Pande, S. Pietruszczak, and H. Schweiger, editors, *Int. Symp. on Numerical Models in Geomechanics, NUMOG VII*, pages 53–58, Graz, 1999. A.A.Balkema.

[41] E. Hoek and E. T. Brown. Empirical strength criterion for rock masses. *Journal of Geotechnical Engineering ASCE*, 106(9):1013–1035, 1980.

[42] G. Houlsby. How the dilatancy of soils affects their behaviour. In *Proc. 10th ECSMFE, Florence*, volume 4, pages 1189–1202. A.A.Balkema, 1991. Theme lecture.

[43] R. J. Jardine, D. M. Potts, A. B. Fourie, and J. B. Burland. Studies of the influence of non-linear stress-strain characteristics in soil-structure interaction. *Géotechnique*, 36(3):377–396, 1986.

[44] F.-S. Jeng, M.-C. Weng, T.-H. Huang, and M.-L. Lin. Deformational characteristics of weak sandstone and impact to tunnel deformation. *Tunnelling and Underground Space Technology*, 17:263–274, 2002.

[45] L. Jing. A review of techniques, advances and outstanding issues in numerical modelling for rock mechanics and rock engineering. *Int. J. of Rock Mechanics and Mining Sciences*, 40:283–353, 2003.

[46] M. Karakus and R. Fowell. FEM analysis for the effects of the NATM construction technique on settlement above shallow soft ground tunnels. In *GeoEng*, Melbourne, 2000. CD-ROM.

[47] Y.-S. Kim, F. Tatsuoka, and K. Ochi. Deformation characteristics at small strains of sedimentary soft rocks by triaxial compression tests. *Géotechnique*, 44(3):461–478, 1994.

[48] D. Kolymbas. The misery of constitutive modelling. In D. Kolymbas, editor, *Constitutive Modelling of Granular Materials*, pages 11–24, Horton, 2000. Springer.

[49] D. Kolymbas. Tunnelling mechanics. In D. Kolymbas, editor, *Tunnelling Mechanics (Eurosummerschool, Innsbruck 2001)*, volume 5 of *Advances in Geotechnical Engineering and Tunnelling*, pages 1–85. Logos Verlag, Berlin, 2002.

[50] D. Kolymbas and E. Bauer. Soft oedometer – a new testing device and its application for the calibration of hypoplastic constitutive laws. *Geotechnical Testing Journal*, 16(2):263–270, 1993.

[51] D. Kolymbas, M. Mähr, and I. Herle. Tail gap grouting. In *Urban underground space: a resource for cities*, Torino, 2002. Associazione Georisorse e Ambiente.

[52] D. Kolymbas and W. Wu. Recent results of triaxial tests with granular materials. *Powder Technology*, 60:99–119, 1990.

[53] D. Kolymbas and W. Wu. Introduction to hypoplasticity. In D. Kolymbas, editor, *Modern approaches to plasticity*, pages 213–223. Elsevier, 1993.

[54] R. Kondner. Hyperbolic stress-strain response: cohesive soils. *Journal of the Soil Mechanics and Foundations Division ASCE*, 89(SM1):115–143, 1963.

[55] H. Konietzky, R. Hart, and D. Billaux. Mathematische Modellierung von geklüftetem Fels. *Felsbau*, 12(6):395–400, 1994.

[56] P. V. Lade. Elasto-plastic stress-strain theory for cohesionless soil with curved yield surfaces. *International Journal of Solids and Structures*, 13:1019–1035, 1977.

[57] J. Lanier, C. Di Prisco, and R. Nova. Étude expérimentale et analyse théorique de l'anisotropie induite du sable d'Hostun. *Rev. Franç. Géotech.*, (57):59–74, 1991.

[58] K. Lee and R. Rowe. Deformations caused by surface loading and tunnelling: the role of elastic anisotropy. *Géotechnique*, 39(1):125–140, 1989.

[59] H. Mroueh and I. Shahrour. A full 3-d finite element analysis of tunneling-adjacent structures interaction. *Computers and Geotechnics*, 30:245–253, 2003.

[60] T. Ogawa and K. Lo. Effects of dilatancy and yield criteria on displacements around tunnels. *Canadian Geotechnical Journal*, 24:100–113, 1987.

[61] J. Ohde. Zur Theorie der Druckverteilung im Baugrund. *Bauingenieur*, 20:451–459, 1939.

[62] R. Peck. Advantages and limitations of the observational method in applied soil mechanics. *Géotechnique*, 19(2):171–187, 1969.

[63] D. Pickering. Anisotropic elastic parameters for soil. *Géotechnique*, 20(3):271–276, 1970.

[64] D. Potts and T. Addenbrooke. A structures influence on tunnelling-induced ground movements. *Geotechnical Engineering, ICE*, 125:109–125, 1997.

[65] D. Potts, K. Axelsson, L. Grande, H. Schweiger, and M. Long, editors. *Guidelines for the use of advanced numerical analysis*. Thomas Telford, 2002. COST C7 WGA report.

[66] D. M. Potts and L. Zdravković. *Finite element analysis in geotechnical engineering — Application*. Thomas Telford., London, 2001.

[67] A. Powderham. An overview of the observational method: development in cut and cover and bored tunnelling projects. *Géotechnique*, 44(4):619–636, 1994.

[68] A. Puzrin and J. Burland. Non-linear model of small-strain behaviour of soils. *Géotechnique*, 48(2):217–234, 1998.

[69] T. Ramamurthy. Shear strength response of some geological materials in triaxial compression. *Int. J. of Rock Mechanics and Mining Sciences*, 38:683–697, 2001.

[70] C. Sagaseta. Analysis of undrained soil deformation due to ground loss. *Géotechnique*, 37(3):301–320, 1987.

[71] I. S. Sandler, F. L. DiMaggio, and G. Y. Baladi. Generalized cap model for geological materials. *Journal of Geotechnical Engineering Division ASCE*, 102(GT7):683–699, 1976.

[72] H. Schuller and H. F. Schweiger. Application of multilaminate model for shallow tunnelling. *Felsbau*, 17(1):44–47, 1999.

[73] H. F. Schweiger. Results from two geotechnical benchmark problems. In A. Cividini, editor, *Proc. 4th European Conf. Numerical Methods in Geotechnical Engineering*, pages 645–654. Springer, 1998.

[74] H. F. Schweiger, M. Kofler, and H. Schuller. Some recent developments in the finite element analysis of shallow tunnels. *Felsbau*, 17:426–431, 1999.

[75] S. Sharan. Elastic-brittle-plastic analysis of circular openings in Hoek-Brown media. *Int. J. of Rock Mechanics and Mining Sciences*, 40:817–824, 2003.

[76] B. Simpson. Engineering needs. In M. Jamiolkowski, R. Lancellotta, and D. Lo Presti, editors, *Proc. 2nd Int. Symp. on Pre-failure Deformation Characteristics of Geomaterials, Torino*, volume 2, pages 1011–1026, Lisse, 2001. Swets & Zeitlinger.

[77] T. Sitharam, J. Sridevi, and N. Shimizu. Practical equivalent continuum characterization of jointed rock masses. *Int. J. of Rock Mechanics and Mining Sciences*, 38:437–448, 2001.

[78] M. Souley, F. Homand, and A. Thoraval. The effect of joint constitutive laws on the modelling of an underground excavation and comparison with in situ measurements. *Int. J. of Rock Mechanics and Mining Sciences*, 34:97–115, 1997.

[79] C. Tamagnini and G. Viggiani. On the incremental non-linearity of soils. Part I: theoretical aspects. *Rivista italiana di geotecnica*, 36(1), 2002. 44-61.

[80] C. Tamagnini, G. Viggiani, and R. Chambon. A review of two different approaches to hypoplasticity. In D. Kolymbas, editor, *Constitutive modelling of granular materials*, pages 107–145. Springer, 2000.

[81] F. Tatsuoka and Y. Kohata. Stiffness of hard soils and soft rocks in engineering applications. Report No.242 5, Institute of Industrial Science, University of Tokyo, 1995.

[82] C. Thornton and J. Lanier. Uniaxial compression of granular media: Numerical simulations and physical experiment. In Behringer and Jenkins, editors, *Powders and Grains*, pages 223–226. A.A.Balkema, 1997.

[83] P. A. Vermeer and N. Ruse. Die Stabilität der Tunnelortsbrust in homogenem Baugrund. *Geotechnik*, 24(3):186–193, 2001.

[84] A. Verruijt and J. R. Booker. Surface settlements due to deformation of a tunnel in an elastic half plane. *Géotechnique*, 46(4):753–756, 1996.

[85] W. Wu. Rational approach to anisotropy of sand. *International Journal for Numerical and Analytical Methods in Geomechanics*, 22:921–940, 1998.

[86] W. Wu and P. O. Roony. The role of numerical analysis in tunnel design. In D. Kolymbas, editor, *Tunnelling Mechanics (Eurosummerschool, Innsbruck)*, volume 5 of *Advances in Geotechnical Engineering and Tunnelling*, pages 87–168. Logos Verlag, Berlin, 2001.

[87] O. C. Zienkiewicz. The era of computational mechanics: Where do we go now?. *Meccanica*, 36:151–157, 2001.

[88] O. C. Zienkiewicz, C. Humpheson, and R. W. Lewis. Associated and non-associated visco-plasticity and plasticity in soil mechanics. *Géotechnique*, 25(4):671–689, 1975.

SCL Tunnel Design in Soft Ground
-insights from monitoring and numerical modelling

Chris Clayton[1], Alun Thomas[2], Pierre van der Berg[3]

[1] University of Southampton, United Kingdom
e-mail: c.clayton@soton.ac.uk

[2] Mott MacDonald, United Kingdom
e-mail: aht@mm-croy.mottmac.com

[3] Jones and Wagener, South Africa
e-mail: vdberg@jaws.co.za

Abstract:

The typical design process for sprayed concrete lined (SCL) tunnels in the UK is reviewed. Methods of field monitoring and 3-dimensional numerical modelling are examined in the light of recent experience, development and research on SCL tunnelling in the UK.

1. Introduction

The Euro Summerschool on Rational Tunnelling is intended to provide access to "the latest thinking and processes" associated with tunnelling. This paper has therefore been written with a view to the future rather than the past.

Soft ground tunnelling in the UK has traditionally been based on closed-form analytical solutions, experience and empiricism, using hand dig or TBMs, and segmental linings. The introduction of sprayed concrete lined (SCL) tunnelling into the UK has not been without difficulty - the collapse of tunnels at Heathrow Airport [15] and the consequent delays to Heathrow Express and the Jubilee Line Extension caused major problems for the tunnelling industry. Despite this, there is a considerable awareness of the economic potential for SCL (also termed NATM) tunnelling, and of the important role that tunnelling is likely to have as a part of future urban infrastructure – SCL tunnelling features prominently in the design of the proposed Crossrail project in London. Most of the UK's major cities are built on soils rather than rocks. Although this paper draws on experience of tunnelling in the London Clay Formation at Heathrow, to the west of London, the lessons are broadly applicable.

An ability to understand and analyse the interaction between a shotcrete tunnel shell and the ground around it is central to the rational and safe exploitation of the SCL tunnel technique. Whilst empirical design techniques and simple hand calculations provide a sensible background to the design of tunnels, they must be seen as potentially conservative and limiting, especially when introducing new construction techniques or complex tunnel geometries. These approaches may be unsafe if they ignore / overlook potentially critical aspects of behaviour – e.g. heterogeneity in the ground, adjacent structures and compensation grouting. Realistic insights into the behaviour of SCL tunnelling require the use of a wide range of techniques, including advanced analysis and monitoring.

2. UK design

This section provides a brief introduction to the design of SCL tunnel linings in soft ground in the UK. A number of issues are of importance in urban tunnel design:-

- Tunnel and face stability needs to be ensured, not only for the safety of tunnel workers, but also to prevent damage to overlying buildings;
- Ground movements need to be sufficiently small that they can be tolerated by overlying buildings and adjacent infrastructure (e.g. other tunnels);
- The sprayed concrete shell should be of sufficient strength and thickness to ensure that it can sustain the ground loading that can be expected at all stages of its life;
- The shell geometry should be easy to construct, in order to minimise the potential for poor construction quality.

To assess the overall stability of a tunnel, expected conditions are compared with a critical stability ratio (N_{TC}), which for SCL tunnels (where the tunnel pressure is zero) is equal to the total vertical stress at the tunnel axis divided by the undrained shear strength. Values of N_{TC} have been obtained from experiment or case studies (e.g. Broms and Bennermark [7], Peck [30], Mair [22], Kimura and Mair [18]), from plasticity solutions (Davis *et al.* [13]), and from numerical modeling (Sloan and Assadi [34]). Based on a range of data, Mair [23] proposed a design chart that has been widely used for lined tunnel headings with thin clay cover.

The UK design approach is based on Eurocode 7. A moderately conservative estimate of the undrained shear strength can be obtained from the results of the site investigation. This value is reduced by a factor of 1.4 and used in the N calculation. The estimated N value must be less than the design line. A similar approach can be applied to granular materials using the analytical solutions proposed by Leca and Dormieux [19].

As with tunnel stability, a range of methods are available for calculating ground movements around tunnels. In familiar soil conditions surface settlements are probably best estimated using empirical methods. Martos [25] first proposed the use of a Gaussian distribution curve to describe the transverse settlement profile above mine workings, and many others (notably Peck [30]) have shown that this does indeed provide a good fit with field observations. Attewell and Woodman [2] extended this approach to allow the prediction of the full surface settlement profile.

Calculation of surface settlements requires *a priori* that the surface settlement trough width (which depends upon the parameter *i*) and the volume loss (which defines the maximum settlement) are known or estimated. In familiar ground conditions the parameter *i* is an approximately linear function of tunnel axis depth, and can be determined from O'Reilly and New's [28] survey of UK tunnelling data. Approximately, it equals one-half of the tunnel axis depth, and is insensitive to tunnel construction method and (for tunnels with cover of more than one diameter, diameters smaller than 5m and depths not exceeding 30m) tunnel diameter.

It is common practice to use the Relative Volume Loss (the volume of surface settlement trough divided by the volume of excavated tunnel, normally expressed as a percentage) to define the surface settlement profile. Mair and Taylor [24] state that volume loss depends principally on ground conditions and upon the tunnelling method, including the details of tunnel construction, whilst Macklin [21] has proposed a relationship between volume loss and stability number. The literature (O'Reilly and New [28], Attewell et al. [3], New and Bowers [27]) suggests that in stiff over-consolidated clays such as the London Clay relative volume loss will be between 1% and 2%.

Whilst empirical methods of support design are available for rock tunnels (Barton *et al.* [4], Bieniawski [5]) soft ground tunnels are generally designed using a combination of analytical solutions (Muir Wood [25], Curtis [11], Panet and Guenot [29]) and 2-D numerical modelling. Recent guidance is provided in the Institution of Civil Engineers' recommendations [17], and the forthcoming British Tunnelling Society's Lining Design Guide [6]. Analytical methods are relatively simple to implement, and provide valuable estimates of the stress in the lining and its likely deformation. But they typically assume plane strain and axi-symmetry, and are developed for circular tunnels with full-face excavation in homogeneous ground, making no allowance for stress redistribution ahead of the face. 2-D numerical models (such as the popular Hypothetical Modulus of Elasticity approach) must rely on empirical correction factors to allow for arching and stress redistribution ahead of the face, and the three-dimensional nature of the tunnelling process.

They can therefore be similarly criticised as unlikely to capture the influences on real tunnels, but in addition are much more complex to implement and require proper validation (HSE [15]). 3-D numerical modelling increases the complexity of the computation once again, and therefore at this time is probably out of the reach of all but the most sophisticated design practices.

Lining design is normally based on a fairly simple approach. The loads within the lining in the hoop direction are estimated either using "closed-form" analytical models or 2D numerical models. Moderately conservative soil parameters are used in the design models. These loads are then increased by a partial factor of safety (e.g. 1.40), according to standard concrete design codes, and compared with the capacity of the lining, adjusted with a material factor of safety. Interaction diagrams are used for the comparison of loads and capacity. On the basis of engineering judgement the loads in the temporary case before the closure of the ring are often not checked because they are short-term. In the case of these soft ground tunnels the ring is usually closed within half a tunnel diameter of the face. However most tunnel collapses occur close to or at the face. 3D numerical models have revealed the complete "stress history" of the tunnel lining – from the face back to the closed ring. Simple analyses often predict loads that exceed the strength of the young shotcrete.

More problematic than the design of basic tunnel linings is the design of tunnel junctions. Here there are no empirical design charts or analytical solutions to guide the designer. Crude estimates can be made using an analytical solution for a hole in an infinite flat elastic plate under stresses. Numerical analyses often predict high concentrations of compressive stresses at the axis level of the opening in the tunnel and concentrations of tension above and below the opening. The application of normal design codes then leads to very heavy reinforcement around the tunnel junction. Experience suggests that the high predicted stresses are not present around real tunnel junctions.

Since the Heathrow collapse the distinction in the UK between the primary (sprayed concrete) lining, which was treated as temporary, and the secondary (cast concrete) permanent lining has been abandoned. Primary linings must now be designed with the same rigour and factors of safety as secondary linings.

3. Monitoring

In the UK there is some reluctance to label SCL tunnels in soft ground as 'NATM' because of our design and construction philosophy. Sprayed concrete lined (SCL) tunnels in the UK are 'fully designed' – that is to say, the details (e.g. thickness of the shell) and the construction sequence, as well as any anticipated variations / contingencies, are predetermined. Advance lengths and support measures may be varied on site by the designer, depending upon the observed ground conditions at the time of construction, but these variations are expected and allowed for. It is not expected that construction will stray beyond the boundaries of what has been designed unless the results of monitoring suggest that the design or construction method has for some unforeseen reason proved inadequate. Monitoring is therefore used by the tunnel designer to check that the tunnel is performing in line with the design assumptions. If this is not the case then either the design assumptions are incorrect, in which case the design must be revised, or construction is not being carried out to the design.

In contrast, where compensation grouting is used, surface or building settlement measurements provide feedback on its effectiveness, and allow progressive grouting and adaptation of the timing, distribution and volume of grout injection. This process permits the movements of overlying buildings as a result of tunnelling to be reduced to acceptable values. Monitoring is used to control the compensation grouting process, in an observational approach more usually associated with NATM.

There appears to be widespread scepticism in the UK industry about the value of all but the simplest monitoring. In this section we examine experience at Heathrow Express Terminal 4 station, with a view to determining:-

- The potential value of the monitoring data;
- Which instruments have performed well, in terms of survival rates and quality of data;
- The observed pattern of behaviour for some tunnels.

3.1 The potential value of monitoring data

We have already noted two important functions of monitoring, namely as part of a 'design-as-you-construct' system (such as NATM), and for checking that fully designed tunnel systems are performing satisfactorily. The first approach, whilst potentially leading to economy, can be risky, particularly for construction of shotcrete shells in soft ground where convergence is not necessarily associated with large reductions in ground loading. The second is often seen as wasteful, particularly since tunnel monitoring is expensive and can be disruptive. If things go well during construction there can appear to be little in the way of benefit in compensation for the cost, time and inconvenience involved in monitoring. The worst of all outcomes is when extensive instrumentation is installed and monitored, but data are not adequately evaluated in time to prevent problems. This has been suggested to have been the case for the Heathrow CTA collapse [15].

There is also the considerable danger when constructing SCL tunnels, that monitoring instrumentation will fail to provide the necessary warning that ground conditions are deviating from what was expected. The UK Health and Safety Executive's review of the safety of NATM tunnels [16] found that the most frequent reported causes of tunnel collapse could be attributed to unexpected ground conditions at the tunnel heading, i.e. at a stage before most instrumentation would have been installed.

Instrumentation can also be used to give confidence in the performance of new construction materials or techniques, to support less conservative design (as in the Observational Method [31]), and to provide data for the development of new prediction and design methods. And it provides essential data for the validation of analyses, for example using finite element or finite difference methods.

Where monitoring is used to verify that the tunnel is performing as designed it is important to focus the efforts of the monitoring to achieve this goal. The danger with modern real-time automatic monitoring is that the construction team can be overwhelmed with irrelevant data. The monitoring system should be designed to be simple, relevant and robust.

During construction the results from the monitoring should be judged against a set of criteria determined from the design. These criteria are often termed "Key Performance Indicators" (KPIs) and typically take the form of a hierarchy of limits on parameters such as surface settlement or in-tunnel deformations. If a key parameter – for example, lining deformation – exceeds a "Trigger level", this means that the tunnel is not performing as well as expected. This will "trigger" a set of predefined actions, such as a review by the construction team of the causes and potential remedial measures or an increase in the frequency of monitoring. If the parameter exceeds the "Limit level", then corrective actions must be taken. Beyond this there may be a final "Alarm level" and if this is exceeded, the tunnel is evacuated and an emergency plan is implemented. Limits for KPIs are set for absolute values and trends and are often related to the relative position of the active tunnelling face. A robust management system is required to ensure that the KPIs are reviewed at a daily meeting involving all the relevant people and that corrective actions are taken if levels are exceeded. This was one of the key lessons from the collapse at Heathrow in 1994.

This system of reviewing monitoring data represents the umbilical cord that connects the growing construction with its design. Information flows both ways through the system to ensure healthy construction progress. In this way informed decisions can be made on corrective actions to address inadequate performance of the tunnel or on improvements in the support measures to suit the prevailing ground conditions. It is good practice in all construction work for designers and constructors to maintain an open dialogue but this is crucial to the success of SCL tunnelling.

3.2 Performance of different types of instruments

A very large range of instrumentation and monitoring techniques is available for tunnelling, allowing the measurement of:

- Displacements and convergence of the tunnel;
- Changes in ground levels and building levels above the tunnel;
- Vertical and horizontal displacements in the soil around the tunnel;
- Stresses in the shotcrete shell, and between the tunnel shell and the surrounding soil;
- Pore pressures in the soil.

All of these have been used, with varying degrees of success, in recent tunnel construction in the UK. In addition, one should not neglect the important contribution that good qualitative engineering observation, by inspectors / engineers, can make if the process is properly structured and managed.

A survey of instrumentation has been carried out (Clayton et al.[9], [10], van der Berg et al. [37]) using the data obtained from the station tunnels at Heathrow's Terminal 4 (Figure 1). A typical section through the instrumentation is shown in Figure 2.

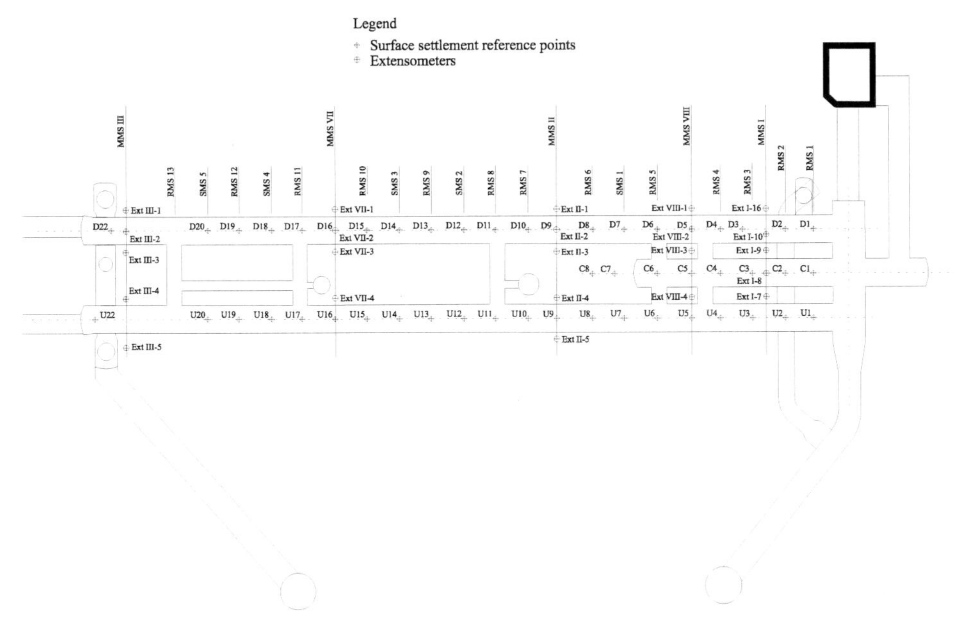

Figure 1: Plan view of Heathrow Terminal 4 station showing the platform tunnels and concourse tunnl instrumentation layout.

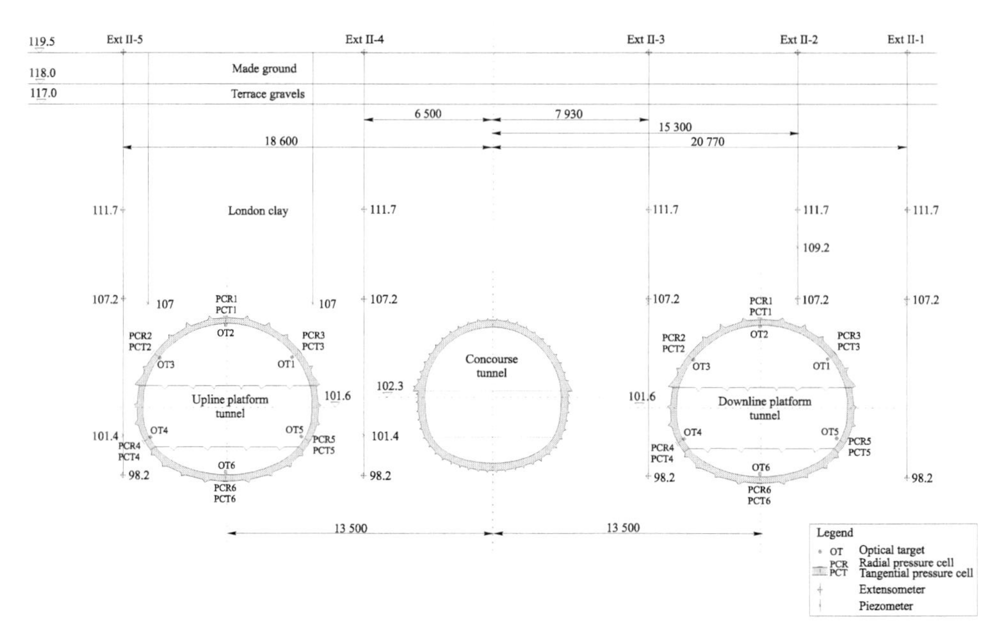

Figure 2: Cross section through Heathrow Terminal 4 station and concourse tunnels showing instrumentation

The conclusions from this work were that:

- Although tape extensometers have traditionally been used to monitor tunnel convergence, in recent years (as at Terminal 4) total stations and optical surveying have been dominant. Both methods suffer from the fact that measurements cannot be made until some distance back from the face, although optical surveying is potentially the more powerful in SCL tunnelling in this respect, since it does not rely on the existence of pairs of measuring points. For the monitoring of tunnel convergence, tape extensometers offer much greater precision and accuracy than total stations and optical targets (Dunnicliff and Green, [12]). However, given the pace of work in a tunnel heading, there are practical and safety problems associated with such systems.

- Whilst optical systems continue to improve, experience at Terminal 4 was, in hindsight, disappointing. Although in ideal conditions a positional precision of +/-1 mm was achieved, under most working conditions the actual precision was found to be as poor as +/-3 mm over distances not exceeding 30m, meaning that convergence as large as 6mm might under unfavourable conditions be difficult to detect. Although a well-constructed lining should have maximum displacements of similar magnitude as this, optical convergence measurements can still be expected to warning of a defective, and therefore more compliant, lining (for example see the HSE report on the failure at Heathrow's Central Terminal Area [15].

- Procedures for monitoring surface settlements are well established. The precision and repeatability of precise levelling at Terminal 4 was generally of the order of +/-0.5 mm, and was seldom worse than +/-1.0 mm. Given that trigger levels for compensation grouting may be of the order of 10 mm, it was found that this method of monitoring was more than adequate for its intended purpose.

- Single or multi-point extensometers can give excellent precision in terms of vertical changes in distance between (for example) the surface and a monitored point in the ground below. But in practice their ability to measure absolute positional change is limited to that of precise levelling, since this is required to determine the level of the top of the instrument.

- Vertical and horizontal displacements can be measured along vertical, horizontal and inclined lines using inclinometers, electro-levels and chain deflectometers. We have found inclinometers and chain deflectometers, placed ahead of the advancing face, to be useful in uniform ground for assessing the contribution of face take to surface settlement [39]. There has also been quite widespread use of electro-levels in the UK, for example in monitoring the effects of tunnelling on existing adjacent tunnels and buildings. Our experience is that when inclinometers are used with care (for example, with the same instrument used by the same operator, with exactly the same routine for each set of readings) the repeatability of measurements of horizontal movements down a 25m vertical borehole can be very good, and certainly better than +/2 mm. If the instruments are left in the holes between readings, chain deflectometers can be expected to have rather better repeatability than this. Change in horizontal and vertical position can be measured with a repeatability better than 1mm. The grouting of instrumentation casing for inclinometers and chain deflectometers needs some care, in order to ensure good fixity of the measuring system in the soil.

- In the UK stress is generally considered extremely difficult to measure. Although it is somewhat easier to measure the stress at the boundary between construction and soil than within soil or within shotcrete (Clayton et al.[10]), lack of attention to detail during installation frequently seems to lead to poor performance of both boundary (radial) and embedded (tangential) pressure cells.

- For boundary (radial) cells to work the cell should be significantly stiffer than the surrounding soil – Clayton and Bica. [8] give guidance on the design of such cells to provide an adequate cell action factor. When monitoring shotcrete shells it is also necessary to limit bedding and the disturbance to the soil against which the cell is to operate - practical details such as the provision of box-outs, and the use of wet grout as bedding are important.

- Embedded (tangential) cells are affected by numerous factors. The cell action factor will be made ineffective if blind zones during shotcreting allow the formation of cavities beside the sensing face ("shadowing"), crimping results and temperature effects must be taken into account, and shotcrete shrinkage has been shown to lead to significant increases in the measured stress [10].
- Finally, the measurement of pore pressure is not without its complications. SCL tunnels proceed at a relatively rapid rate, and therefore piezometer response time is an important issue. Vibrating wire and pneumatic piezometers can provide a good response time if properly de-aired and placed below the water table, but only provided that the stress relief and any changes in groundwater regime caused by tunnelling do not lead to low or negative pore pressures. It is unlikely that a hydrostatic groundwater regime will exist during (and perhaps after) tunnelling, and in this case the proper interpretation of (necessarily limited) piezometer data will require some estimation of both anisotropy and flow regime.

3.3 Instrument survival

If monitoring is to play a key role in tunnel construction then it is essential that sufficient instruments are placed in and around the tunnel to provide the necessary data, that the data are collected sufficiently often, and that interpretation of the data is fast and expertly carried out. In order to ensure that sufficient instruments are installed, the designer must make estimates of the required instrument spacing, and modify this in the light of expected losses. Most modern instrumentation is rugged, and losses are generally small, except when cabling is severed. Although they may survive, stress measurements often remain unconvincing. Convergence points are vulnerable to plant movements.

A survey by van der Berg [40] gives an overall indication of the survival of the optical targets used for tunnel convergence monitoring in the tunnels at Heathrow Terminal 4. Convergence monitoring sections were installed every 10m down the station tunnels, with 3 optical targets in the top heading, 2 in the bench and 1 in the invert:

- Of the 68 targets installed in the top heading, meaningful results were obtained at 45 (approximately 66%);

- Of the 40 targets installed in the bench, meaningful results were obtained at 21 (approximately 50%);
- Of the 16 targets installed in the invert only 2 gave meaningful results. It is almost impossible to protect targets in this position from the effects of plant moving on the backfilled invert;
- At only 1 of the 24 monitoring sections was it possible to use the results from all the targets. If invert targets are excluded, there were still only 2 satisfactory sections. Only 7 sections (<30%) had 4 surviving targets.

The HSE report on the collapse of NATM tunnels at Heathrow [15] states that Monitoring is vital to determine the behaviour of the lining and ground. Unsatisfactory trends must be identified sufficiently early to enable corrective steps to be taken. ... Direct measurement of movement on a tunnel's cross-section is the most vital (method of measurement) and provides the most readily interpretable data'. In the light of the above, such low survival rates must be of great concern.

3.4 Observed behaviour

Based on the complete monitoring data set for the station platforms, and for special instrumentation installed on the concourse tunnel, general trends of movement have been established for SCL tunnels in London Clay at Heathrow Terminal 4. These are described in more detail in van der Berg *et al.* [38, 39, 40, 41]. The patterns observed are in line with those seen on other recent projects in the London Clay, and in similar ground elsewhere. The typical sequence of tunnel excavation is illustrated in Figure 3. A layout of the tunnels and a cross section through a monitoring section are shown in Figures 1 and 2. The clay is heavily overconsolidated, with a coefficient of earth pressure at rest estimated as considerably above unity. Examples of data obtained are given below.

Displacements of the top heading result from three components, namely convergence, ovalisation, and settlement. The position of the optical targets is shown in Figure 2, and typical values in Table 1. The data show that the top heading typically underwent a settlement of some 8 mm, the remaining 1.5 mm being due to ovalisation (in the vertical direction) and convergence. The lining in the bench, and as far as can be deduced in the invert, showed very little deformation.

These movements are very small in comparison with those typically allowed in Alpine NATM tunnels. But it is essential to restrict tunnel convergence in soft ground, not only to protect overlying structures, but also because the rate of soil stiffness reduction with increasing strain means that ongoing movements may be associated with increasing, rather than decreasing, loads on the shell. Trends in convergence readings tend to be important – movement should be expected to cease within about 2 diameters after the closure of the ring, provided that the shell has been well constructed.

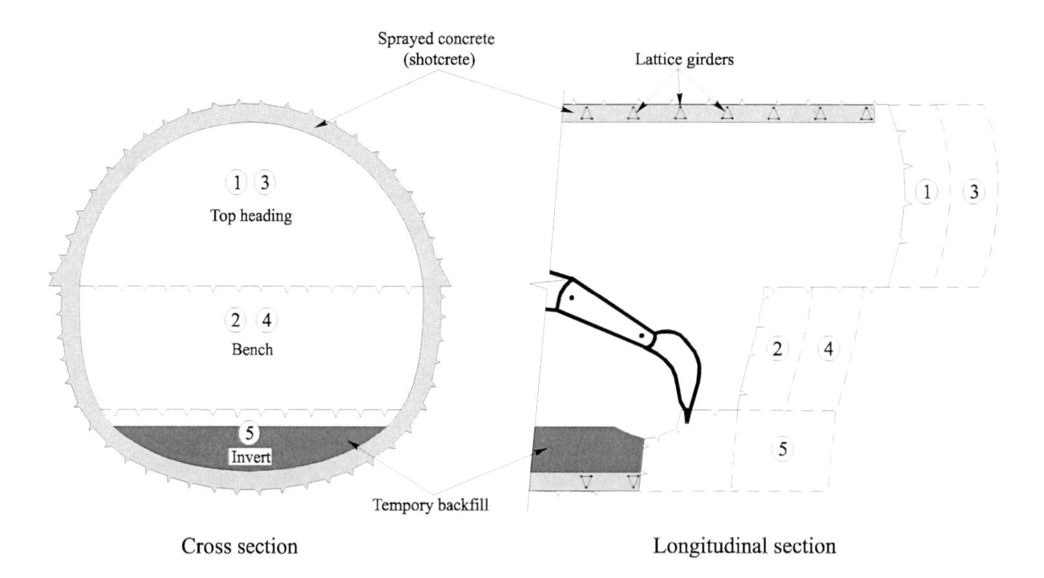

Figure 3: Tunnel excavation sequence

Target	Vertical displacement (mm)	Transverse displacement (mm)	Longitudinal displacement (mm)
2	-8.4	0	2.8
3	-9.8	-4.8	-0.6
1	-9.5	4.6	-0.5
5	-1.5	-0.8	-0.2
4	-2.4	1.8	-0.5

Table 1: Typical displacements measured at optical targets – Heathrow Terminal 4 Station tunnels

Horizontal displacements in the direction of tunnelling ('longitudinal displacements', contributing to face take) were measured during the construction of the concourse tunnel, using inclinometers and deflectometers. Examples of the data obtained are shown in Figures 4 and 5. Integration of the data and extrapolation showed that the maximum horizontal ground displacement at the tunnel face was of the order of 25-30 mm. Face take accounted for 0.45% of the excavated tunnel volume. The volume loss estimated from surface settlements was about 0.8% of tunnel volume.

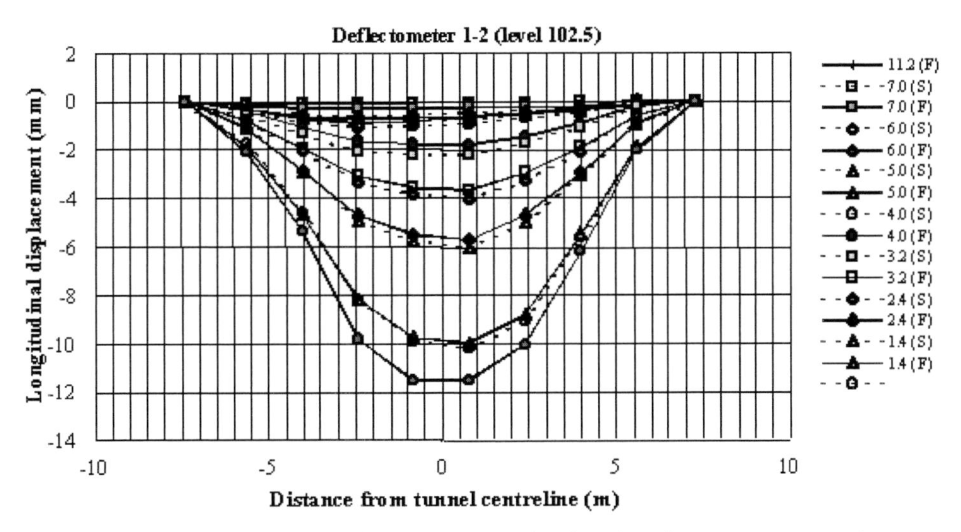

Figure 4: Longitudinal movement (face take) caused by the advancing concourse tunnel, as measured by a deflectometers chain.

4. Numerical modelling

Despite the apparent optimism displayed by many analysts, numerical modelling is not straightforward. However, whilst some practising engineers believe that computer analyses will never be able to provide reasonable predictions of tunnel behaviour, we see it as an essential part of future design. The simple analytical and empirical techniques currently used in design leave many important questions unanswered.

In the UK, initial interest in the application of geotechnical numerical modelling to tunnelling seems to have concentrated on the prediction of ground movements. At the time this work started realistic 3-D modelling was impractical, so many analyses were carried out in two-dimensional plane strain. It proved very difficult to simulate the shape of the surface settlement trough observed above actual tunnels (Gunn [14], Addenbrooke *et al.* [1], with numerical profiles typically having a much wider spread even when relatively sophisticated ground models were used. Given the three-dimensional nature of SCL construction, typically with top-heading, bench and invert sequences, but sometimes further subdivided into a side gallery and enlargement, the use of 2-D analysis alone might seem an assumption too far.

In most early geotechnical analyses the step-wise construction sequence of the shell was not modelled, and the settlement trough resulted from a combination of upward movement (as a result of unloading by excavation) and unsupported convergence (but in an undrained elasto-plastic material). Various empirical correction factors or modelling techniques have been used in order to get realistic predictions from 2D numerical models (e.g. the stress-reduction approach (β-value), the Hypothetical Modulus of Elasticity or a reduction in the K_0 value adjacent to the tunnel). These may obviously be of limited applicability if the tunnel geometry and construction differ significantly from those of the tunnels originally used to develop the methods.

4.1 3-D Numerical modelling – effects of the ground model

In the 1990s we carried out 3-D analyses using the finite element package ABAQUS, since by that time it had become feasible (using a very high specification computer) to incorporate both reasonable levels of construction detail and non-linear soil behaviour. Our objectives were to:

- Examine whether, as we expected, 3-D analyses would really produce significantly different (and hopefully better) results than 2-D analyses;
- Provide preliminary guidance on boundary positions and mesh discretization;
- Determine whether it was reasonable, given the factors above and the very large number of options available for the constitutive modelling of soil, to expect objective and realistic determinations of ground movements around tunnels.

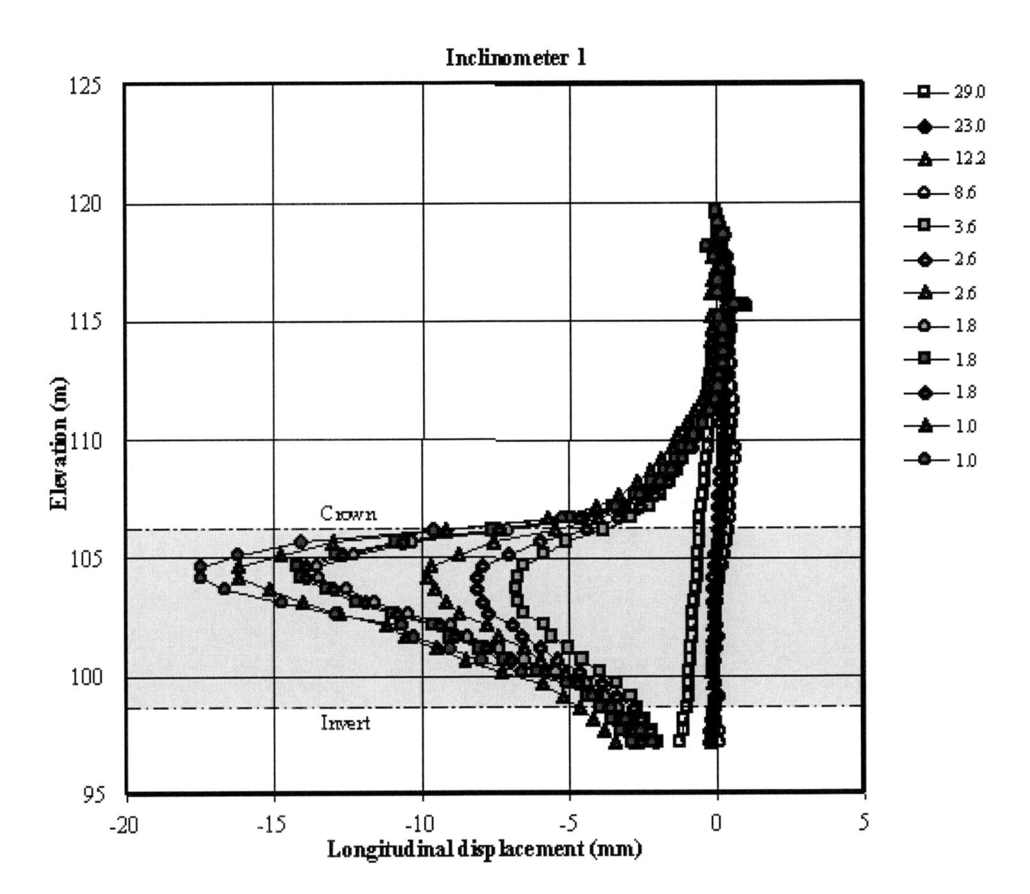

Figure 5: Longitudinal movement (face take) caused by the advancing concourse tunnel, as measured by an inclinometer

In these analyses the construction of the shotcrete shell of an 8 m diameter tunnel, with its axis 17.2 m below ground level in a material with similar parameters to London Clay, was modelled. The shotcrete was taken to be linear elastic, with stiffness independent of time, but nonetheless a number of important lessons were learnt from these numerical experiments. Figure 6 shows the 3D model that was used.

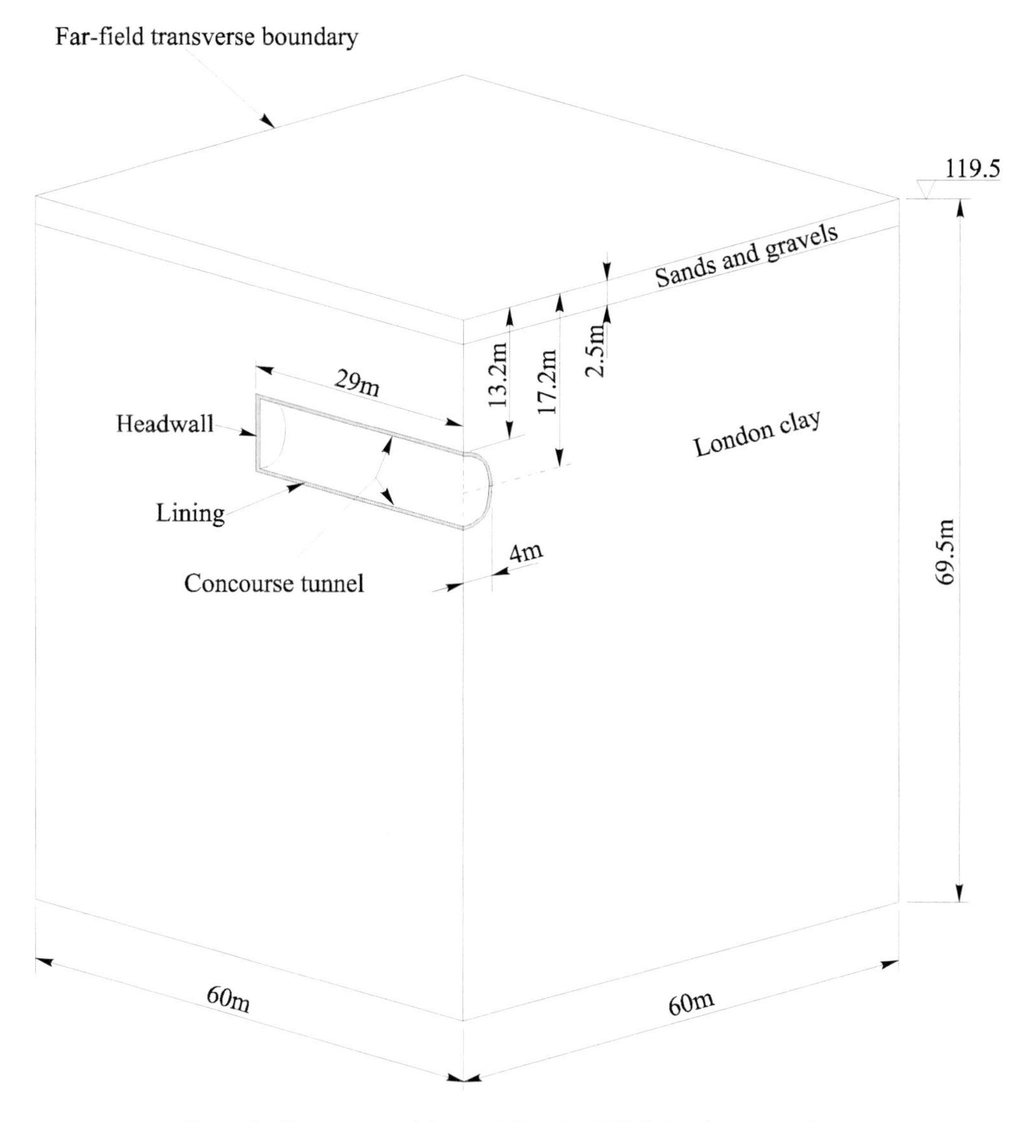

Figure 6: Geometry used for the 3-D ABAQUS finite element model

As might be expected, the type of finite element employed could have a significant influence on the magnitude of predicted surface settlements, with volume loss varying by as much as 50%. For the three different element types that were used, however, it was concluded that provided the mesh was sufficiently refined the difference in prediction could be reduced to an insignificant level. Figure 7 shows the influence of different element types in a 2D analysis.

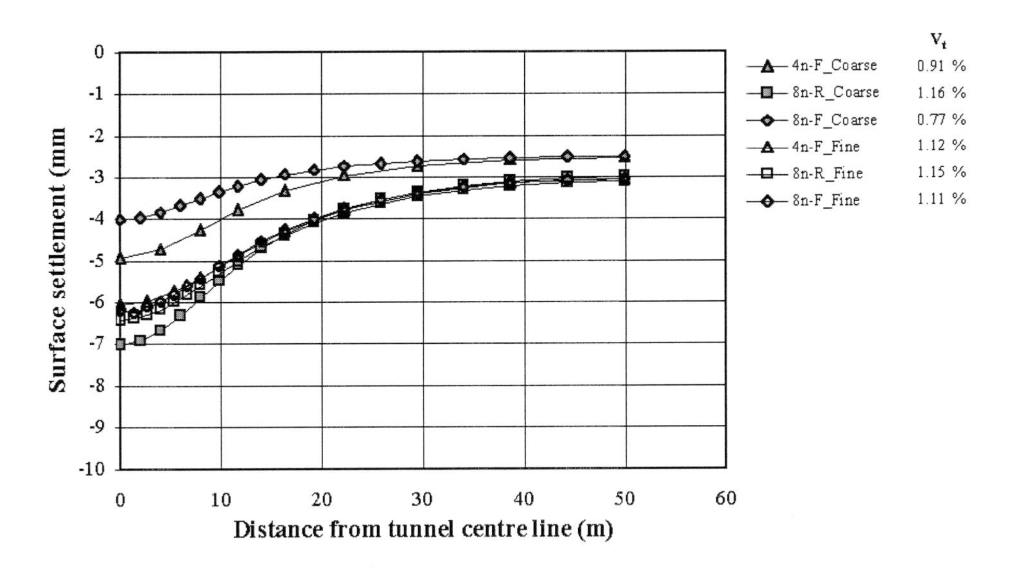

Figure 7: Effect of element type on predicted surface settlements in 2-D finite element analyses

Boundary positions were found to have a very significant effect on surface movements, based on analyses of a tunnel in a linear isotropic homogeneous soil. An initial analysis was carried out with a 60m long mesh 49.5m deep, but when significant longitudinal strains were predicted near the bottom of the mesh, its overall depth was increased by 20m to 69.5m. For the 8m circular tunnel it was found that to reduce surface settlements at the transverse boundary to approximately zero required the boundary to be placed some 150m to 200m from the tunnel centreline. Despite this, the predicted local volume loss was approximately the same with the far transverse boundary at only 30 m to 60 m from the tunnel centreline, supporting the guidelines suggested by Gunn [14] – see Table 2.

Mesh dimension	Van der Berg 1999	UK [1]	Gunn 1993
Transverse	13 R	13 R	3 times depth to axis
Ahead of face	10 R	12 R	-
Behind face	4 R	12 R	-
Depth	13 R	10 R	3 times depth to axis

Table 2: Recommendations for numerical modelling mesh sizes

[1] Typical values used by UK designers

As should be obvious, initial 'excavation' of a 3-D tunnel through the vertical boundary of a finite element mesh produces progressively changing surface settlements, until a steady state is reached some distance from the starting boundary. For both a 60m-long mesh and a 100m-long mesh the results indicated that after an advance of about 4 tunnel radii the vertical near boundary from which excavation commenced no longer influenced the results. However, for the 60m-long mesh the far vertical boundary soon affected (and reduced) the longitudinal displacements above the tunnel. This suggests that realistic ground movements are only calculated for a simple advancing tunnel when the face is 2-3 tunnel diameters away from the longitudinal boundaries.

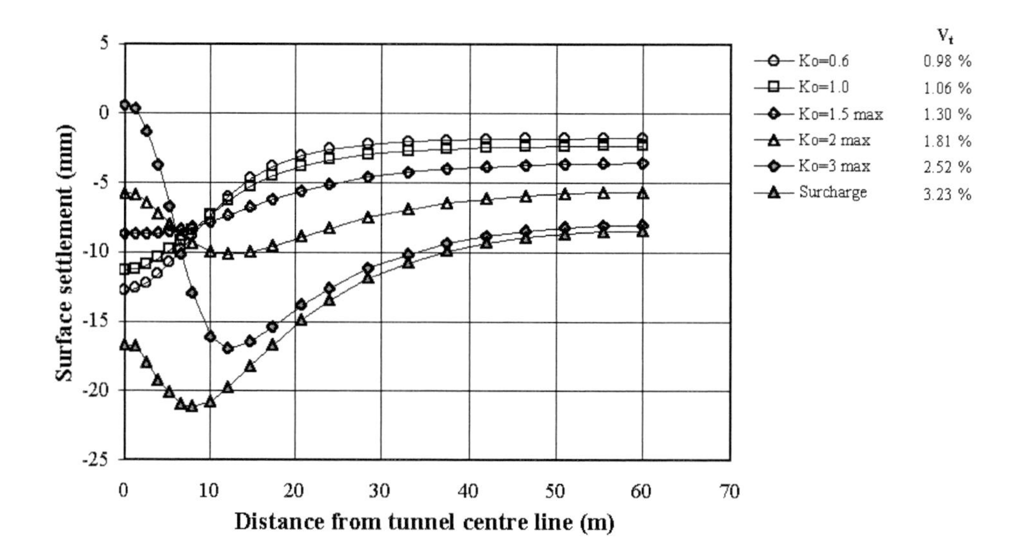

Figure 8: Effect of Ko on predicted surface settlements and calculated volume loss for 2-D analyses (note: An 85 kPa surcharge was also applied to model the effects of an overlying car park at Heathrow Terminal 4)

Figure 8 shows the very significant effect that the coefficient of earth pressure at rest (Ko) can have on the surface settlements predicted above a tunnel modelled in two dimensions. It should be remembered that when modelling a tunnelling problem the initial stresses in the soil is the loading when the tunnel is excavated. One would therefore expect that for the different initial stress conditions (loading conditions) the predicted ground displacements would differ.

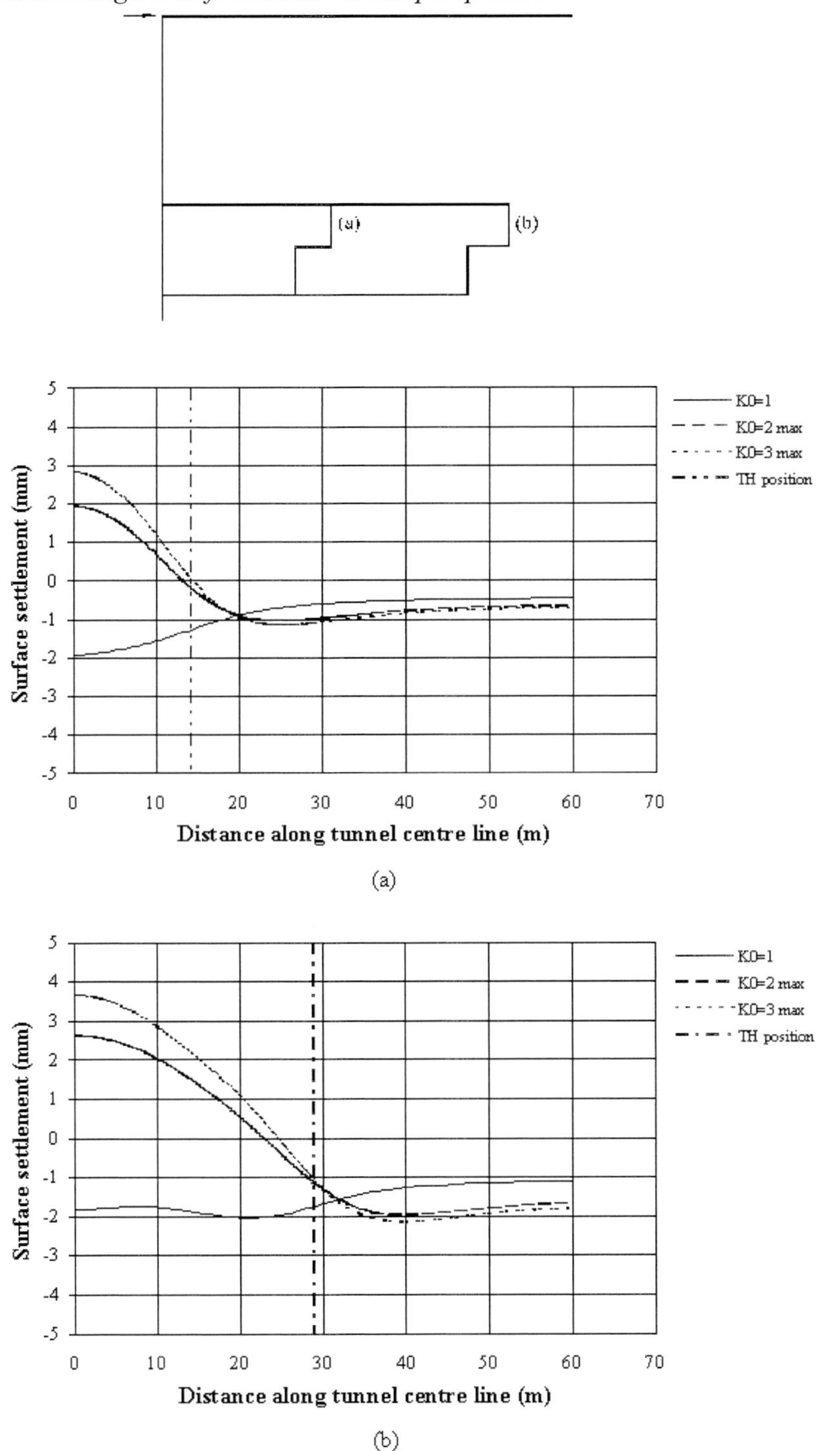

Figure 9: Effect of stress conditions and distance from starting boundary for settlements predicted along the tunnel centreline for 3-D analyses

In this case (see Figure 6) there is only 13.2m of ground above the 8m diameter tunnel. Construction of the tunnel was modelled, albeit in 2D, in stages (top heading, bench and invert), with excavation of each stage being followed by placement of the shotcrete lining. The increase in horizontal stresses results in the rotation of the displacement vectors from predominately vertical (low Ko) to horizontal (high Ko). This results in an increase in the settlement predicted in the far-field. With an increase in Ko the settlement of the tunnel crown increases, but there is an area above the tunnel where the surface settlement profile shows a heave effect. This can possibly be explained by the fact that with a high Ko value the stress state above the tunnel is close to passive failure. With the excavation of the tunnel, a zone of soil above the tunnel fails resulting in the heave effect.

These effects are repeated in Figure 9, which not only shows the influence of distance from the starting boundary, but also the range of surface settlement predictions that can result from the use of different soil models in the same basic 3D analysis. As before, low values of Ko lead to predictions of settlement, whilst higher values predict heave above the tunnel.

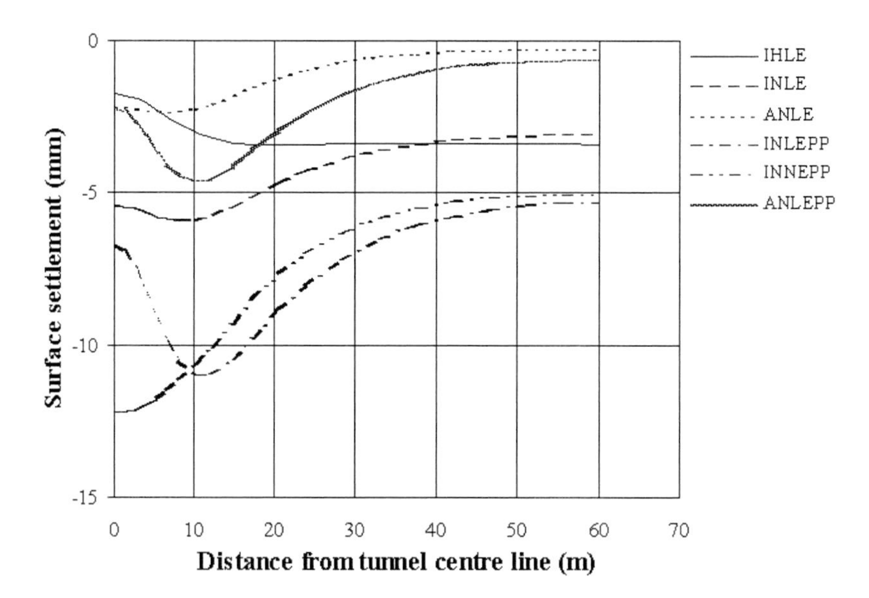

Figure 10: Effect of soil model on predicted transverse settlement trough from 2-D analyses (I = isotropic, H = homogeneous, A = anisotropic, N = non-homogeneous, NNE = non-homogeneous non-linear elastic, LE = linear elastic, PP = perfectly plastic)

Figure 10 illustrates the effect of the material model on the calculated surface settlements in 2D using a best estimate of in-situ stress conditions. From the results it is evident that the inclusion of non-linear elastic behaviour provides a reasonable estimate of settlement trough, whilst the inclusion of anisotropy reduces the settlements predicted at the far boundaries.

Modelling a tunnel in 2-D severely restricts the way in which construction detail can be modelled. As an extreme example, a 2-D wished-in-place lining can not model ground movement and therefore stress distribution ahead of the face (i.e. the stress reduction factor $\beta = 0$), and therefore moves upwards as a result of excavation unloading. 3-D analyses produce significantly different results from 2-D, primarily due to the inclusion of face take and the ability realistically to model the construction sequence. Using the high quality soil data and construction data available, in a 3-D finite element analysis with an isotropic non-homogeneous non-linear elastic perfectly plastic constitutive model for the soil, it was impossible to obtain realistic longitudinal and transverse settlement profiles. Factors that may have contributed to this disappointing (but not unexpected) result could be:-

- The difficulty in obtaining realistic estimates of the initial in-situ horizontal stress;
- The fact that anisotropy of stiffness and history effects were not modelled - Simpson *et al.* [33];
- The possible need to introduce a low value of the independent shear stiffness parameter, G_{vh} (Lee and Rowe [20], Addenbrooke *et al.* [1];
- The use of a soil model which, although sophisticated, remains unrealistic. One important factor to emerge from comparisons of monitored and predicted movements of the shotcrete shell is that the top heading was observed (as noted above) to undergo a significant vertical translation. Numerical modelling was unable to replicated this movement, which may be associated with tensile loading and fissure opening in the clay;
- The use of a simple linear elastic model for the shotcrete. The fact that approximately 25% of the settlement above the tunnels was observed to occur after invert closure suggests that either fissures are re-closing with time, or that creep and shrinkage of the shell is significant.

Although it might seem from the above discussion that numerical modelling has little to offer the tunnel designer, this is not so. The 3-D analyses conducted showed that it was possible to produce reasonable predictions of the ground movements ahead of the tunnel, and since this is an important contributor to relative volume loss, such estimates are valuable. Figure 11 shows a comparison of the 3D analyses and the inclinometer results, while Figure 12 compares the results

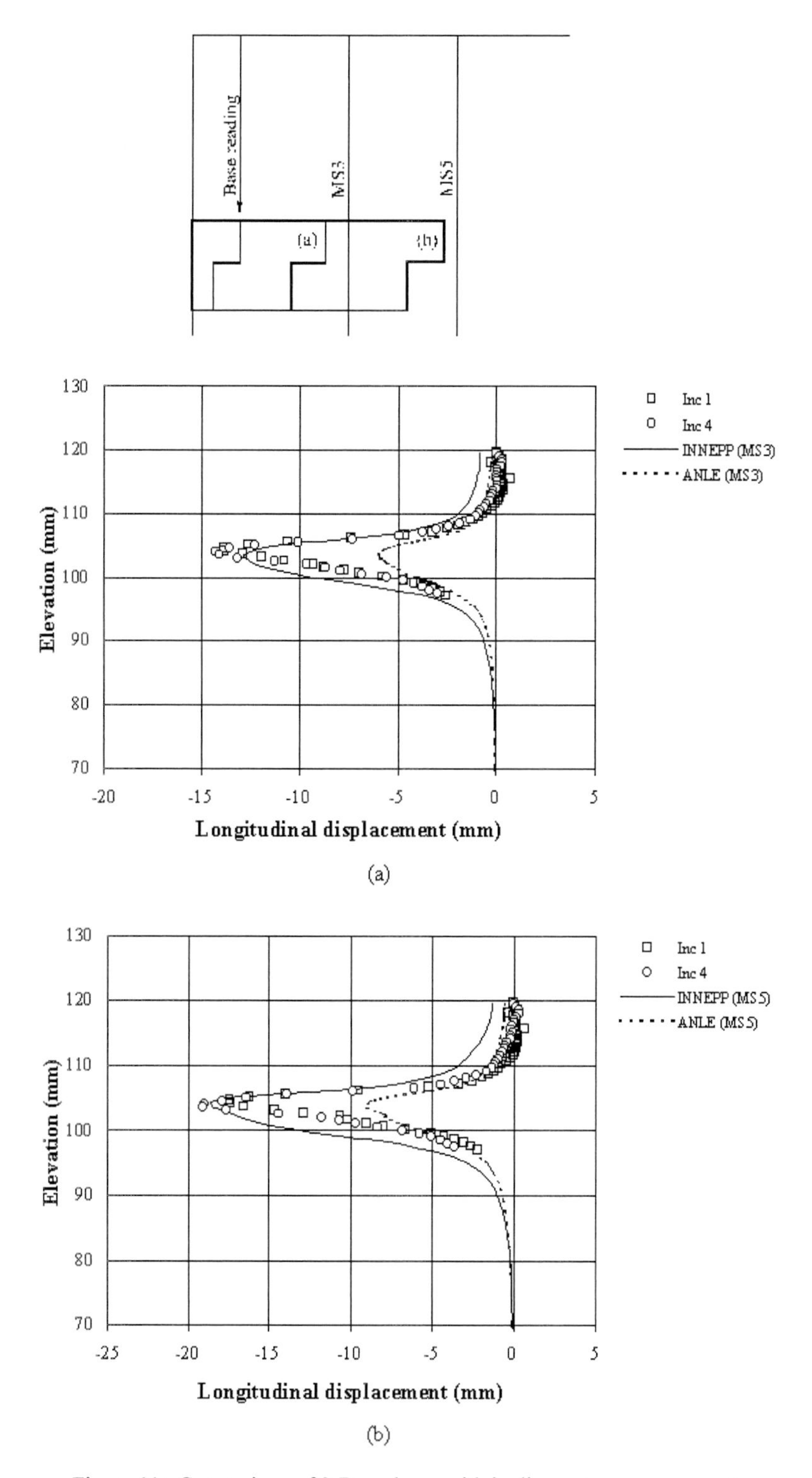

Figure 11: Comparison of 3-D analyses with inclinometer measurements

with the deflectometer measurements. For the case described here, 3-D analysis predicted radial movements and movements into the tunnel face (so-called "face-take") as 0.33% and 0.3% respectively.

Field monitoring suggested that the relative volume loss for the concourse tunnel was 0.8%, of which approximately 0.45% could be attributed to face take. In unfamiliar ground conditions numerical modelling should be able to provide estimates for the relative volume loss, which is required as input for the empirical estimation of surface settlements. Furthermore, estimates of movements induced on nearby tunnels or structural foundations should be of value.

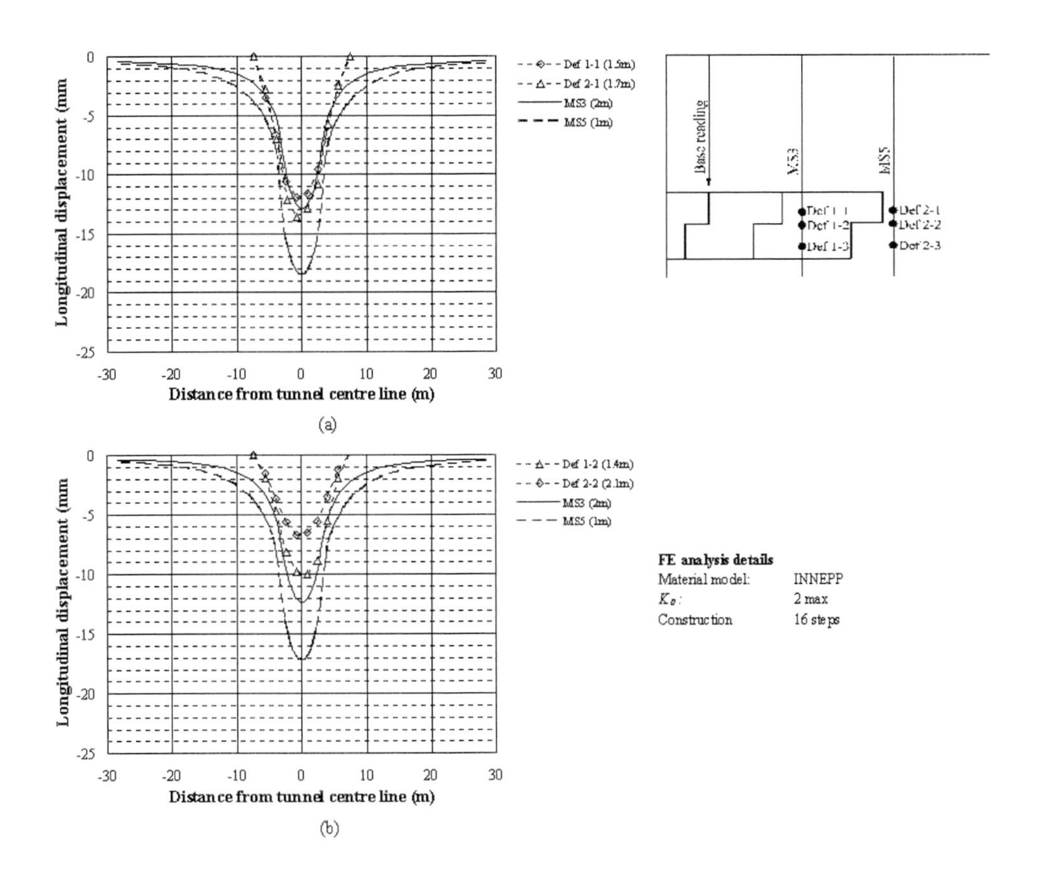

Figure 12: Comparison of 3-D analyses with chain deflectometers data

4.2 3-D Numerical modelling – effects of sprayed concrete model

In the past three years we have investigated the use of numerical analyses to estimate the loads in sprayed-concrete tunnel linings. The major question to be answered was whether the constitutive model chosen to represent the behaviour of the shotcrete has a significant impact on the normal stresses and bending moments that are predicted, since the ultimate objective is to be able to obtain practical design information (such as axial forces and bending moments). A series of 3-D numerical experiments were therefore carried out using the finite difference program FLAC.

Sprayed concrete is a complex material, and a numerical model for use in SCL tunnelling needs to take account both of long-term and early-age behaviour. Unfortunately relatively little is known about the early-age behaviour of such materials (at least partly because of the inherent difficulties of testing a material that is evolving rapidly), and the development of such knowledge is not helped by the fact that shotcrete technology is still changing rapidly. However, the following facets of shotcrete behaviour need to be considered and evaluated before a simple constitutive model can be used in practice:-

- Strength and stiffness increase rapidly during the early life of the shotcrete. At this stage the applied stresses are high relative to strength, and since the invert has not been closed in a typical SCL heading, the tunnel is at its most vulnerable;

- Unlike soil, cementitious materials have significant strength in tension, and this may need to be considered. Furthermore sprayed concrete tunnel linings in soft ground are almost always reinforced with steel bars or fibres to increase the tensile capacity;

- Shotcrete can reasonably be modelled as an elasto-plastic material, although there is not universal agreement as to the best yield criterion to use, or whether work-hardening or perfect plasticity is more appropriate. Pre-yield, isotropic elasticity is often assumed. The stress-strain curve for concrete is non-linear;

- Sprayed concrete is likely to exhibit much greater creep in tunnels than is experienced by conventional cast concrete, because of the high levels of early loading, and its higher cement content;

- It also suffers more shrinkage, because of its higher cement content, and the larger surface to volume ratio in a tunnel lining.

- There are likely to be imperfections in the shell. Junctions between panels are unlikely to be as strong as the shotcrete itself. As the investigations at Heathrow showed, junctions between panels must be designed with constructability in mind, and rebound must be carefully removed in order to avoid a laminated invert.

- Since shotcrete tunnels are constructed in panels, the details of construction are likely to affect the stress level and distribution in the lining, given the changing properties of shotcrete with time, and the shrinkage and creep that it undergoes;

Where necessary one can draw on existing information for conventional cast concrete to assist the assessment above.

3-D numerical modelling carried out on a similar tunnel geometry to that used by van der Berg [40] suggests that the hoop forces in a typical SCL tunnel will be significantly affected by the assumed stiffness of the lining (Figure 13). The models referred to in Figure 13 are identified in Table 3. As the strains in the shell increase the load drops, as previously suggested by Peck *et al.* [32] and Soliman *et al.* [35]. Time dependency of stiffness, non-linearity of stress-strain behaviour, and creep can all produce larger strains for a given load compared with a constant high stiffness shotcrete model, and lead to lower estimates of hoop load, due to a favourable interaction between the structure and the ground.

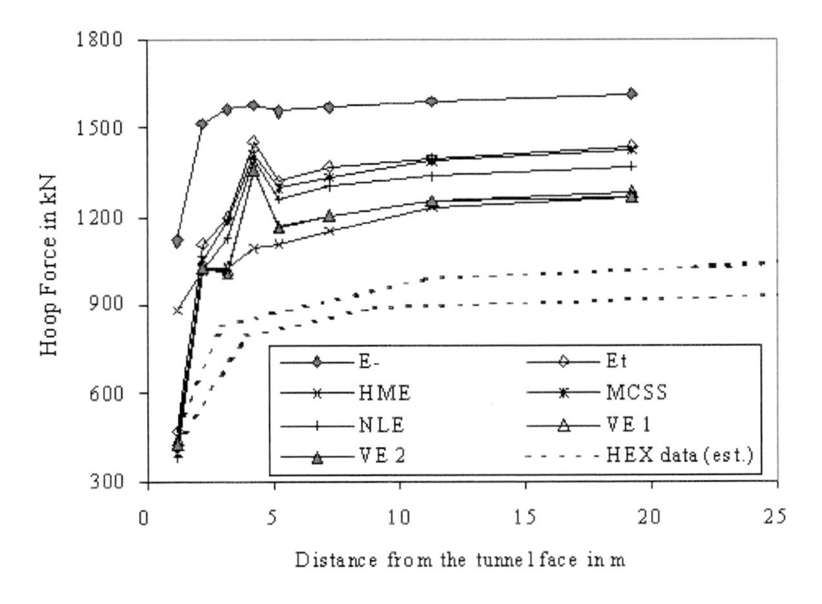

Figure 13: Influence of constitutive model for shotcrete on predicted hoop axial force in the crown

Abbreviation	Description
E-	Linear elastic, constant stiffness = 28 day value
Et	Linear elastic, age-dependent stiffness
HME	Hypothetical Modulus of Elasticity (HME)
MCSS	Strain-hardening plasticity model (Mohr Coulomb)
NLE	Nonlinear elastic model - Kotsovos & Newman (1978)
VE 1	Visco-elastic "Kelvin" creep model - stress independent
VE 2	Visco-elastic "Kelvin" creep model - stress dependent
VE 3	Visco-elastic "Kelvin" creep model after Yin (1996)
JR	MCSS model but strength reduced by 50% on radial joints
JL	MCSS model but strength reduced by 50% on longitudinal joints
J	MCSS model but strength reduced by 50% on radial & long. Joints

Table 3: Constitutive models used for shotcrete

Analyses showed that hoop bending moments were even more strongly affected by the constitutive model for the shotcrete. A softer shotcrete response led to favourable redistribution of stresses, and considerably lower predicted bending moments.

Longitudinal loads and bending moments seem rarely to be considered in tunnel design. Given that the top heading is cantilevered off the completed tunnel ring, and that there is considerable inward movement at the tunnel face, it might be expected that the stress regime would be complex. Numerical modelling suggests that longitudinal forces are significantly affected by the assumed shotcrete behaviour, but that they will be small if the initial stiffness is low. Bending moments will be high, with the extrados of the top heading in tension as a result both of vertical ground loading and ground movement towards to the face. Cracking on the extrados will remain undetected, and potentially this could lead to durability problems if the shotcrete shell were to be used below ground water level in a single-pass shell.

Finally, the way that a tunnel is constructed can be expected to have a significant influence on the loads it must carry. As an example, 3-D numerical modelling has been carried out to examine the effect of advance length (AL), advance rate (AR) and distance to ring closure (RCD). Thomas [36] has proposed combining all these variables into a single Sequence Factor

$$S = \mathrm{RCD/AR} \;.\; \mathrm{AL/R} \;.\; \mathrm{E_x/E_{28}} \qquad (1)$$

where E_x/E_{28} is the ratio of the Young's modulus of the sprayed concrete at ring closure to the 28 day value, and R is the tunnel radius. Results from numerical modelling suggest (Figure 14) that both hoop forces and bending moments increase with an increasing Sequence Factor.

4.3 Numerical modelling – the future?

Objective numerical modelling, using best estimates of ground conditions and construction sequences, has shown how difficult it is to predict ground movements (e.g. the surface settlement trough) at a distance from an advancing tunnel. Given the difficulty of obtaining some key parameters (e.g. the coefficient of earth pressure at rest, Ko) and the fact that critical details of tunnel construction may remain unknown at the time of such an analysis, it is our view that finite element and finite difference analysis should only be used for this type of "best shot" predictions when the results can be calibrated against good quality data for a similar tunnel in similar ground conditions. The added complexity of 3-D analysis is not justified without proper validation and calibration against high-quality monitoring data.

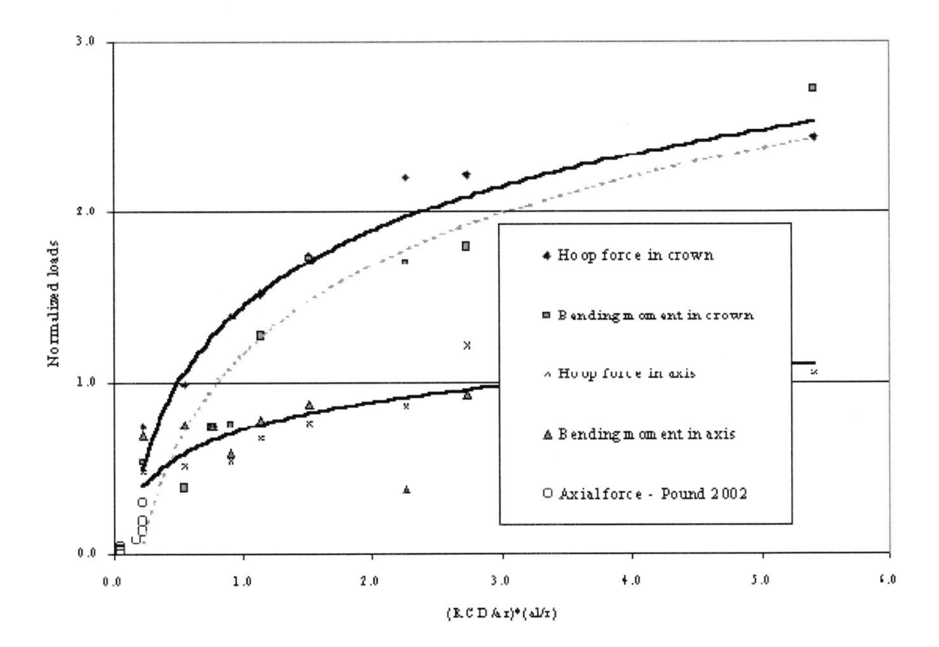

Figure 14: Effect of construction sequence (in terms of Sequence Factor) on hoop loads and bending moments (loads normalized with respect to results from linear elastic, age dependent model (Et) and Sequence Factor values corrected for tunnel radius and stiffness at ring closure)

3-D numerical modelling of ground movements near to the tunnel face would appear at this stage to be worth further investigation, particularly since estimates of face take are important to the designer, and are difficult to obtain by other means.

3-D analyses involving realistic constitutive models of both soil and shotcrete have shown that the predicted loads and bending moments in a sprayed-concrete lining are significantly affected by the shotcrete model, but are less significantly affected by the soil model. Although in some cases (e.g. around junctions) there is no other means of estimating the detailed distribution of loads, "best shot" analyses to determine stresses in the shotcrete shell during design need to used with caution, given the lack of high quality monitoring data for the stresses in shotcrete shells against which to judge their results. However, parametric studies (i.e. studies where parameters are varied to indicate the sensitivity of the predictions to the assumed shotcrete and soil conditions) have provided and will continue to provide valuable insights into those factors that affect the stresses in the lining, particularly at early age (in the heading) and during the construction of junctions. As an example, Figure 15 shows the results of a simulation of the Heathrow Terminal 4 concourse tunnel construction, with the shotcrete modelled as having time dependency of stiffness. Utilization factors in the top heading typically exceed 0.5 towards the leading edge of each ring. The utilization factor is defined as the current (deviatoric) stress divided by the current (deviatoric) strength. Sensitivity analyses and parametric studies using modelling of this type can potentially provide the designer with useful information to augment experience and precedent practice.

5. Conclusions

This paper has briefly reviewed the design of SCL tunnels in soft ground. It has examined the results of high-quality field instrumentation and from advanced 3-D numerical modelling related to soft-ground SCL tunnelling in the London Clay Formation at Heathrow. Both activities are extremely demanding, and should be undertaken with great care. There is abundant evidence that without skill and care both can produce poor and misleading results. However, even the experienced engineer needs carefully to examine such data, validating them against experience, precedent practice, common sense, and the work of others.

Despite the difficulties that we have identified, it is hard to imagine a future for SCL tunnelling in soft ground that does not make use of the results both monitoring and 3-D numerical modelling to support future designs. Whilst it is conceivable that some future tunnels may be designed and constructed without the benefit of these tools, principally because of lack of expertise and commitment on the part of designers and constructors, failure to analyse and monitor will reduce confidence in construction, and is likely to inhibit future development of promising new tunnelling methods.

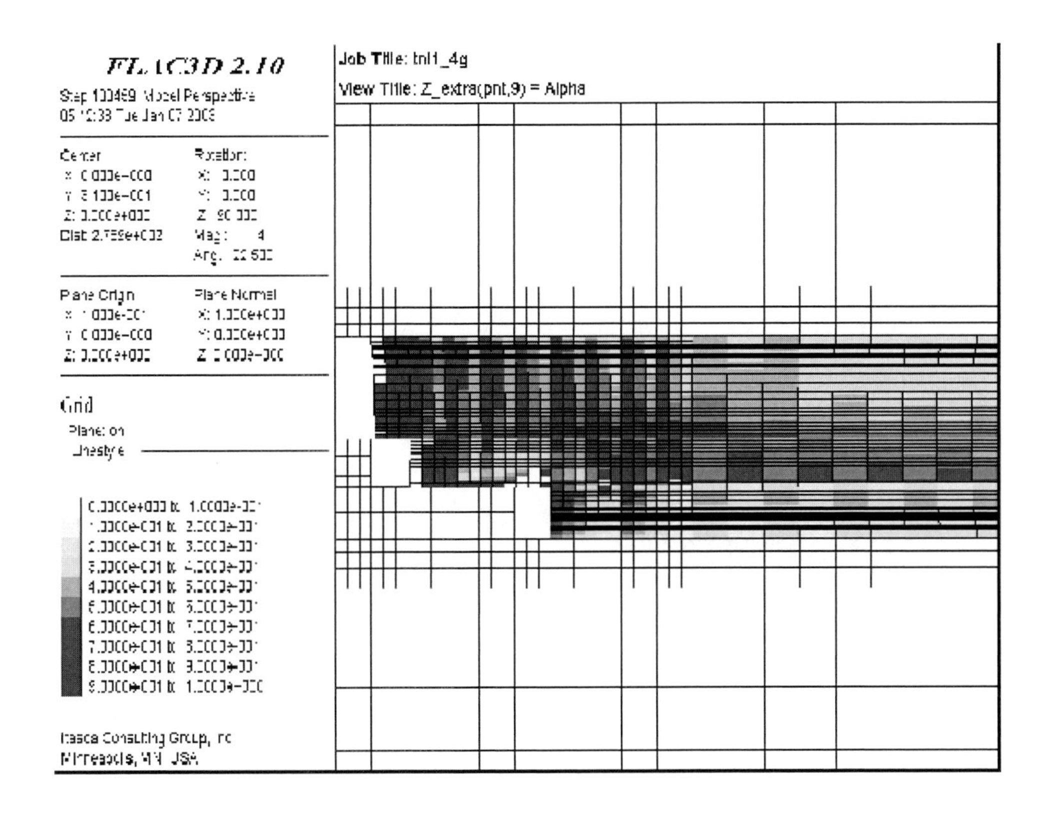

Figure 15: Shotcrete stress utilization factors predicted using a time-dependent linear-elastic stiffness model for shotcrete

Bibliography:

[1] Addenbrooke, T.I., Potts, D.M. and Puzrin, A.M.: The influence of pre-failure soil stiffness on the numerical analysis of tunnel construction, Geotechnique, Vol. 47, No. 3, pp. 693-712, 1997

[2] Attewell, P.B. and Woodman, J.P.: Predicting the dynamics of ground settlement and its derivatives caused by tunnelling in soil, Ground Engineering, Vol. 15, No. 8, pp. 13-22 and 36, 1982

[3] Attewell, P.B., Yeates, I. and Selby, A.R.: Soil movements induced by tunnelling, Blackie, Glasgow, 1986

[4] Barton, N., Lien, R. and Lunde, J.: Estimation of support requirements for underground excavations. Proc. 16th ASCE Symp. On Rock Mechanics, pp. 163-177, 1975

[5] Bieniawski, Z.T.:Rockmechanics design in mining and tunnelling, Balkema, Rotterdam, 1984

[6] British Tunnelling Society: Tunnel lining design guide, In draft, 2003.

[7] Broms, B.B. and Bennermark, H.: Stability of clay at vertical openings, ASCE J. of Soil Mechanics and Foundation Engineering, Vol. 93, SM1, pp. 71-94, 1967

[8] Clayton, C.R.I. and Bica, A.V.D.: The design of diaphragm-type boundary total stress cells, Geotechnique, Vol. XLIII, No. 4, pp. 523-536, 1993.

[9] Clayton, C.R.I., Hope, V.S., Heymann, G. van der Berg, J.P. and Bica, A.V.D.: Instrumentation for monitoring sprayed concrete lined soft ground tunnels in built-up areas, Proc. Inst. Civil Engnrs, Geotechnical Engineering, vol 143, pp.119-130, 2000

[10] Clayton, C.R.I., van der Berg, J.P., Heymann, G., Bica, A.V.D. and Hope, V.S.: The performance of pressure cells for sprayed concrete tunnel linings, Geotechnique, Vol. 52, No. 2, pp. 107-116, 2002

[11] Curtis, D.J.: Discussions on Muir Wood, The circular tunnel in elastic ground, Geotechnique, Vol. 26, No. 1, pp. 231-237, 1976

[12] Dunniciff, J. and Green, G.E.: Geotechnical instrumentation for monitoring field performance, Wiley, New York, 577 pp., 1988

[13] Davis, E.H., Gunn, M.J., Mair, R.J. and Seneviratne, H.N.: The stability of shallow tunnels and underground openings in cohesive material. Geotechnique, Vol. 30, No. 4, pp. 397-419, 1980

[14] Gunn, M.J.: The prediction of surface settlement profiles due to tunnelling, Proc. Conf. on *Predictive Soil Mechanics*, Thomas Telford, London, pp. 304-316, 1993

[15] Health and Safety Executive: The collapse of NATM tunnels at Heathrow Airport, HSE Books. 110 pp. 2000

[16] Health and Safety Executive: Safety of New Austrian Tunnelling Method (NATM) Tunnels – A review of sprayed concrete lined tunnels with particular reference to London clay, HSE Books, 86 pp. 1996

[17] Institution of Civil Engineers: Sprayed concrete linings (NATM) for tunnels in soft ground, Thomas Telford, London, 1996

[18] Kimura, T. and Mair, R.J.: Centrifugal testing of model tunnels in soft clay, !0th Int. Conf. On Soil Mechanics and Foundation Engineering, Vol. 1, pp. 319-322, 1981

[19] Leca, E. and Dormieux, L.: Upper and lower bound solutions for the face stability of shallow circular tunnels in frictional material, Geotechnique, Vol. 40, No. 4, pp. 581-606, 1990

[20] Lee, K.M. and Rowe, R.K.: Deformations caused by surface loading and tunnelling: the role of anisotropy, Geotechnique, Vol. 39, No. 1, pp. 125-140, 1989

[21] Macklin, S.R.: The prediction of volume loss due to tunnelling in overconsolidated clay based on heading geometry and stability number, Ground Engineering, April, pp. 30-34, 1999

[22] Mair, R.J.: Centrifugal modelling of tunnel construction in soft clay, Ph.D. Thesis, University of Cambridge, 1979

[23] Mair, R.J.: Developments in geotechnical engineering research: applications to tunnels and deep excavations. Proc. Instn of Civil Engineers, London, Vol. 93, pp. 27-41, 1993

[24] Mair, R.J. and Taylor, R.N.: Bored tunnelling in the urban environment, Proc. 14th Int. Conf. on Soil Mechanics and Foundation Engineering, Vol. 4, pp. 2353-2386, 1997

[25] Martos, F.: Concerning an approximate equation of subsidence trough and its time factors, International Strata Control Congress, Leipzig, pp. 191-205, 1958

[26] Muir Wood, A.M.: The circular tunnel in elastic ground, Geotechnique, Vol. 25, No.1, pp.115-127, 1975

[27] New, B.M. and Bowers, K.H.: Ground movement model validation at the Heathrow Express Trial Tunnel, Tunnelling '94, pp. 301-329, 1994

[28] O'Reilly, M.P. and New, B.M.: Settlements above tunnels in the United Kingdom – their magnitude and prediction. Tunnelling '82, Inst. Of Mining and Metallurgy, London, pp. 197-204, 1982

[29] Panet, M. and Guenot, A.: Analysis of convergence behind the face of a tunnel, Tunnelling '82, Institution of Mining and Metallurgy, pp. 197-204, 1982

[30] Peck, R.B.: Deep excavations and tunnelling in soft ground, Proc. 7th Int. Conf. On Soil Mechanics and Foundation Engineering, Vol. 3, pp. 225-290.

[31] Peck, R.B.: Advantages and limitations of the Observational Method in applied soil mechanics, Geotechnique, Vol. 19, pp. 171-187, 1969

[32] Peck, R.B., Hendron, A.J. and Mohraz, B.: State of the art in soft ground tunnelling, Proc. Rapid Excavation and Tunnelling Conf., American Inst. Of Mining, Metallurgy, and Petroleum Engineers, New York, pp. 259-286, 1972

[33] Simpson, B., Atkinson, J.H. and Jovicic, V.: The influence of anisotropy on calculations of ground settlements above tunnels, Geotechnical Aspects of Underground Construction in Soft Ground, ed. Mair, R.J. and Taylor, R.N., Balkema, Rotterdam, pp. 591-594, 1996

[34] Sloan, S.W. and Asadi, A.: Stability of shallow tunnels in soft ground. Predictive Soil Mechanics, Thomas Telford, London, pp. 644-663, 1993

[35] Soliman, E., Duddeck, H. and Ahrens, H.: Effects of the development of stiffness on stresses and displacements of single and double tube tunnels, Tunnelling and Ground Conditions, ed. Abdel Salem, pp. 549-556, 1994

[36] Thomas, A.H.: Numerical modelling of sprayed concrete lined (SCL) tunnels. PhD Thesis, University of Southampton, UK, 2003

[37] van der Berg, J.P., Clayton, C.R.I. and Hope, V.S.: An evaluation of the role of monitoring during the construction of shallow NATM tunnels in urban areas, North American Tunnelling '98 Conference (NAT'98) on 'New Horizons: Building our Future', Newport Beach, California, 21-25 February (ed. Ozdemir, L.), pp. 251-257, 1998

[38] van der Berg, J.P., Clayton, C.R.I., Powell, D.B. and Savill, M.: Monitoring of the Concourse Tunnel at Heathrow Express Terminal 4 station constructed using the NATM, Proc. World Tunnel Congress '98 (ed. Negro, A. and Ferreira, A.A.), Sao Paolo, 25-30 April 1998. Balkema, Rotterdam, vol. 2, pp. 1163-1168, 1998

[39] van der Berg, J.P. and Clayton, C.R.I.: Monitoring ground movement ahead of a tunnel heading in London clay, Proc. Int. Symp. on 'Geotechnical Aspects of Underground Construction

in Soft Ground', Tokyo (ed. O. Kusakabe, K. Fujita, Y. Miyazaki), Tokyo, pp.173-178. Balkema, Rotterdam, 2000

[40] van der Berg, J.P.: Measurement and predictions of ground movements around three NATM tunnels, Ph.D. Thesis, University of Surrey, 1999

[41] van der Berg, J.P., Clayton, C.R.I. and Powell, D.B.: Displacements ahead of an advancing NATM tunnel in the London clay, accepted for publication by Geotechnique.

Ground reinforcing and steel pipe umbrella system in tunnelling

Daniele Peila[1], Sebastiano Pelizza[2]

[1] Department of Georesources and Land, and Istituto di Geologia Ambientale e Geoingegneria- CNR (Italian National Research Council), Politecnico di Torino, Italy
e-mail: daniele.peila@polito.it

[2] Department of Georesources and Land, Politecnico di Torino, Italy

Abstract

The various types of ground reinforcing techniques that are used in tunnelling construction are discussed and analyzed in this work and are classified on the basis of their action in tunnelling. Specific attention has been paid to forepoling which is also known as the steel pipe umbrella system. A specific care has been given to the design methods of this techniques.

1. Introduction

Tunnelling is characterized by the tendency to "mechanise" or rather "industrialise" the excavation as much as possible in order to guarantee lower construction times, lower total costs and safer and more hygienic working conditions. This can be attained both through the systematic use of high productivity excavation and mucking machines (therefore large machines) during a conventional advancement or full face tunnelling machines TBMs (Tunnel Boring Machines) and Shields [18, 29, 35, 43, 44].

When adverse geotechnical conditions are met and the free span and self supporting times are too short and not compatible with an industrial activity the technological possibilities are:

- to reduce the size of the excavation sections, increasing the number of working attacks;
- to improve the rock mass quality or reinforce it;
- to pre-support the excavation;
- to apply a pressure to the tunnel face.

Already back in the Seventies the great importance of ground reinforcement techniques was brought to light, although it was then mainly limited to the tunnel boundary (in radial directions) and only behind the face. This technique did not permit advancement at full section in "difficult" conditions, but usually required proceeding at partialised sections [5, 11, 12, 21, 23, 35].

The evolution of reinforcement techniques, which has occurred with the systematic use of interventions ahead of the face has instead led to enlarging sections, thus allowing works to be carried out with large free span.

2. Classification of reinforcement and improvement techniques

The classification of the various ground reinforcing techniques for tunneling refers both to the performance and technological typology of the reinforcement techniques or to the action that the intervention is designed for. When referring to the technological typology, the following main reinforcement groups can be ideintified [28, 34, 35]:

- improvements;
- reinforcements;
- pre-supports;
- drainage.

On the other hand it is possible to identify [35, 43, 44] the following main actions of the various methods:

- to modify the convergence-confinement laws of the tunnel boundary;
- to modify the entity of the radial displacements at the moment of applying traditional support structures, that is, in correspondence or just behind the excavation face (it is here intended that the structures placed inside the void, after the excavation, are traditional supports and have the purpose of guaranteeing stability until the final lining is placed);
- to guarantee the stability of the excavation face;
- to guarantee stability of the free span close to the face;
- to guarantee stability of local portions of the rock mass in particular geological and geotechnical conditions;
- to permit the control of the water flow.

Each type of reinforcement therefore plays a different role (or several roles at the same time) in the stress-strain control at the boundary of the tunnel or ahead of the face during or after the excavation stage.

Figure 1 gives a global overview of the various ground reinforcing applications used in tunnels on the basis of the proposed classification.

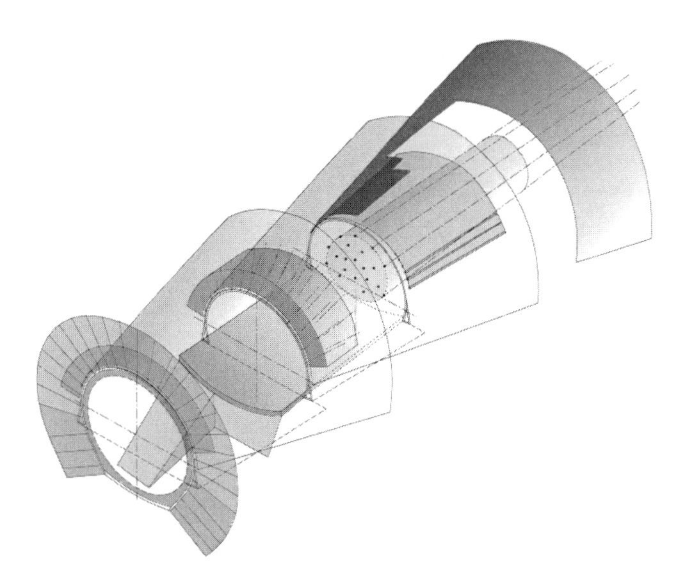

Figure 1: Possible locations of the various reinforcement applications within a tunnel (courtesy of Geodata S.p.A., Torino)

To describe both the variations of the convergence-confinement behaviour law and the reduction of the radial displacements at the face in an efficient way, it is useful to refer to the convergence-confinement method. On this basis the actions of the reinforcement techniques are synthesed in figure 2: the convergence-confinement curve of the reinforced tunnel has a different trend from that which refers to undisturbed conditions and is such that it facilitates the stability of the void, while the action at the face causes a reduction of the displacements at the moment the supports are applied ($u_0 \rightarrow u_0^*$) [35].

A short description of the various techniques is given in the following. **Improvements** act by improving (in term of allowing a better mechanical performance) the properties of the soil around the excavation and can be divided as:

- *grouting (at a low pressure)*: a lasting intervention which produces its action with the injection of grout mixes at a 20-40 bar pressure into the soil The mixes are able to fill the voids of the soil thereby reducing the permeability and improving the soil geotechnical parameters. The grout mixes can be either cement or chemically based (figure 3);
- *jet grouting*: a lasting intervention which is carried out by injecting a cement based grout mix into the soil at a high pressure (over 300 bars). Different injection layouts have been developed. The final result is a column of treated soil with better geotechnical characteristics than the ground itself;

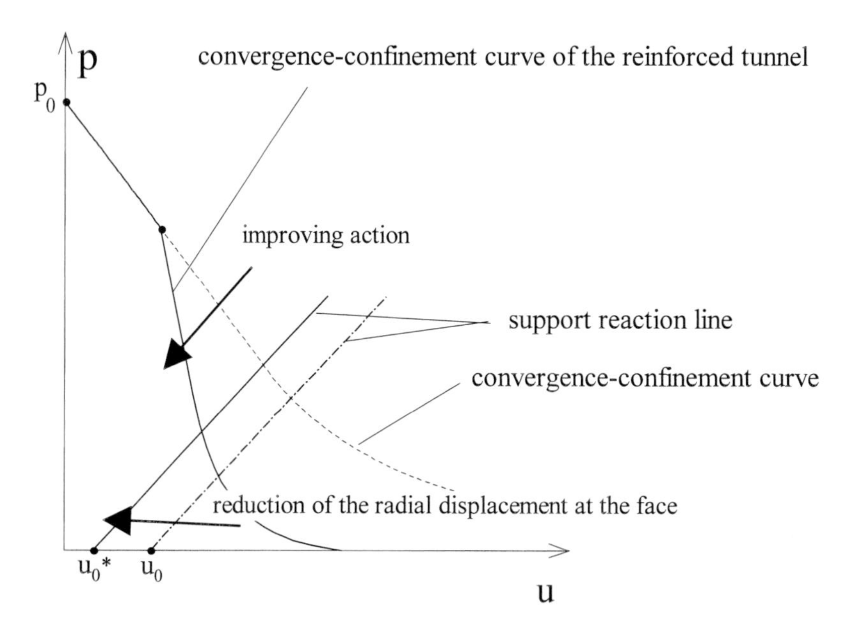

Figure 2: Synthesis of the actions that are carried out by reinforcements in a tunnel through an analysis with the convergence-confinement method. The convergence-confinement curve of a tunnel is the connection between the radial pressure (p) that acts on the tunnel perimeter and its radial displacement (u): with a progressive diminishing of the pressure (starting from p_0), the radial displacements increase. The convergence-confinement curve is a function of the physical and mechanical parameters of the ground, of the stress state at the depth of the excavation and the radius of the tunnel. In the presence of some reinforcements, such as radial reinforcements at the tunnel boundary, the ground parameters improve in the reinforced zone and the convergence-confinement curve varies, coming closer to the axis of the displacements. Other reinforcement interventions carried out ahead of the excavation face induce a reduction of the radial displacements of the tunnel perimeter in correspondence to the face. As a consequence, the reaction line of the supports moves towards the left [34, 35].

- *freezing*: a temporary intervention which can be applied when the soil is wet. The aim of this technique is to create an area of frozen ground, which is more resistent than the original ground and which has no flowing water inside it.

Ground reinforcements are instead lasting interventions which are carried out by introducing structural elements that are more resistent and rigid than the soil (i.e. bolts, steel cables, steel or fiberglass pipes, etc.) and which are directly connected to the ground with the purpose of obtaining a better (in terms of tunnel excavation) global behaviour of the reinforced ground. By using a resistent element of notable geometry and well known characteristics, the various applications offer the maximum adaptability to the soil variations (i.e. changing dimensions and number of the structural elements). The use of well defined reinforcing elements guarantees the correct carrying out of the designed work, while the durability of the reinforcing elements is one of the problems if the intervention must be long lasting [32, 39].

There are several ground reinforcing schemes which can be used according to the structural requirements; in technical literature it is possible to find the following main examples:

- *transversal intervention from the surface:* "comb", "mantles" and "fan voulting" layout of the reinforcing elements (steel or fiber-glass pipes or steel cables) (figure 4).

- *radial underground interventions:*

  a) from the tunnel after the advancement of the face: bolts (fully grouted or pre-tensioned), steel cables, steel, glass-fiber, carbon-fiber pipes or bars;

  b) preventively from a pilot tunnel: steel cables, steel or fiber-glass pipes or bars (figure 5);

- *transversal underground interventions*: from side tunnels or pilot tunnels steel cables or bars, steel or fiber-glass pipes (figure 6);

- *longitudinal underground interventions*

  a) tunnel face reinforcing: fiber-glass pipes or bars (figure 7, 8);

  b) pipe umbrella: steel pipes or bars (figure 9).

Pre-reinforcement techniques are generally obtained through the "creation" of a structure ahead of the tunnel face to improve the stability in the tunnel span (they can also be obtained by using a layout of ground improving techniques):

- *cellular arch or the use of microtunnelling*: used for short and large underground excavations: before the underground excavation is carried out a supporting structure is made with many small section tunnels or microtunnels. This technique can be applied if the overburden is so thin that not allow the use of other supporting techniques or it is necessary to control to a high extent the possible subsidence of the surface (figure 10).

- *precut*: a lasting intervention which is carried out by excavating a preventive "tile cut" around the tunnel using a chain-saw blade machine. The cut is filled with a concrete lining of high mechanical characteristics (reinforced with steel fiber) (figure 11).
- *jet grouting arch*: an arch of sub-horizontal jet-grouting columns is made at the crown of the future tunnel, while subvertical columns are often placed in the walls of the tunnel. The arch acts by supporting the soil during the excavation and by homogenizing the stresses which will act on the final supports (figure 12);

Figure 3 : Example of ground reinforcing using radial grouting carried out from a pilot tunnel where longitudinal jet grouting columns were used as ground reinforcement (Milan metro works) (Courtesy of Rodio S.p.A.)

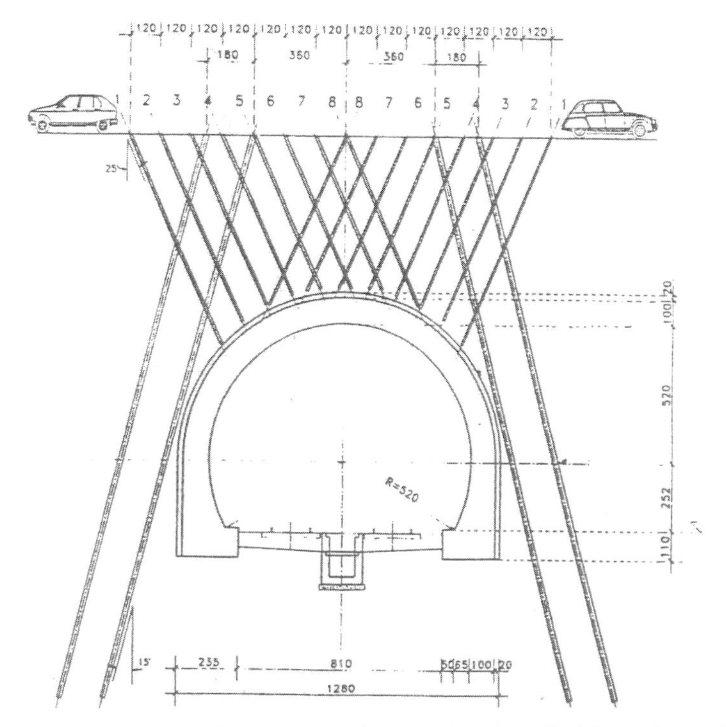

Figure 4 : Example of ground reinforcement with steel pipes installed from the surface [16]

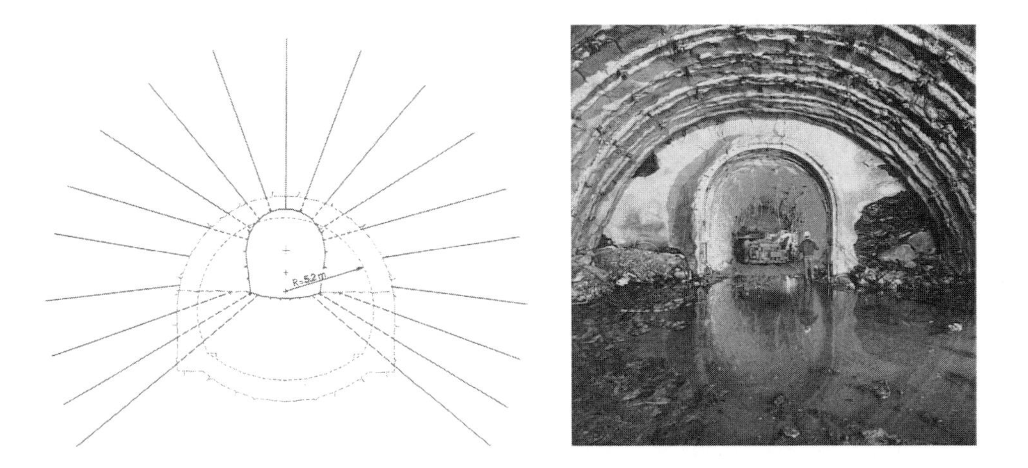

Figure 5: Example of radial ground reinforcing with fully grouted cables installed from the pilot bore [14, 15]

		Type of action					
Methods		method which modify the coinfinement-convergence curve	method which modify the value of the radial displacement at the moment of installing the supports	method which control tunnel face stability	method which control the length of the free span	method which control the local stability conditions	method which control the water inflow
Improvement	injection at low pressure	intervention around the tunnel	intervention on the core ahead of the tunnel face	intervention on the core ahead of the tunnel face	intervention around the tunnel	underpinning of steel arches	reduction of ground permeability
	jet-grouting	massive intervention around the tunnel	improvement of the ground ahead the tunnel face	improvement of the ground ahead the tunnel face	-	underpinning of steel arches	reduction of ground permeability
	freezing	massive intervention around the tunnel	improvement of the ground ahead of the tunnel face	improvement of the ground ahead of the tunnel face	massive intervention around the tunnel	-	massive intervention around the tunnel
Reinforcement	bolting	radial bolting below the tunnel face	radial bolting (from pilot tunnel)	-	-	local intervention on unstable rock blocks	-
	fully grouted cables	radial bolting below the tunnel face	radial bolting (form pilot tunnel, from a near tunnel, from the surface)	-	radial bolting (form pilot tunnel, from a near tunnel, from the surface)	-	-
	micropiles	radial bolting below the tunnel face	-	-	from the surface	underpinning of steel arches	-
	longitudinal bolting with VTR elements	-	longitudinal bolting of the core ahead of the tunnel face	longitudinal bolting of the core ahead of the tunnel face	-	-	-
presupports	mechanical precutting	-	concrete shell	-	concrete shell	-	-
	pretunnel	-	final concrete lining ahead the tunnel face	-	final concrete lining ahead the tunnel face	-	-
	forepoling	-	-	-	longitudinal pipes or bars	-	-
	jet grouting umbrella	-	longitudinal columns with or without reinforcement	-	longitudinal columns with or without reinforcement	-	-
	ground reinforement with fibre glass ("coronella")	shell of ground around the tunnel injected and reinforced with fiberglass elements	shell of ground around the tunnel injected and reinforced with fiberglass elements	-	shell of ground around the tunnel injected and reinforced with fiberglass elements	-	shell of ground around the tunnel injected and reinforced with fiberglass elements
drainage	-	-	-	all types	-	-	all types

Table 1: Classification of the various ground reinforcing methods used in tunnel construction [translated from 35].

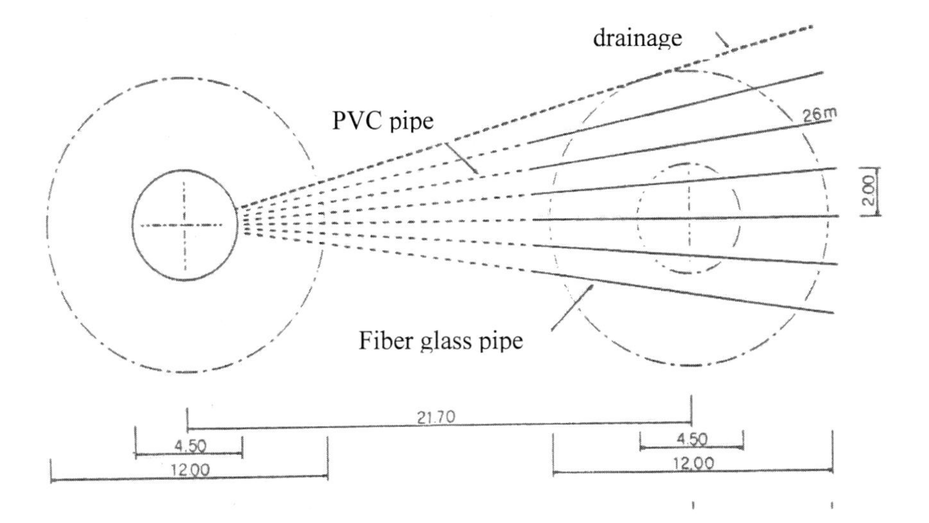

Figure 6 : Example of ground reinforcement with transversal fiber-glass pipes installed from a tunnel nearby [27]

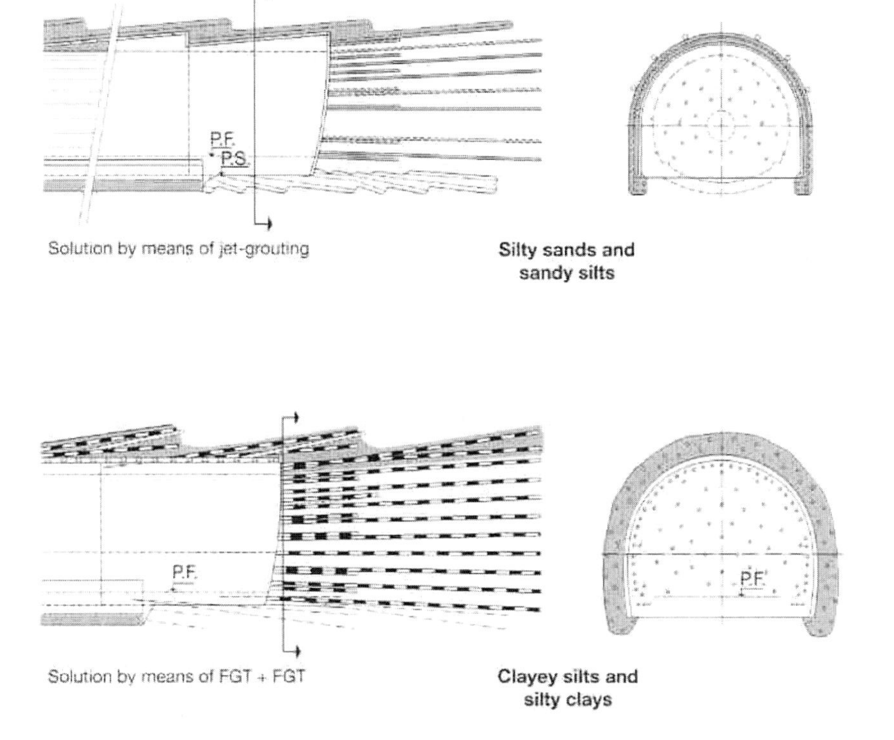

Figure 7: Example of some scheme of ground reinforcement using longitudinal fiber glass reinforcement [23]

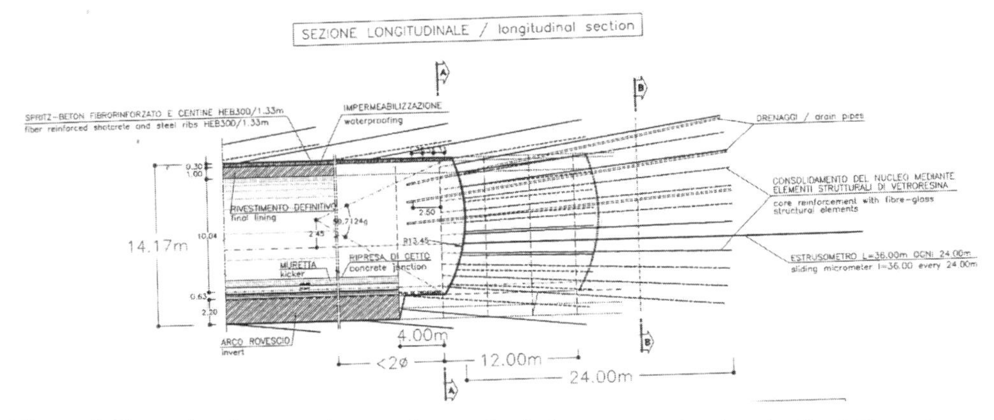

Figure 8: Example of some scheme of ground reinforcement using longitudinal fiber glass reinforcement in Tartaiguille tunnel (France) [22]

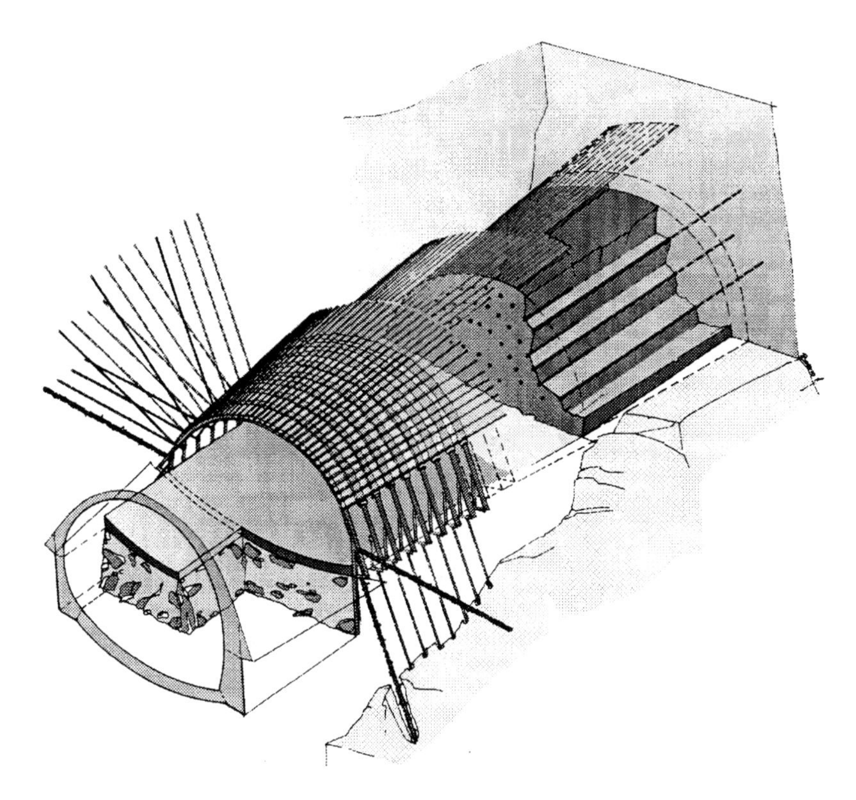

Figure 9: 3D view of the steel pipe umbrella technical and face reinforcement (Courtesy of Geodata S.p.A., Torino)

Figure 10 : View of the tunnel built to cross Highway 285 (Atlanta – USA). It is possible to see the crown of the microtunnels which were filled of concrete before carring out the excavation of the tunnel (Herrenknecht News, 1997).

Figure 11: Precut method and of the chain-saw machine

- *steel pipe umbrella or forepoling:* an umbrella of steel pipes or bars with a truncated conic shape is made on the crown of the future tunnel;
- *ground reinforcement using fiber-glass around the tunnel*: the ground is reinforced around the tunnel with a truncated conic shape, using grouted fiber glass pipes ("coronella") [21, 22, 34];
- *pretunnel*: a concrete lining (up to 1.5 m in thickness) is installed along the tunnel perimeter. This pre-lining may be used as the final lining or is integrated with the final lining. The cut is made with a special cutting boom. This technique has also been applied to widen existing tunnels which are kept in operation [22].

TYPE OF INTERVENTION	FIELD OF APPLICATION				
	Cohesive terrain	Sandy/gravely terrain	Terrain with boulders	Fractured rock	Complex Formations
grouting	⦿1	●			
jet-grouting	⦿2	●	⦿3		
freezing		●	●		
dewatering	⦿4	●	●	●	
fibreglass elements	●	⦿5		⦿6	●
pilot tunnel	●	●	●	●	●
precut	●	●			
pre-tunnel	●	●		●	
umbrella-arch	●	⦿7	●	●	●

Table 2. Fields of application for different interventions that can be performed from within the tunnel, ahead of the tunnel face (modified from [9]).

Key: ● Applicable. ⦿ Applicable with special intervention: 1 – chemical grout; 2 - two or three-fluid jet grouting; 3 - steel rebar or pipe reinforced jet grouting; 4 – active dewatering (vacuum pump required); 5 – additional grouting; 6 – high resistance element; 7 – additional grouting.
The interventions listed in this table can be combined in order to guarantee safe tunnelling conditions in almost all geotechnical conditions. Grouting, jet-grouting, freezing and dewatering can be normally be applicable also when tunneling under water table. The other interventions when the tunnel is under the water table must be combined with impermeabilization techniques.

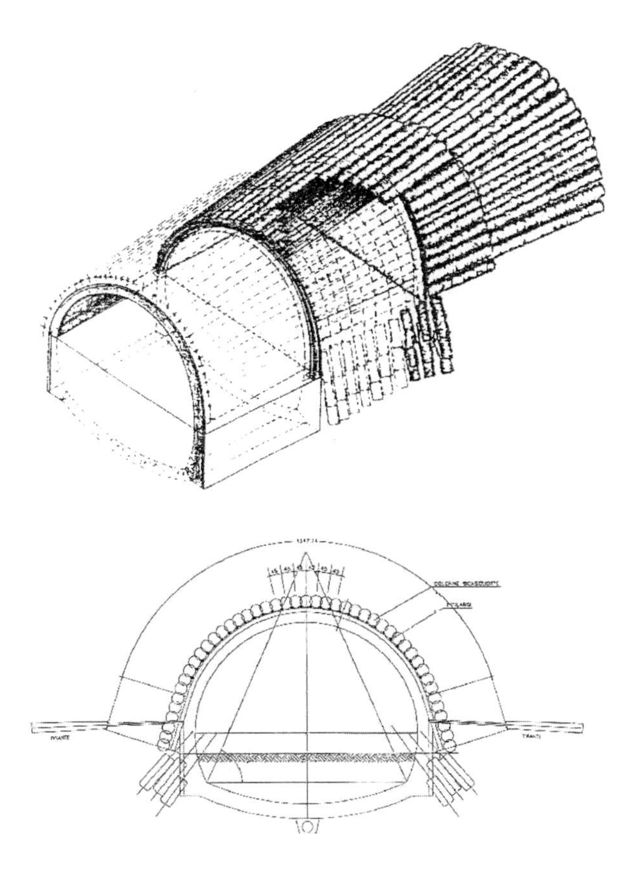

Figure 12: Example of the jet grouting arch [10]

The fields of application of the previously described techniques are summarized in Table 2 while the comparison among them is given in Table 3 [44]. Figure 13 reports the classification proposed by Hoek [18] for face excavation and use of support for large tunnels.

3. Steel pipe umbrella

3.1 Introduction

The steel pipe umbrella or forepoling method is a pre-reinforcement technique which is obtained by installing steel pipes or bars ahead of a tunnel face to provide a stiff structure able to mantain stable the advancement step for the time necessary to install the support [1, 2, 3, 7, 8, 9, 24, 25].

When steel pipes are used they are installed with a 5°-10° dip (with reference to the horizontal) in such a way as to form an umbrella with a truncated cone shape and which allows the overlapping of two adjacent umbrellas. When using this method, it is possible to cover advance lengths of 12-13 m, 9- 10 m of which are of excavation, because it is necessary to have an overlap between two sets of pipes to guarantee the stability of the face. This technique allows safe excavation even in poor geotechnical grounds and at a low overburden but it cannot be used in a ground where large boulders are present [33, 50].

The evolution and diffusion of this method has increased thanks to the technological developments that have been used which have led to the construction of bigger and more efficient installation machines (figure 14). Sometimes in small tunnels or when the rock mass properties are suitable it is possible to use short steel bars which can be installed using a conventionl drilling machine ("Jumbo"). In these cases in technical literature the intervention is described as "forepoling" (figure 15).

When steel bars are used their inclination is usually greater and length of the intervention is lower than when steel pipes are used. In this last case self drilling bolts have been applied with good results as was done for the excavation of some of major rail links in Germany [17, 24, 25, 26] and also in some tunnels in Italy, France and Greece.

Methods	Flexibility	Feasibility	Durability	Ease of Carrying Out the Work	Speed of Carrying Out the Work	Field of Application	Controls
Jet grouting arch	From medium to high (it is possible to reinforce the columns with pipes)	Medium (difficult to know the condition before excavation)	High	From medium to difficult (need for trained workers)	Medium (to be alternated with excavation)	Various soils (difficult in soil with boulders or clay)	Difficult
Pipe umbrella	High	High (elements are placed securely in the ground)	Not yet well known (possible to use non-corrodibl materials)	Easy	High (to be alternated with excavation)	Various soils (from moraine to sands)	Possible
Radial ground reinforcement	Very high	High (elements are placed securely in the ground)	Not yet well known (it is possible to use non-corrodibl materials)	Easy (using either cables or bars)	From medium to high: 1. To be alternated with excavation; 2. Preventive from a pilot tunnel (or from an adjacent tunnel)	Various soils (from moraine to hard rock)	Possible
Face reinforcement	High (can be coupled with other methods)	High	Timely intervention	Easy	From medium to high (to be alternated with excavation)	Poor soil or clay (and/or with boulders)	Possible
Grouting	Medium (requires suitable ground)	From medium to high (necessary to carry out tests)	High	From medium to easy	Low; normally preventive: • from a pilot tunnel; • from the surface; • from advancing face; • from an adjacent tunnel	Sand, gravel	Possible to difficult
Mechanical precutting	Low (often coupled with face reinforcing)	High	High (as a concrete lining)	Difficult	Medium (to be alternated with excavation)	Clay or cohesive soil	Possible
Freezing	Low	Medium	Timely intervention	Difficult	Medium	Saturated soil	Possible to difficult
Drainage	High	High	From medium to high	From medium to easy	High	Soil with water	Difficult

Table 3 Comparison of soil and rock improvement methods in underground works [43]
Key: flexibility: adaptability of the method to the variations of the geological and geotechnical conditions; *feasibility*: guarantee that the result is in compliance with the desired one; *durability:* decay in time of the resistance characteristics of the reinforced elements or of the reinforced soil; *carrying out*: easeness of construction related to the dimensions, type and costs of the machines and the need of specialized workers;*control*: if it is possible to monitoring the intervention

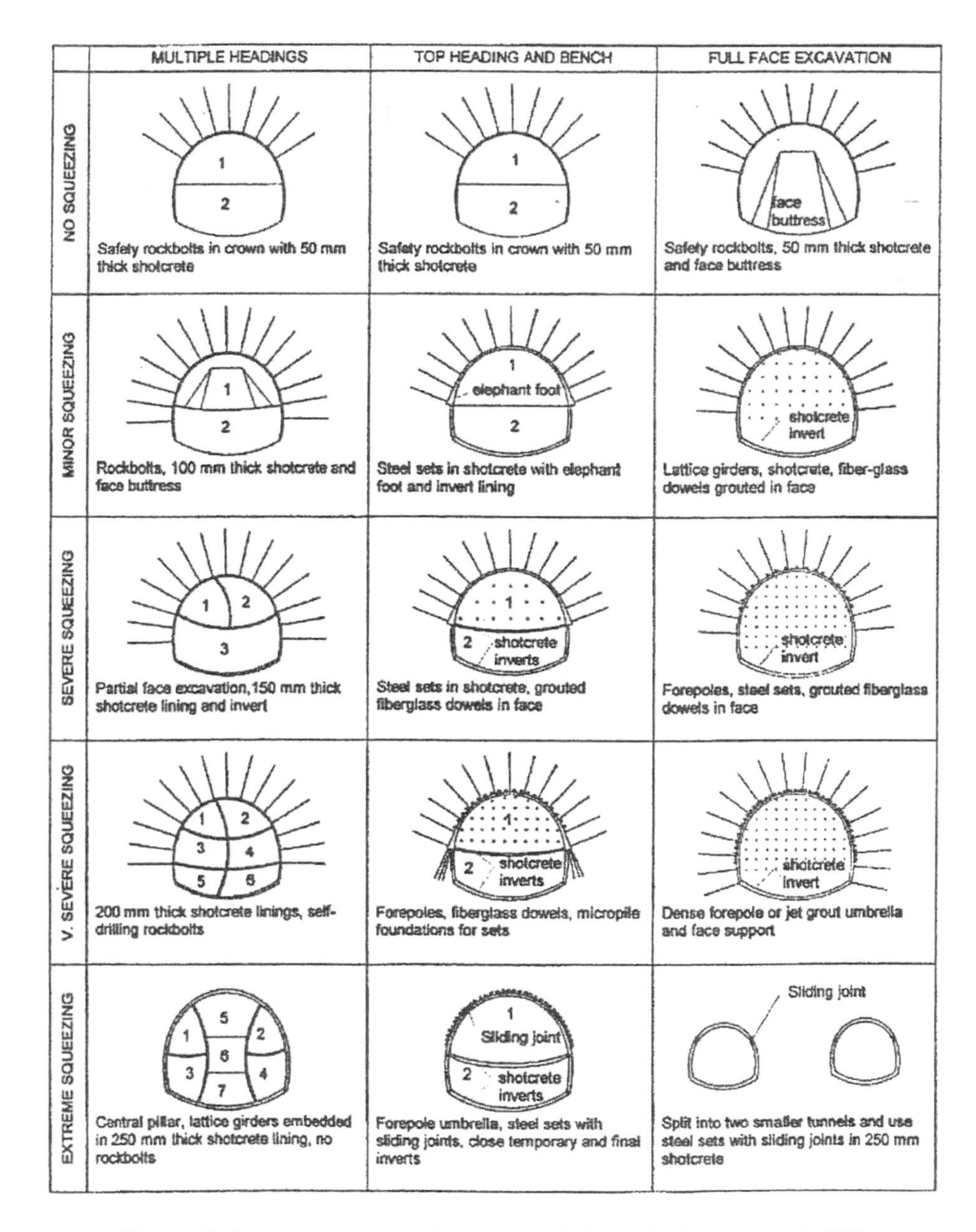

Figure 13: Face excavation and support techniques for large tunnels [18]

Figure 14: Examples of machine for the installation of the steel pipe umbrella (courtesy of Rodio S.p.A. and Soilmec brochure)

Generally speaking, this technique has been used for [1, 2, 8, 9, 33, 35]:

- the construction of shallow tunnels in soft ground where a good control of the ground displacements for reducing the surface subsidence is necessary, this been necessary to avoid damage to any infrastructures and buildings on the surface and also prevent the start or the activation of slope instability phenomena [42];
- for the construction of tunnel portals which are frequently problematic zones (figure 16) [47];
- for the construction of tunnels through weak ground with a high overburden or through fault zones (figure 17) [4, 44];
- for crossing zones where the tunnel has collapsed (figure 18) .

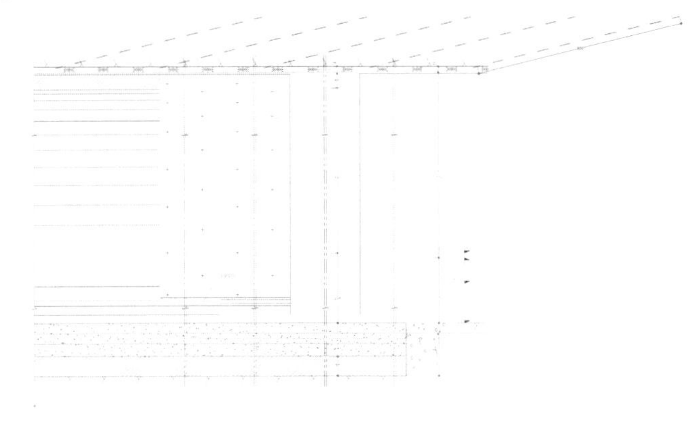

Figure 15: Example of a forepoling scheme (Courtesy of Atlas Copco S.p.A.)

Figure 16: Examples umbrella-arch construction at a tunnel portals

3.2 Carrying out procedure

The typical procedures for the steel pipe umbrella technique are [8, 9, 33, 50] (figures 19, 20):

- positioning and blocking of the guiding steel set machine;
- perforation (the diameter is usually between 140 and 200 mm) and placing of the metallic tubes (external diameter between 70 and 180 mm and thickness between 4 and 10 mm);
- formation of a buffer seal at the face edge of the pipe;
- injection stage of the pipes;
- excavation advancement in stages for a length equal to the distance between the steel sets;
- installation of a new steel set and of a shotcrete lining.

Points 1-4 and 5-6 represent two different operative cycles: the first refers to the pre-reinforcement intervention, the second to the tunnel excavation and support intervention.

The main stages if an excavation heading and benching is chosen are:

- *heading excavation*

Generally with a free span of 0.75-1.5 m and with the installation of the steel arches, shotcrete (mesh or fiber reinforced and with a thickness of 15-20 cm).

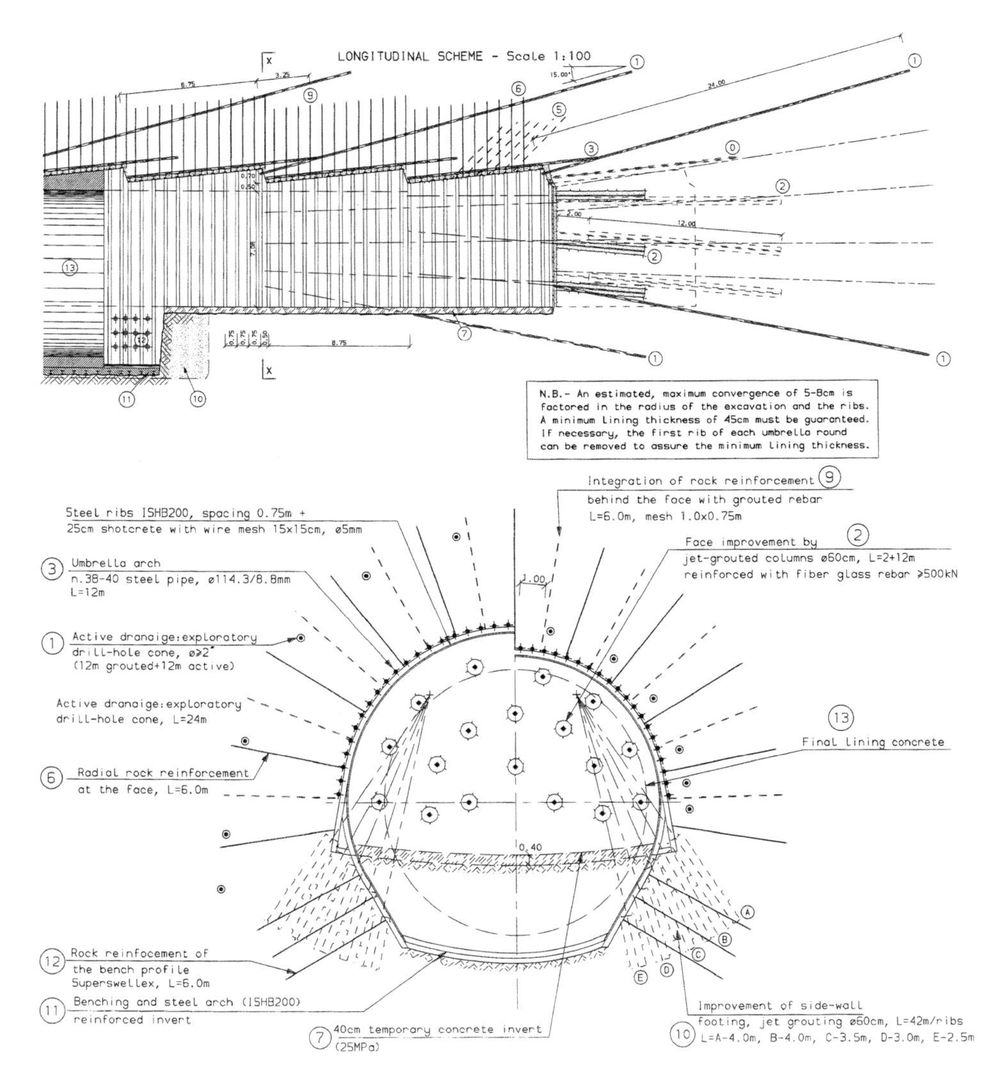

Figure 17: Section and longitudinal tunneling scheme that was adopted to cross the Daj Khad shear zone (Headrace tunnel of Nathpa Jhakri Hydroelectric Project - India) [9].

- *bench excavation*

Normally this operation is carried out alternatively on the two sides of the tunnel after the reinforcement of the feet of the steel ribs by micropiles or jet grouting columns. After the excavation the legs under the steel ribs are installed. Finally, if necessary the reverse arch is excavated and cast.

In same cases the excavation has been carried out full face but this procedure requires the use of systematic tunnel face reinforcement, in other cases multiple drifts attach was used [25, 26].

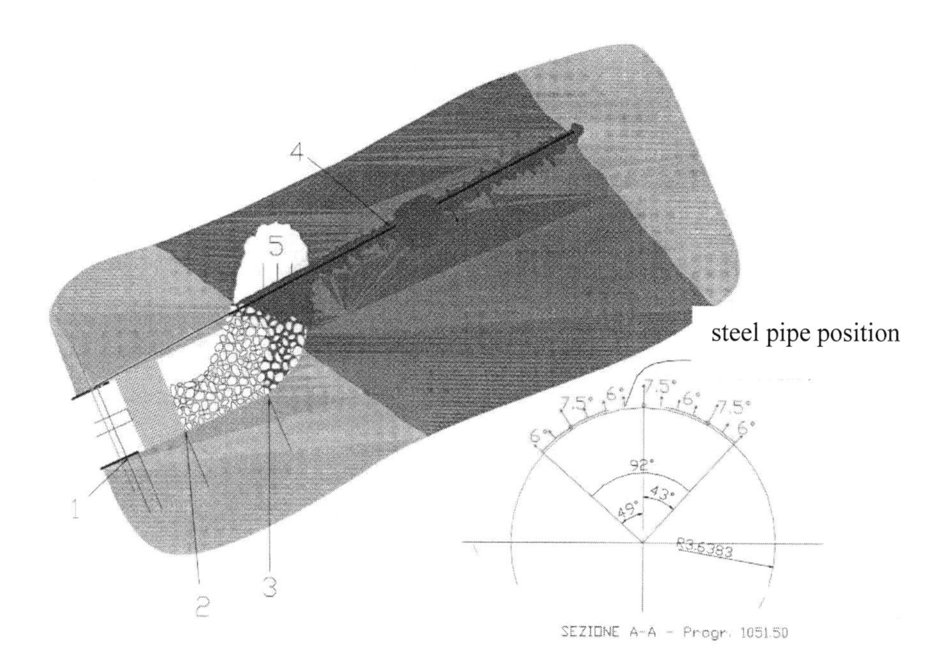

Figure 18 : Example of steel pipe umbrella used to cross a zone which has collapsed in a tunnel excavated with a TBM [6]. Key : 1-TBM; 2- position of the TBM head during the execution of the remedial works; 3- TBM head position when the roof instability occurred; 4- Steel pipe umbrella; 5-Instability area

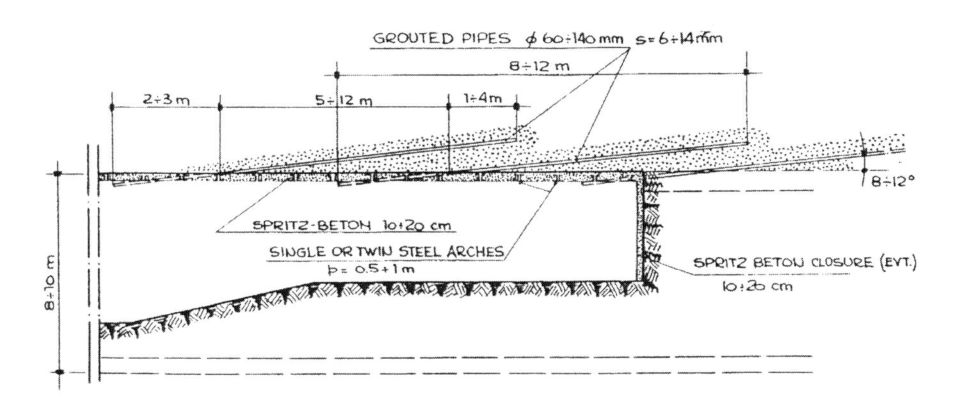

Figure 19: Cross-session of a tunnel excavated with steel pipe umbrella [2]

The geometry of the intervention, and more in particular the diameter, the thickness and the transversal spacing of the pipes, depend on: the depth of the tunnel, the distance between the steel sets, the type of steel set and its foundation and the stiffness of the ground ahead of the face.

Where necessary (depending on the bearing capacity of the ground and the trasmitted loads) the feet of the steel ribs can be underpinned with micropiles or bolts and their size must be enlarged (figures 22, 22).

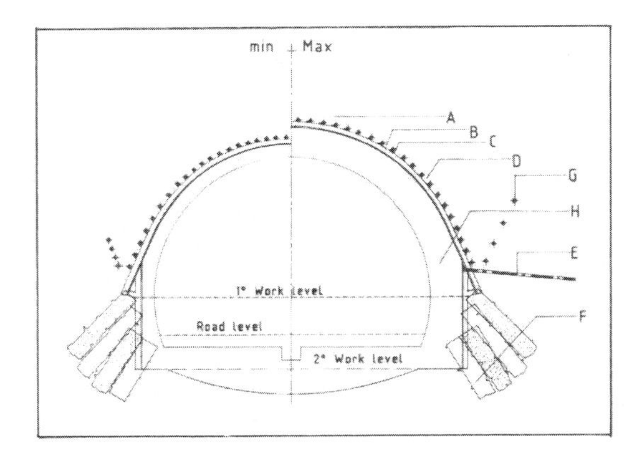

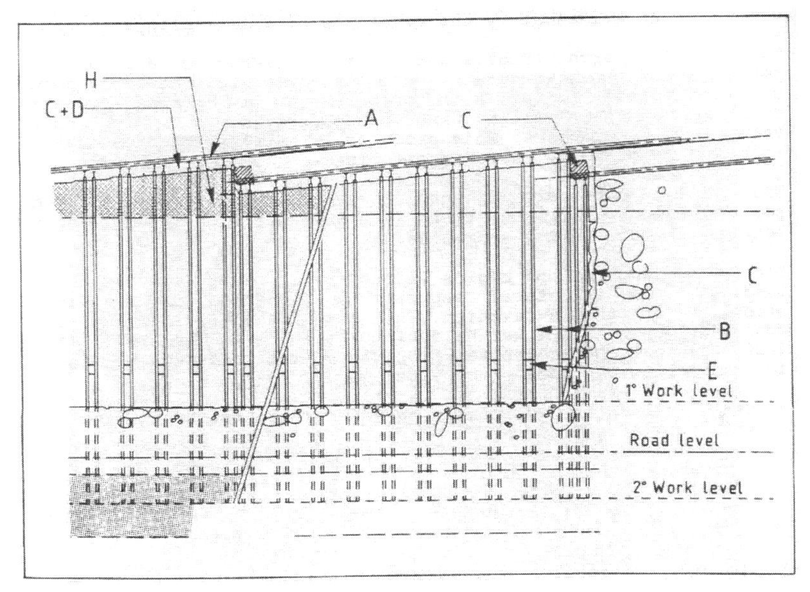

Figure 20: Cross-section of Lonato tunnel (Italy) where steel pipe umbrella has been used [8]. The excavation has been carried out heading and bench.
Key: A: steel pipes; B: twin steel arches; C: shotcrete; D: electro-weldes mesh; E: micropile; F: jet grouting column; G: drainage pipe, H: concrete lining

Figure 21: Detail of the installation of the steel arch under the steel pipe umbrella used for crossing a fault zone (D. Peila)

Figure 22: View of a tunnel face where a steel pipe umbrella with the enlarged steel arch foot is evident. On the right there is the detail of the enlarged foot of a steel arch (S. Pelizza)

Due to the great importance of the contact between the pipes and the support structures, which are generally steel arches and shotcrete, a "bullflex hose" has sometimes been used to assure quick and secure contact. With this technique it is possible to assure that there is no gap between the pipes and the steel arches therefore reducing any possible settlements (figure 23).

3.3 Drilling techniques

For the installation of streel pipe umbrella the need of drilling relatively long, properly aligned boreholes in a truncated cone arrangement along the perimeter of the tunnel prompted drilling equipment constructors to design and develop machinery that would enable perfectly aligned holes to be drilled quickly and economically for a considerable distance ahead of the excavation.

The machines were designed so that a long horizontal mast supported by two telescopic columns, could be mounted onto a highly manoeuvrable hydraulic base carrier. The most recent machines have two independent masts to reduce the drilling times. A number of experimental projects were carried out in the early eighties and from these the length of 12 m was determined to be optimal for the boreholes [1, 2, 8, 9].

For many years now this length has been adopted as the length of steel pipe umbrella; consequently the machines, which are designed for "one-pass" drilling over all the length, are built with long, 14 m stroke masts. There are however designs today that call for the boreholes to be as long as 18 m and machine manufacturers are re-designing some of their larger rigs to cope with this new requirement.

Since the two telescopic columns are mounted onto ring gears that permit their rotation, the mast can rotate through an arc greater than 180°. Therefore once the rig has been set up correctly it is possible to drill all the boreholes without moving the machine.

Numerous drilling techniques are used in forepoling and they are chosen on the basis of the formation of the ground that has to be drilled. The most important examples are briefly described in the following [9].

Figure 23: Bullflex hose installed between the steel arch and and the pipes (S. Pelizza). The bullflex is a tube of plastic fiber with some valves to allow it to be filled with the same type of concrete that is used for shotcrete. Its filling allows an optimal contact between the steel arches and the pipes and gives a precharge to the steel arch.

3.3.1 Open hole drilling

When the formation to be drilled is stiff and the borehole will remain open for a short period of time, open hole drilling methods are used. The boreholes usually range in diameter from 104 mm to 220 mm. Where the formation is stable, continuous flight augers, drag bits, tricone roller bits or down-hole hammers can be used to drill. Efficient flushing or removal of cuttings from the borehole is one of the most important factors that governs the speed and efficiency of all drilling operations and requires special attention. When choosing the most appropriate flushing fluid it is important to ensure that its application is appropriate to the drilling method, the nature of the ground and the tunnel environment. The flushing fluid must also not give rise to any significant overbreak and consequent instability of the borehole or the face and it must not pollute the working environment.

Air is the most commonly used substance since it is suitable for drilling in most types of rock, for stiff clays and well compacted, stiff or cemented sands but it not recommended for drilling in soft clays, loose sands or where soils are saturated. When drilling underground, there is the problem of the large amounts of dust it creates, therefore in tunnel environment air is used only in conjunction with special dust suppression systems.

Water can be used for flushing in all formations. It is commonly used when drilling through soft soils, or when drilling under the water table and at a depth but cannot be used when drilling with D.T.H. hammers.

Drilling muds are used when the soils have a high permeability, such as in sands and gravels and where there is danger of instability of the borehole due to loss of the flushing fluid through the borehole wall.

Polymers are being used increasingly to replace water and bentonite. They are usually made from biodegradable materials and are therefore environmentally much more friendly. In the tunnel environment polymers are used frequently when air is chosen as the flushing fluid. Small quantities of a polymer are injected into the air stream, and when it exists from the tool, it helps suppressing dust and, in difficult formations, it provides short-term stability of the borehole. In contrast to the use of bentonite, only small quantities of polymer are needed during the drilling process and the expensive treatment and recycling systems, which are necessary when using bentonite, can be avoided.

Foam is also used in conjunction with air flushing in those circumstances where air alone cannot be used as a flushing fluid and where water or bentonite cannot be used because of the type of drilling tool being used or because of the nature of the soil.

3.3.2 Drilling with casings (liners)

When borehole stability is not guaranteed or when water, bentonite, polymers or foam cannot be used to provide sufficient support (e.g. when there are cavities in the rock) a steel casing (lining) can be installed in the borehole to guarantee its stability. Air or any other fluid can then be used for flushing. Two methods are generally used:

"Drive drilling method": in soft formations.

An open-ended length of steel tube, fitted with a cutting crown, or occasionally a casing fitted with a “lost tip”, is directly connected to the rotary head by a blind coupling bell and is driven to the required depth without the soil being removed. The tube is kept full of water under pressure; in this way it displaces the soil as the cutting crown or “lost tip” advances.

“Duplex drilling method”: in stiff formations.

A duplex drill string is made up of an inner rod or auger, fitted with a pilot bit, and an outer casing fitted with a cutting crown. Both the rod and casing are connected to the rotary head by an open coupling bell or by a bayonet type connector; both are then made to rotate and are driven simultaneously. The flushing fluid is pumped to the pilot bit through the inner rod and the fluid and cuttings then return to the surface through the annular space between the rod and the casing.

When drilling in very hard formations and in fractured rock or through an overburden that contains boulders, rotopercussive methods are generally used to install the casings. Three basic procedures are used in this case:

Pilot bit and casing

This system is used when drilling through medium hard formations or soft rock. A rotopercussion pilot bit is mounted onto the drill rod or onto a D.T.H. hammer and a suitable cutting crown is fitted to the casing shoe.

Eccentric systems

Two eccentric bit systems exist:

“Top drive” system. The drill string is made up of an outer casing (without a casing shoe) that is connected to the drill rod by a bayonet connector. The inner drill rod is directly connected to the shank adaptor of a drifter or to the rotary head and to a D.T.H. hammer. The drill bit is made up of two elements: a central pilot bit and an eccentric element, which extends beyond the radius of the casing when rotated in the clockwise direction. When the eccentric bit extends, it underreams the borehole drilled by the pilot bit to a diameter slightly larger than the casing’s outer diameter. The casing is pushed into the enlarged borehole by the rotary head. When the bit is made to rotate in the reverse direction, the eccentric bit retracts and can be withdrawn leaving the casing in place.

“Bottom drive” system. The drill string is essentially the same as in the “Top drive” system with the exception that the casing is not connected to the drill rod or to the rotary head; it has a special shoe with a machined inner edge that coincides with an impact shoulder that has been machined into the drilling bit. In this system the casing is drawn into the enlarged borehole by the ring bit.

Eccentric drilling systems are rarely used when drilling horizontally for forepoling in tunnels as they do not always ensure efficient flushing and consequently good productivity, particularly where ground formations are not homogeneous. Furthermore, the eccentric cutting bits, in many instances, cause the borehole to deviate from the desired alignment, a condition that is not in line with good forepoling practice.

Concentric systems

Two basic concentric bit systems exist: the Centrex and the Symmetrix systems. A third system, called the Concentrix system, is essentially the same as the Symmetrix system but is marketed by a different company.

In these systems a special shoe is welded to the casing. The casing shoe has machined shoulders on the inside that correspond to machined shoulders on the pilot bit and it accommodates a special cutting ring bit that is fitted externally with tungsten carbide inserts. The ring bit is able to rotate independently of the casing shoe. Impact energy from the external drifter or from the D.T.H hammer is transmitted to both the pilot and ring bits and both are made to rotate by the rotary head:

Centrex. Developed by Sandvik Rock Tools. In this system, percussive energy from the D.T.H. hammer or drifter and rotation from the rotary head are transmitted to both the pilot and ring bits to break up the rock. The casing does not rotate. It is drawn into the ground as the pilot and ring bits advance. Upon completion of drilling, the drill rods, hammer and pilot bit are unlocked from the casing shoe and are retrieved. The Centrex system is designed in such a way that it is also possible to unlock the pilot bit from the ring bit and to continue drilling without tripping out.

Symmetrix. Developed by Rotex Oy (Finland), is similar to the Centrex system with the exception that the ring bit has an internal bayonet coupling that locks onto the pilot bit so that they advance and rotate together. When drilling is completed, the drill rods, hammer and pilot bit are retrieved by a slight reverse rotation which unlocks the bayonet coupling. In this system, if further drilling without the casing is required, the pilot bit must be replaced, after tripping out, with a standard D.T.H. hammer bit whith a smaller diameter so that it will pass through the ring bit.

Concentric systems tend to be more expensive than traditional duplex drilling systems due to the cost of the special casing shoe and drive systems, but they are nevertheless quite popular when forepoling in difficult formations as they offer very good alignment of the borehole and casing and good productivity.

4 Design procedures

Over the recent years, in spite of the significant technological progress, the development of numerical design methods that are able to take into account all the parameters involved in the design of a steel pipe umbrella has not undergone such a deep development due to:

- the high number of parameters that are involved;
- the impossibility of correctly considering the influence of all the parameters through empirical methods (based on experience acquired on similar grounds);
- the inadequacy of analytical methods to simulate the tunnel face effect;
- the complexity of three-dimensional numerical method applications;
- the difficulties of interpreting the effect of the overlapping of the umbrellas and their connection to the steel sets, of the stiffness of the steel sets in comparison to the vertical loads and of the influence of the mechanical characteristics of the ground at the excavation face;
- the influence of the position of the excavation face on the stresses inside the pipes.

Therefore the design of a steel pipe umbrella is still based on empirical considerations or simplified schemes. A short description of these approaches is given in the following.

4.1 Empirical design approaches

To determine the *required steel pipe section*, the pipe is considered to be a continuous beam on two or multiple supports (steel arches) embedded in the ground ahead of the excavation face. The concrete which fills and surrounds the pipe is not normally considered in the calculation.

The computation is carried out for the most critical phase, which is just before the installation of the steel rib as the free span is the longest.

The two most used calculation schemes are summarized in figure 24:
The acting load on the pipe [q] can be evaluated, starting from the value of the maximum vertical stress $q = p_v \cdot i$ where "i" is the spacing between the pipes. One of the problems is the evaluation of the vertical stress that is acting near the face: in many cases, it is empirically assumed that $p_v = 0.50\text{-}0.75$ of the total vertical load before excavation. In some real cases the load proposed by the well known formulations of Terzaghi was used.

The length ahead of the tunnel face which is not acting as support of the pipes (g) is usually empirically chosen and very often the value of 0.5 m is assumed. This value is, obviously, directly linked to the geomechanical properties of the ground and to the presence of tunnel face reinforcements [13, 36, 38, 40, 41]. More detailed research must be developed for a complete and final definition of this length.

Considering that the support action of the of the pipes must be developed for a short period of time before the tunnel support are installed (steel archs and shotcrete), the admissible working stress (S_{adm}) of the steel of the pipes can be close to its field stress ($1.5>F_s>1.1$). Therefore knowing the acting stresses it is possible to chose the type of pipe.

The choice of the *length of the steel pipe* is linked to practical reasons, that is, drillability and the maximum bore hole deviation which limit the length of 15–18 m; while the length of the overlap between two subsequent umbrella is controlled by the behaviour of the ground ahead the tunnel face. In recent years there have been numerous studies on tunnel face reinforcement with longitudinal pipes based on small scale laboratory tests [51], field tests and numerical modelling [31, 37, 38, 41, 51, 52]. The results of these researches suggest that the length of the overlap must not be less that 0.4 times the equivalent diameter of the tunnel.

The *interax between the pipes* is chosen taking into account the fact that the ground must not flow between the pipes. Therefore, the natural cohesion of the ground should be able to control and prevent the occurrence of this phenomenon. Simple calculations can be carried out, considering the stability of the slice of ground onto two nearby pipes.

The described appoach is very simple and its application has been consolidated in time but it neglects some parameters which are very important for the design. It does not consider the real stiffness of the supports (steel arches and tunnel face), the effect of the ground ahead of the face and the own bending stiffness of the steel pipe. In very complex problems (where for example it is necessary to know the induced settelements exactly) it is necessary to use numerical models (mainly tri-dimensional) but better results could be obtained by using a model based on the approach of a beam on multiple supports. Which use could be suggested.

A study based on this approach has been presented by Oreste and Peila [33] where the steel pipe was modelled as a beam on multiple supports and the ground ahead of the face was simulated with a series of independent springs (according to Winkler's ground theory); the steel sets were modelled with springs with their own stiffness (due to the foundation and the arch); the overlapping of two successive umbrellas was schematised with a yielding fixed end which allows only vertical displacements. These authors developed a wide parametric analysis for typical geomechanical and geometric parameters permitting useful qualitative considerations to be made on the bending moments, shear forces and displacements along the steel pipes. They also developed many abacus of the maximum bending moment in the steel pipe, as a function of the applied load thus allowing a simple pre-designing of the steel pipe umbrella.

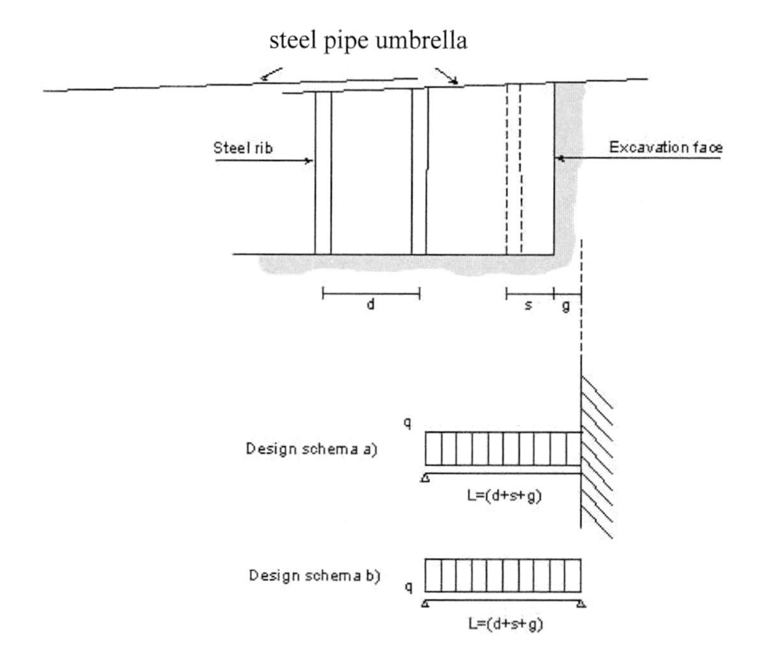

Figure 24: Simplified schemes used for steel pipe umbrella design [33, 35]

From an analysis of their results it is possible to see that:

- the bending moment ahead of the face is not always higher than the continuity moment in correspondence to the last steel arch (contrary to what is usually assumed in the calculation with simplified methods);
- the rigidity of the supports has a great influence on the results, in particular when these values are small (i.e. in a ground of low geotechnical characteristics or with yielding supports) (figure 22);
- the bending stiffness of the pipe has a great influence on both the bending moments and shear forces: an increase of this value causes an increase of the continuity moments in correspondence to the last steel arch and a reduction in the bending moments in the length ahead of the tunnel face;
- the trend of vertical displacements is mainly controlled by the pipe bending modulus and by the distance between the steel arches.

Since the definition of correct values of the support stiffness is very difficult and its values are not constant for all the pipes of the same umbrella (the stiffness which must be considered for the pipe at the centre of the roof is different from the value which must be considerd on a pipe located on the side) it could be suggested to carry out parametric analyses in order to design the steel pipe umbrella on the basis of the worst foreseeable condition.

Some simple procedure for the calculation of these parameters usign the stiffness of the steel arches and of the ground ahead of the tunnel face can be found in [33]; reference values of steel arches stiffness obtained with experimental tests can be found in [20].

A great care should also be taken to evaluate the stiffness of the springs that simulate the ground ahead of the face. Obviously this value should be defined taking into account the effect of the tunnel face stability and the use ot tunnel face reinforcement.

It is very important to point out that the calculation shows that if a support moves down the acting bending moment inside the pipes rises up to very high (and often critical) values. Therefore a great care must be given in practice to have a good and stable foundation of the steel arches.

3.2 Numerical modelling

The use of numerical modelling for the design of tunnel support and reinforcement works, but tri-dimensional numerical models, which are more suitable for the analysis of the real problem of ground reinforcing ahead of the tunnel face have, however, rarely been used to analyse the behaviour of steel pipe umbrella [18, 33].

This procedure has the following problems:

- it supposes that the steel pipes act also in a transversal way by forming a continuous arch which it is not true for steel pipes (but is surely true for jet grouting arch);
- it does not correctly represent the tri-dimensional bending strength of the umbrella;
- it is not completely clear which is the correct value of the radial internal pressure to be applied on the tunnel boundary to simulate the support effect of the tunnel face (particularly in shallow overburden tunnels).

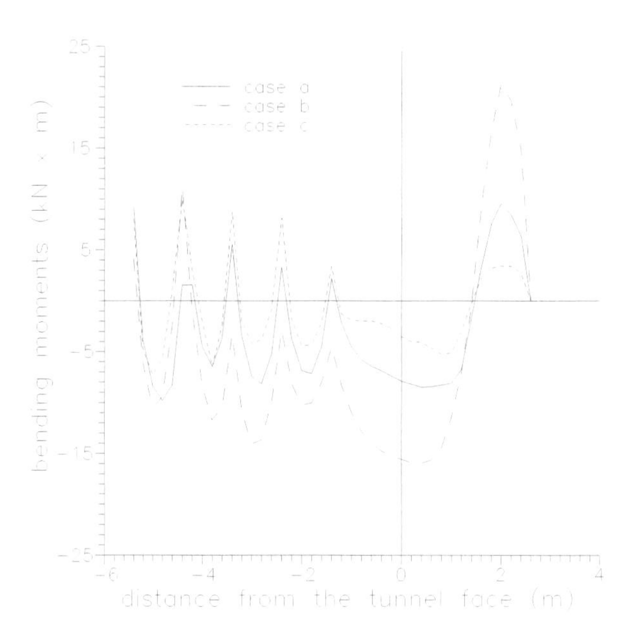

Figure 22. Computed bending moments for 3 different cases of the pipes bending stiffness and different steel arches stiffness [33]

case a): $J=429,6\ cm^4$, $k=2400\ kN/m$;

case b): $J=843,9\ cm^4$, $k=800\ kN/m$;

case c): $J=249,1\ cm^4$, $k=9200\ kN/m$.

These figures confirm the importance of the stiffness on the calculation results.

Uhtsu et al. [49], who have studied this probem in urban areas at a shallow overburden with bi-dimensional and tri-dimensional FEM models, demonstrated that the settlement of the foot of the steel rib is one the major causes of settlements at the tunnel crown and at the surface. These results are in good agreement with those obtained by Oreste and Peila [33] who used tri-dimentional FLAC modelling.

A design procedure for a tunnel where steel pipe umbrella has been used ahead of the tunnel face by a bi-dimesional numerical modelling has been proposed in [18]. In this paper was proposed to model the steel pipe umbrella as an arch of improved material using a weighted average (based on cross-sectional area) of the properties of the pipes, the grout and the rock mass. Then, as usually done in numerical modelling [31, 34], an internal support pressure has been applied to limit the displacement of the tunnel walls (near the face) to a chosen percentage of the total final value. At this stage the supports (steel arches, shotcrete and rock bolts) are applied and the tunnel face advancement is modelled by reducing this internal pressure. As face advances it is proposed to reduce the capacity of the steel pipe umbrella to that of the rock mass due to the fact that the support is no longer available to allow it to act as fully effective shell.

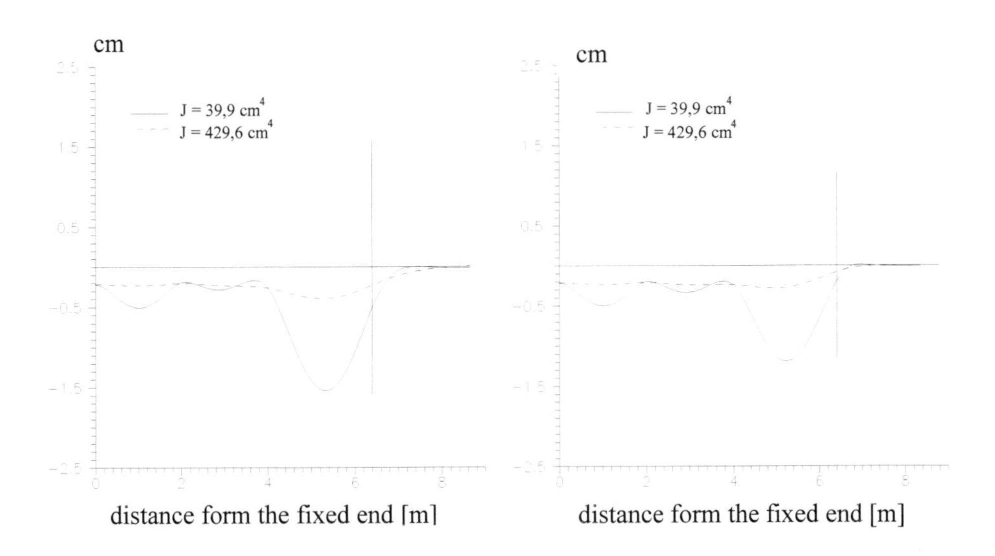

Figure 23: Vertical displacement for various bending stiffens of the pipe for a steel arches stiffeness of 500000 kg/m, for an interax of 1m [33]. Key left: E_{ter}=400 kg/cm^2; right: E_{ter}=2500 kg/cm^2

Oreste at al. [32] have developed many tri-dimensional and bi-dimensional numerical models using FLAC codes to study this last problem and to analyse the influence of tunnel face reinforcement on the behaviour of tunnels at a shallow overburden. These authors have demonstated that the reduction of internal pressure to be applied for a correct bi-dimentional numerical modelling should be lower that the value of 50% (frequently used in practice). They also highligted the great importance of the tunnel face reinforcement on the control of the displacements of the tunnel boundary near and ahead of the tunnel face as can seen in figure 24.

Finally, interesting results on the working behaviour of steel pipe umbrella were, recently, obtained by Takechi et al. [48] who analyzed construction data, laboratory model test data (using decomposed granite) and field measurement data (surface settlements and deflection of steel pipes). These authors have shown that the umbrella-arch is effective for controlling tunnel face stability and surface settlement. In shallow tunnel conditions (overburden is less than 0.5 times the tunnel diameter) the tunnel crown settlement is proportional to the subsidence. The difference between the excavated tunnel crown displacement and the value of the surface settlement ($R_s = S_t - S_s$) decreases starting from an the overburden ratio bigger that than 0.75 times the tunnel diameter. These results are in good agreement with those obtained by Oreste and Peila [33]. They also classified the deformation behaviour of the steel pipes umbrella into three types according to the possible settlements and rotation of the tip of the pipes and their pattern of deflection:

- the steel pipe settlement and rotation at its end is insignificant;
- the tip of the steel pipe settles and rotates slightly but the portion of the steel pipe near the tip moves upwards;
- the tip of the steel pipe rotates and the pipe sags downwards.

On the basis of these concepts, it appears that the use of numerical bidimentional models gives some design problems for the correct modelling of steel pipe umbrella behaviour. In particular more researches must be developed to understand which geomechanical parameters have to be used and which stress release percentage has to be applied near the face to correcly model the tri-dimentional behaviour of the tunnel.

5 Conclusions

The use of reinforcement techniques during the construction of tunnels is by now a widely and systematically used practice through which it is possible to obtain large excavation sections in difficult geomechanical conditions or, as in the case of TBMs for rock excavation, to overcome geological conditions that are complex for a full face machine.

Conventional full section excavation, which today cannot rule out without reinforcement of the rock mass in difficult "geo" conditions, is an extremely complex technological challenge, but allows remarkable advantages to be obtained:

- a high degree of mechanisation of all the operations, thanks to the presence of wide excavation sections;
- high flexibility in adapting the reinforcement during advancement and at the tunnel boundary to the characteristics of the rock mass;
- good control of the excavation wall convergences and of the extrusion and stability of the excavation face;
- good control of any possible subsidences (in shallow tunnels);
- the possibility of constructing the inverse arch close to the excavation face and therefore of installing a closed lining that reacts better to the thrust from the rock.

Figure 24: Vertical displacements of the tunnel roof obtained with tri-dimensional modelling in the presence (continuous line) and without (dashed line) bolting of excavation face and results of the bi-dimensional calculation with a variation of the applied internal pressure before activating the presupports (8m ahead of the face) (granular soil, elasto-fragile behaviour - overburden of two diameters) [32].

However, the loads induced by the supports in tunnels with large diameters and in poor rock conditions with a scarce overburden, can reach such values, because of the limited tolerated convergence, that it is impossible to guarantee the stability of the void, even with heavy and closely fitted steel arches with a thick ring of shotcrete. In these cases, recourse to reinforcement of the rock, even at the tunnel boundary, becomes indispensable.

On the other hand, in the case of a systematic use of reinforcement, the designers have to adopt an elastic and versatile designing approach that is higher than what has been requested up till present, taking into consideration all the methodologies that technology makes available.

This however makes it necessary to evaluate the functioning modalities of the various reinforcements in a correct way: for this purpose measuring and verification calculations should always be performed with the most up to date procedures, in particular with the aid of tri-dimensional modelling methods.

The various alternatives should then be compared, from both the stability point of view (in the short and long term) and from the constructive point of view, that is, in relation to the facility of use and the advantages offered to the workers on the site.

Biblography

[1] Barisone G., Pelizza S. and Pigorini B.: Umbrella Arch Method for Tunnelling in Difficult Conditions - Analysis of Italian Cases, Proc. of IV International Association of Eng. Geol. New Delhi. Vol. IV, Theme 2, pp.15-27, 1982

[2] Barisone G., Pelizza S. and Pigorini B. Italian Experiences with Tunnel Portals in Difficult Ground, Proc. of Eurotunnel '83. Basle, Switzerland, pp. 157-163, 1983

[3] Barisone G., Campo F., Corona G. and Pelizza S.: Rapid Umbrella-Arch Excavation of a Tunnel in Cohesionless Material under an Archeological Site, Proc. of Int. Congress on Progress and Innovation in Tunnelling. Toronto, 1989

[4] Bellini A., De Domenico R. and Da Forno G.: Gli infilaggi metallici a contatto per il superamento con fresa scudata a piena sezione di una frana al fronte in graniti completamente degradati, con forti venute d'acqua, Proc. Congr. on "Soil and Rock Improvement in Underground Works", Milano, pp. 65- 72, 1991

[5] Benedetto W., Scataglini D. and Selleri A.: Evoluzione dei Sostegni in Galleria dalle Armature Classiche alla Vetroresina, Proc. Congr. on "Soil and Rock Improvement in Underground Works", Milano, Vol.1, pp.73-97, 1991

[6] Bethaz E., Fuoco S., Mariani S. Porcari P., Rosazza Bondibene E.: Lo scavo in rimonta con TBM: l'esperienza del cunicolo di Maen, Gallerie e grandi opere sotterranee, 61, pp. 67-80, 2000

[7] Carrieri G., De Donati A., Grasso P.G. , Mahtab A. and Pelizza S.: Ground Improvement for Rapid Advance of Lonato Road Tunnel near Verona, Italy, 8th Annual Conference of Tunnelling Association of Canada. Vancouver, Canada, pp.243-254, 1983

[8] Carrieri G., Grasso P.G., Mahtab A. and Pelizza S.: Ten Years of Experience in the Use of Umbrella-Arch for Tunnelling, Proc. Congr. on "Soil and Rock Improvement in Underground Works", Milano, vol. 1, pp.99-111, 1991

[9] Carrieri G., Fiorotto R., Grasso P., Pelizza S.: Twenty years of experience in the use of the umbrella-arch method of support for tunnelling, Int. Workshop on micropiles, Venice, 2002

[10] Cresta L. and Serra A.: L'uso del jet-grouting nell'esecuzione di gallerie in terreni sciolti, Proc. Congr. on "Soil and Rock Improvement in Underground Works", Milano, vol. 2, pp.145-155, 1991

[11] Dorge G.R and Mouxaux J.: Prevention of Accidents in Difficult Tunnelling Conditions by means of Specialized Techniques such as Grouting, Drainage, Umbrella Arch Methods, Proc. of International Tunnel Symposium. Tokyo, 1978

[12] Dolcini. G., Grandori R. e Marconi M. Importanza della progettazione specifica dei metodi esecutivi e delle relative attrezzature, per l'esecuzione di una galleria in condizioni geologiche difficili ed in tempi programmatici ristretti, Gallerie e grandi opere sotterranee, 1990

[13] Dias D., Subrin D. Wong H., Dubois P. and Kastner R.: Behaviour of a tunnel face reiforced by bolts: comparison beween analystical-numerical models, Proc. IInd Int. Conf. "The geotechnics of hard soils – soft rocks", Napoli, pp. 961-972, 1998

[14] Grasso P., Mahtab A. and Pelizza S.: Reinforcing a rock zone for stabilizing a tunnel in complex formations. Proc. of Int. Congress on Progress and Innovation in Tunneling, Vol.2, Toronto, pp.671-678, 1989

[15] Grasso P., Mahtab A., Pelizza S. and Russo G.: On diverse geothechnical and tunnel construction problems in the la Spezia-Parma rail link in Italy. Proc. Int. Congr. on Tunnel and Underground Works Today and Future, Chengdu. pp. 33-39, 1990

[16] Gusman M.: La linea Bari-Taranto, variante di Castellaneta, attraversamento A14, Proc. Congr. on "Soil and Rock Improvement in Underground Works",Milano, pp. 159-167,1991

[17] Haack A.: Underground Construction in Germany 2000, STUVA, 1999

[18] Hoek E.: Big Tunnels in Bad Rock, Journal of geotechnical and geoenvironmental engineering, september, pp. 726-740, 2001

[19] Iwasaki T., Miura K., Kawakita M. and Sano N.: A long tunnel Project by TBM method, Proc. Int. Congr. Challenges for the 21st century, Oslo, pp. 857-863, 1999

[20] Khn U.H., Mitri H.S. and Jones D.: Full scale testing of steel arch tunnel support, Int. Journal of Rock Mechanics, 33, pp. 219-232, 1996

[21] Lunardi, P., Bindi, R. and Focaracci, A. : Nouvelles orientations pour le project et la construction des tunneles dans des terrains meubles - Etudes et expériences sur le préconfinement de la cavité et la préconsolidation du noyau au front. Proceedings of Congr. Tunnels et micro-tunnels en terrain meuble, Paris, pp. 625-645, 1979

[22] Lunardi P.: La galleria "Tartaiguille", ovvero applicazione dell'approccio ADECO-RS per la realizzazione di un tunnel "impossibile", Gallerie e grandi opere sotterranee, 58, pp. 66-81, 1999

[23] Lunardi P.: The design and construction of tunnels using the approach based on the analysis of controlled deformation in rocks and soils, Tunnels and Tunnelling Int.– Supplement, May, pp.1-30, 2000

[24] Maidl B.: Developements in conventional tunnelling during pas two decades, Underground Construction in Germany 2000, Stuva, pp. 31-41, 1999

[25] Maidl B.: Shotcreting developments for German rail tunnels –part 1, T&T International, May, pp. 31-36, 2001a

[26] Maidl B.: Shotcreting developments for German rail tunnels –part 2, T&T International, June, pp. 25-27, 2001b

[27] Molinari L., Pignatelli M. and Pigorini B.: L'impiego del cunicolo pilota nel consolidamento delle rocce, Proc. Congr. on "Soil and Rock Improvement in Underground Works", Milano, pp. 195-208, 1991

[28] Mitchell J.: Soil Improvement – State of the art report, Proc. X ICSMFE, Stockholm, vol. 4, pp. 509-565, 1981

[29] Muir Wood A.: Tunnelling-Management by design, E&FN SPON, London, 2000

[30] Nishimaki A., Mitarashi Y. and Uematsu S.: Study On the effects of the AFG Method [Proposal of a Simplified AGI' Design Technique], South East Asian Symposium on Tunnelling and Underground Space Development. Japan Tunnelling Association, Bangkok. January, pp.125-132, 1995

[31] Oreste P.P., Peila D. and Poma A.: Numerical Study of Low depth Tunnel Behaviour", Proc. Int. World Tunnel Congress '99, Oslo, pp. 155-162 1999

[32] Oreste P.P. and Peila D.: Radial passive bolting in tunnelling design with a new convergence-confinement model. Int. Journal of Rock Mechanics and Mining Sciences, 33, n. 5, Pergamon Press, Oxford, pp. 443-454, 1996

[33] Oreste P.P. and Peila D.: A new theory for steel pipe umbrella design in tunelling, Proc. Int. Congress "Tunnel and Metropoles", San Paolo, pp. 1033-1040, 1998

[34] Oreste P.P . and Peila D.: I Consolidamenti come mezzo per permettere lo scavo meccanizzato di gallerie, MIR 2000, Ed. Barla, Patron Editore, Torino, 2000

[35] Oreste P.P., Peila D. "Study of the behaviour of tunnel reinforcement techniques using analytical methods", AITES/ITA World Tunnel Congress 2001 "Progress in Tunnelling after 2000", Milano, Vol. 1, Patron Editore, Bologna, pp. 579-587, 2001

[36] Panet M. and Guenot A.: Analysis of convergence behind the face of a tunnel, Proc. Congr. Tunnelling '82, London, IMM, pp. 197-204, 1982

[37] Peila D.: A theoretical study of reinforcement influence on the stability of tunnel face. Geotechnical and Geological Engineering, 12, Chapman & Hall, London, pp. 145-168, 1994

[38] Peila D. and Poma A.: Studio degli Interventi di rinforzo del fronte di scavo di gallerie con elementi longitudinali", Gallerie e Grandi Opere Sotterranee, 45, SIG, pp.39-49, 1995

[39] Peila D. and Oreste P.P.: Axisymmetrical analysis of ground reinforcing in tunnelling design, Computer and Geotechnics, 17, Elsevier Applied Science, Oxford, pp. 253-274, 1995

[40] Peila D., Oreste P.P., Pelizza S. and Poma A.: Study of the influence of sub-horizontal fiber-glass pipes on the stability of a tunnel face. Proc. North American Tunnelling '96, Washington, pp. 425-432, 1996

[41] Peila D., Oreste P.P. and Pelizza S.: Face reinforcement in deep tunnels. Proc. Int. Congr. Underground Construction, Praga, pp. 232-243, 2000

[42] Pelizza S., Barisone G., Campo F. and Corona G.: Rapid umbrella arch excavation of a tunnel in cohesionless material under an archeological site. Proc. Int. Congr. on Progress and Innovation in Tunnelling. Toronto, pp. 885-891, 1989

[43] Pelizza S.: Armatura dei terreni. Proc. Congr. on "Soil and Rock Improvement in Underground Works", Milano, vol. 2, pp. 3-30, 1991

[44] Pelizza S. and Peila D.: Soil and rock reinforcements in tunnelling, Tunnelling and Underground Space Technology, 8, n.3, Pergamon Press,Oxford , pp. 357-372, 1993

[45] Pelizza S., Peila D. and Oreste P.P.: A New Approach for Ground Reinforcing Design in Tunnelling, Proc. Int. Congr. Tunnelling and Ground Conditions, Cairo, pp.517-522, 1994

[46] Pelizza S.: Engineering Risks in Tunnelling, Underground Construction in Germany 2000, STUVA, pp. 103-106, 1999

[47] Pelizza S. and Peila D.: Criteria for technical and environmental design of tunnel portals, Tunnelling and Underground Space Technology, vol. 17/4, Pergamon Press, Oxford, pp. 335-340, 2002

[48] Takechi H., Kavakami K., Orihashi T. and Nakagawa K.: Some considerations on effect of the long-fore-piling-method. AITES-ITA 2000 World Tunnel Congress, Durban, March, pp.491-498, 2000

[49] Uhtsu H., Hakoishi Y., Nago M., Taki H.: A prediction of ground beharing due to tunnel excavation under shallow overlorden with long-length forepilings, South East Asian Symposium of Tunneling and Underground Space Development Japan Tunneling Association, Bangkok, January pp.157-165, 1995

[50] Yamamoto and Pagliacci M.: New tunneling technology for a group of tunnels in Aosta Valley. Tunnels and Underground, Vol.23, No.1, pp.21-28, 1999

[51] Yoo C. S. and Yang K. H.: Laboratory investigation of behaviour of tunnel face reinforced with longitudinal pipes. AITES-ITA 2001 World Tunnel Congress, Milan, pp.757-764, 2001

[52] Yoo C. S. and Shin H.K.: Deformation behaviour of tunnel face pre-reinforced with longitudinal pipes. AITES-ITA 2000 World Tunnel Congress, Durban, March, 2000

Acknowledgements

Special thanks are due to Eng. Oreste, Eng. Oggeri and Dr. Carrieri who have generously shared their ideas and experience as well as to Dr. Fiorotto (Casagrande S.p.A.), Ing. Airoldi (Atlas Copco S.p.A.) and the companies Geodata S.p.A. (Torino) and Rodio S.p.A. which provvided pictures and drawings inserted in the text.
The authors have given the same contribution to the development of this paper.

Part II

Water problems

Problems of TBMs in water bearing ground

Lars Babendererde [1]

[1] Babendererde Ingenieure GmbH, Am Lotsenberg 8, Lübeck – Travemünde, Germany, Email: contact@bab-ing.com, URL: http://www.bab-ing.com

ABSTRACT

Tunnelling in water bearing soft ground is always a challenge even for experienced crews and engineers. The stabilisation of the tunnel face and the water inflow into the excavated space can be mastered by application of mechanised tunnelling methodology. Even though risks can be reduced by TBM tunnelling, unexpected problems may still arise during the tunnelling process. Special methods need to be applied to overcome the problems and to guarantee a successful completion of the tunnelling project. The following four case studies, problematic TBM projects in water-bearing soft grounds, are presented to show how the problems were overcome.

1. Introduction

Water has a more or less significant impact on tunnelling conditions. The softer the ground, the more dramatic its influence. In hard rock, the impact of water is restricted to fault zones with soft consistence.

However, in soft material the ground water determines the tunnelling methodology. If dewatering of the ground is not feasible, tunnelling in water-bearing soft ground can only be performed if the inflow into the excavated space is prevented. To achieve this, the permeability of the ground can be reduced by grouting or ground freezing, or the water inflow can be prevented by application of mechanized tunnelling methodology.

Modern Tunnel Boring Machines, with competent operation procedures, permit tunnelling in water-bearing ground with limited risks. Nevertheless, due to the considerable spectrum of natural ground conditions, even experienced tunnellers will continue to be surprised by unexpected events. An account of some of these events is given below.

2. Jammed by compacted soil in surrounding steering gap.

2.1 Pipe - Jacking Machu Picchu

In 1998 a major landslide occured along the Vilcanota River overflow for the Hydroelectric Power Project Machu Picchu in Peru at the feet of the famous Inka ruins. To re-establish the plant, a 150 m long twin-tunnel of 3.10 m outer diameter had to be driven through water-bearing sedimentary soils in silt, clay and medium dense silty sands. The overburden was 43 to 58 m. A Slurry - TBM in front of a string of pipe sections each 2.5 m long had been jacked from a concrete lined start shaft.

Two intermediate jacking stations, each with a capacity of 20.000 kN, were installed, one behind the machine unit M1 and the other behind pipe section No 6.

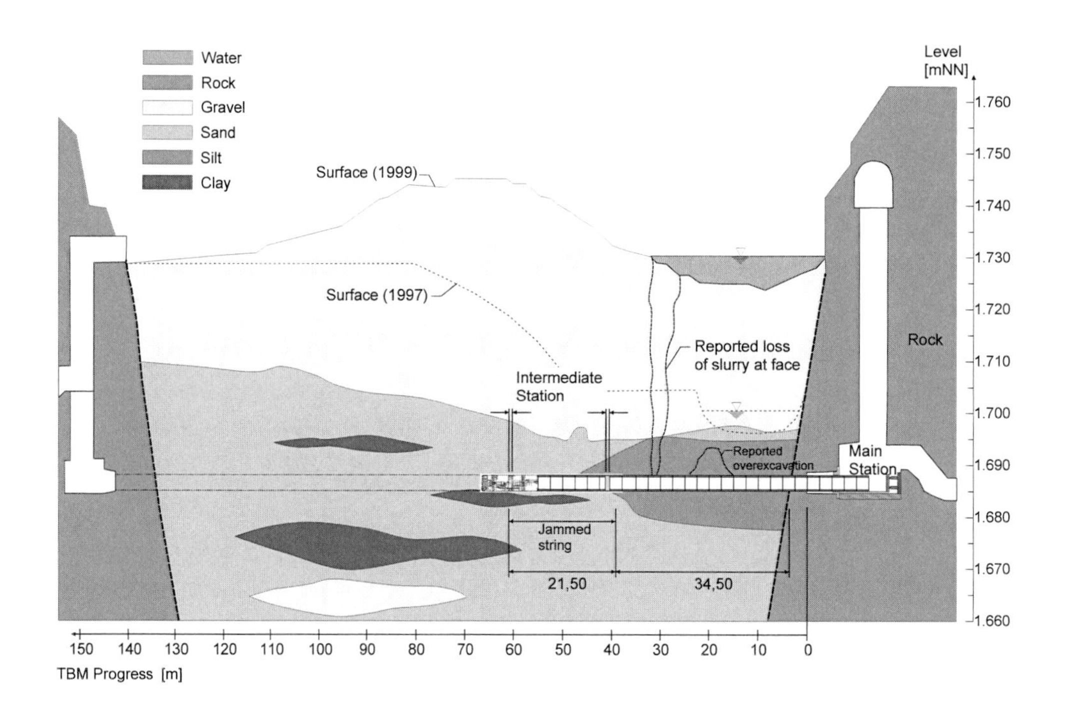

Figure 1: Project layout

After 66 m advance the drive came to a halt because of excessive friction around the perimeter of the forward string section. The 34.5 m long back part remained movable when pushed from the main jacking station in the start shaft to the intermediate station at section No 6. However, the 21.5 m long forward section between the intermediate station behind the machine unit M1 and the pipe section No 6 was jammed by soil compacted in the steering gap around the pipe sections.

The steering gap, usually 2 - 3 cm thick, is injected with bentonite slurry to reduce the friction while jacking the pipe string through the ground.

In the blocked string section, the ground consisted of fine to medium graded medium dense sand, whereas cohesive soil had been encountered in the rest of the string. The permeability of the sand reached k - values of 10-4 m/s.

Because of the relative high permeability, the bentonite slurry disappeared in the ground without creating the intended lubrication. Instead of the lubricant, fine sand filled the steering gap and blocked this section of the string. All endeavours to move the string failed. Attempts to flush out the sand by water injection were not successful.

Figure 2: Special bentonite mixture

Finally a special bentonite suspension, containing polyanionic cellulose, foaming agent and water soluble cutting oil was viscous and stable enough to keep the steering gap open. This suspension was injected in a controlled sequence under high pressure through a narrow pattern of outlets in the string.

Figure 3: Injection outlet pattern

In addition to the creation of stable lubrication, the pore water pressure in the soil around the string was increased by the high injection pressure, thus reducing the shear strength in the sand to the stage of liquefaction.

After two months of interruption, the pipe jacking project could be successfully terminated.

2.2 Slurry-TBM under the Albert - Channel in Belgium

For a life line, the Albert - Channel in Belgium had to be under-passed with 18 m cover to the channel bed by a slurry - TBM with a boring diameter of 3.72 m. The tunnel was lined by grouted reinforced concrete segments.

In 1998 after passing through different water-bearing layers, the TBM entered into a calareous formation. Here the thrust forces increased continuously to a level which stopped the TBM after 30 m, unfortunately exactly under the channel.

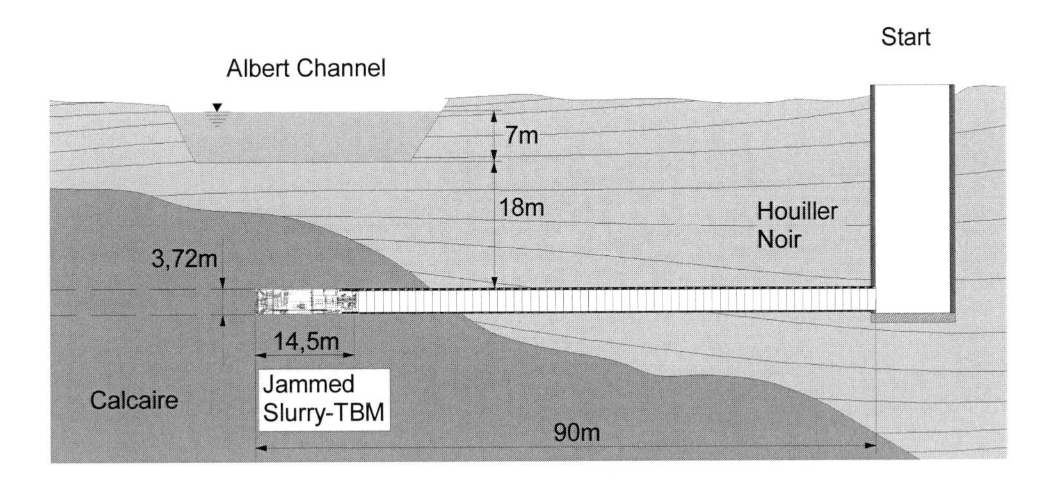

Figure 4: Project layout

The approx. 14.5 m long TBM was completely jammed by finely ground limestone particles which entered the steering gap around the shield of the TBM. This ground limestone passed through the gap between the cutter head and the soil. It was transported by the slurry due to the pressure difference between the support pressure at the face and the ground water pressure. Joints in the limestone facilitated the flow. The grinded limestone settled in the steering gap.

Also grinded limestone penetrated through the 3 - 5 cm wide gap in the cone crasher between cutter head and cutting edge of the shield.

On this TBM the so called back scrapers were missing. These are small scrapers attached to the back side of the cutter head rim to lift the finely grinded material from the bottom and to create local turbulence in the slurry. Otherwise the suction forces at the inlet of the transportation line are not strong enough for dislocation.

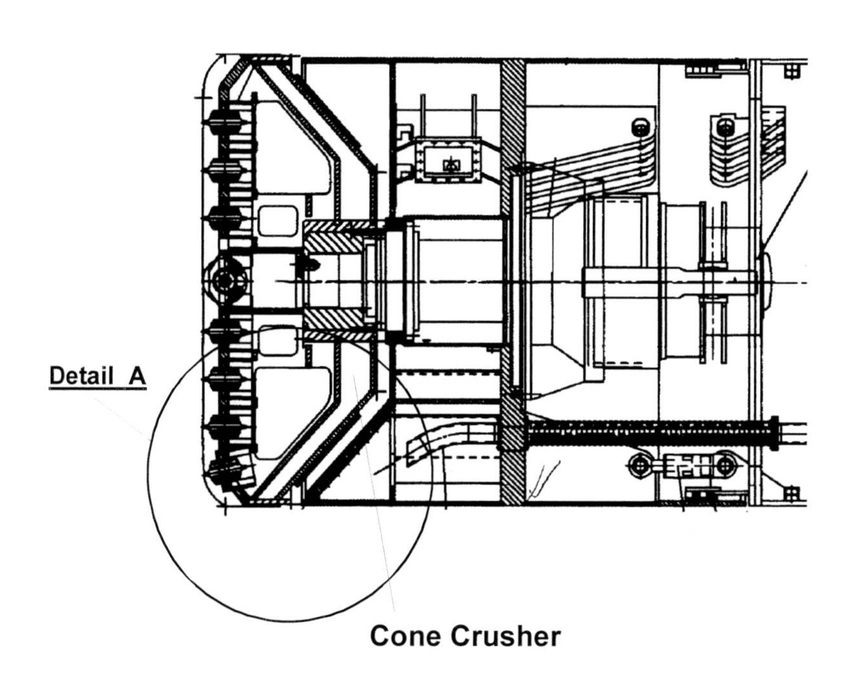

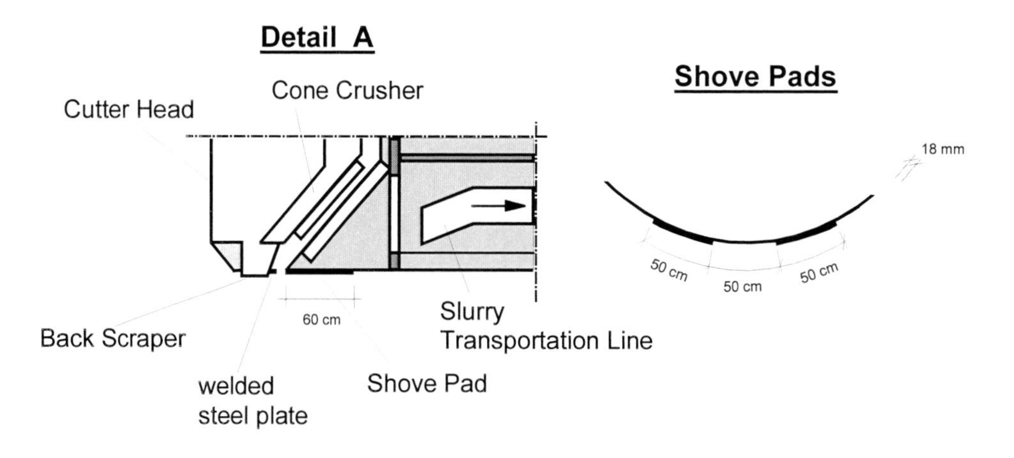

Figure 5: Cutter head

The grinded material finally jammed the TBM. It was concentrated at the invert area and was compacted under the shield.

Thie TBM was relieved by systematic flushing with water through several additional outlets drilled through the skin of the shield and through the 7 injection outlets installed in the tail shield section.

The pressure of the injected water was 0.5 bar higher than the pressure in the working chamber at the front of the TBM. 200 m^3 water per hour were pumped continuously in circulation during several days to relieve the TBM. After adding back scrapers to the cutter head, and with a constant water flow from the tail shield area into the working chamber at the front, the drive continued without further problems.

Additionally, the width of the gap between cone crusher and edge of the shield was reduced by welded steel plates.

The constant water flow has been provided by injection through the 7 injection outlets at the shield tail. The over pressure relative to the slurry pressure in the working chamber was 0.3 bar.

3. Fluctuating support pressure in EPB - TBMs

For a Metro project, 6.3 km tunnels will be excavated by EPB - TBMs with a boring diameter of 8.7 m.

The rock mass through which the tunnels have to be excavated comprises of granite. The conditions vary from fresh to moderately weathered high strength granitic rock to completely weathered soil-like materials. It has to be realized that granite boulders ("core stones") of variable size are embedded in soil-like weathered material and that this results in mixed face conditions for the tunnel drive.

The permeability of the granitic mass is very heterogeneous and varies between $k = 10^{-4}$ to 10^{-6} m/s.

The geological and hydrogeological conditions lead to the decision to use an EPB-TBM.

Figure 6: Heterogene geology

Figure 7: Core samples of heterogene geology

At the end of 2000, on the beginning of the drive, the EPB - TBM was operated in closed mode and partly in a semi-closed mode, i.e. is with a half filled working chamber and compressed air support applied to the upper part of the face.

During the first 400 m of the drive, 3 collapses happened, the last one caused destruction of an overlying house in which a person died.

This accident questioned the method of operation and it was stipulated that the EPB - TBM only be operated with closed mode with strict control of the face support pressure.

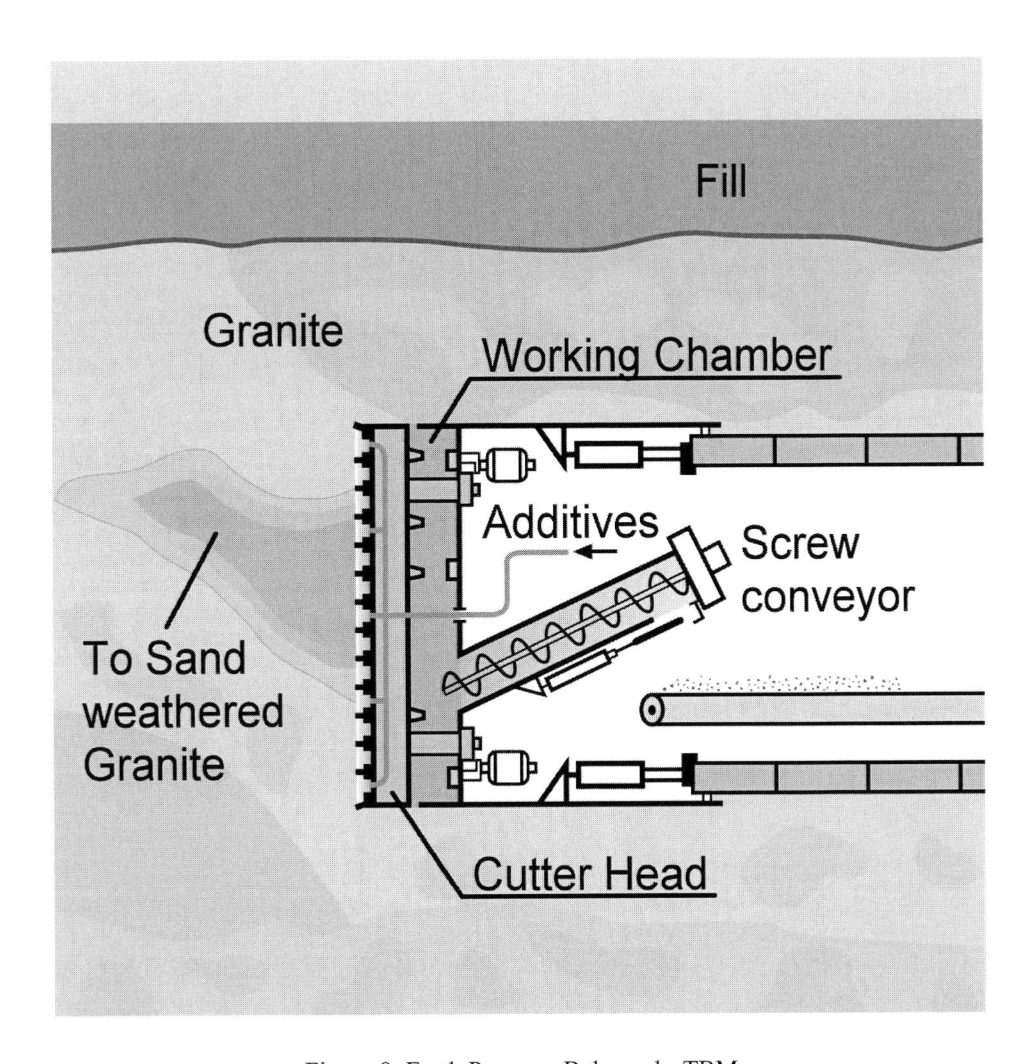

Figure 8: Earth Pressure Balanced - TBM

The face support pressure of EPB - TBMs is controlled by measuring the pressure at the bulkhead with pressure cells, approximately more than 1.5 m from the face.

In closed mode operation, the working chamber is completely filled with conditioned excavated material, the earth paste. The earth paste is pressurized by the advancing forces induced by the advance jacks via the bulkhead. The pressure level is controlled by the progress of the excavating cutter head in relation to the discharging screw conveyor.

To measure the mass within a complete filling of the working chamber, the density of the earth paste in the working chamber is controlled by pressure cells on the bulkhead at different levels. This method satisfies the demand of preventing a sudden instability of the face caused by a not completely falled working chamber. But does this guarante a reliable face support pressure?

Pressure measurement at the bulkhead, 1.5 m behind the face, provides only partial information about the support pressure at the face. The support medium, the earth paste created from excavated ground, conditioned by a suspension with different additives, must have the physical properties of a viscous liquid. However, the shear resistance in is viscous liquid reduces the support forces which can be transferred onto the face. The shear resistance of the earth paste depends on the excavated ground and the conditioning, which is a complex and sensitive procedure. Consequently, the shear resistance of the support medium often varies considerably.

Pressure measurement at the cutter head hase been shown to be not reliable. Therefore, the fluctuation of the face support pressure exceeds 0.5 bar. This fluctuation might be acceptable in an homogeneous soil. However, in mixed ground, as found at the Metro project mentioned above, the variable support pressure entails the danger of a creeping over-excavation.

An additional "active support system" can smooth down the support pressure fluctuations.

This system positioned on the back-up train consists of a container filled with pressurised bentonite slurry linked to a regulated compressed air reservoir. The bentonite slurry container is connected with the crown area of the working chamber of a EPB - TBM.

If the support pressure in the working chamber drops down under a predetermined level the "active support system" automatically injects pressurised slurry as long as the pressure level, loss in the working chamber is not compensated. This "active support system" additionally installed to the EPB - TBM works in the same manner like in a Slurry - TBM.

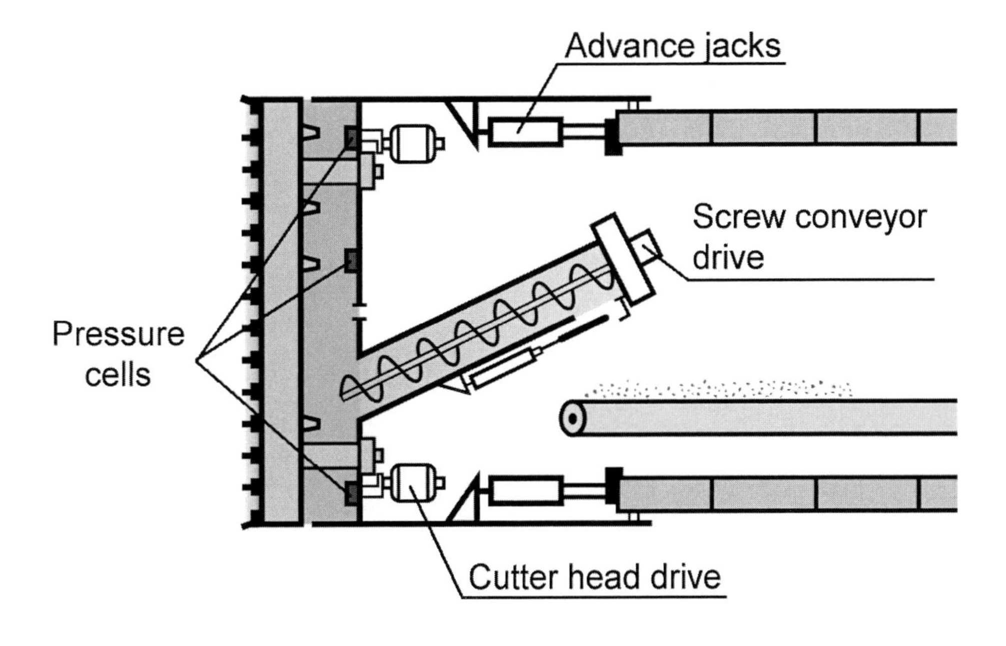

Figure 9: Measurement devices for face support pressure

Automatic pressure control system reduces the range of fluctuations of the face support pressure down to 0.2 bar.

The automatic controll is especially required if the excavated ground is conditioned with foam, which consists mainly of air bubbles. The air bubbles diffuse after some time, reducing the volume of the support medium and consequently lowering the pressure level.

The additional active support system was applied for first time at the described project. Its installation in the EPB - TBM drive of ∅ 8.7 m was essential to underpass old houses with a cover of 3 m, without any mitigating measures and with surface deformations of less than 5 mm. Simultaneously the Active Support System has been linked with the steering gap around the shield, filled with bentonite slurry, providing a reliable filling of the gap under predetermined constant pressure.

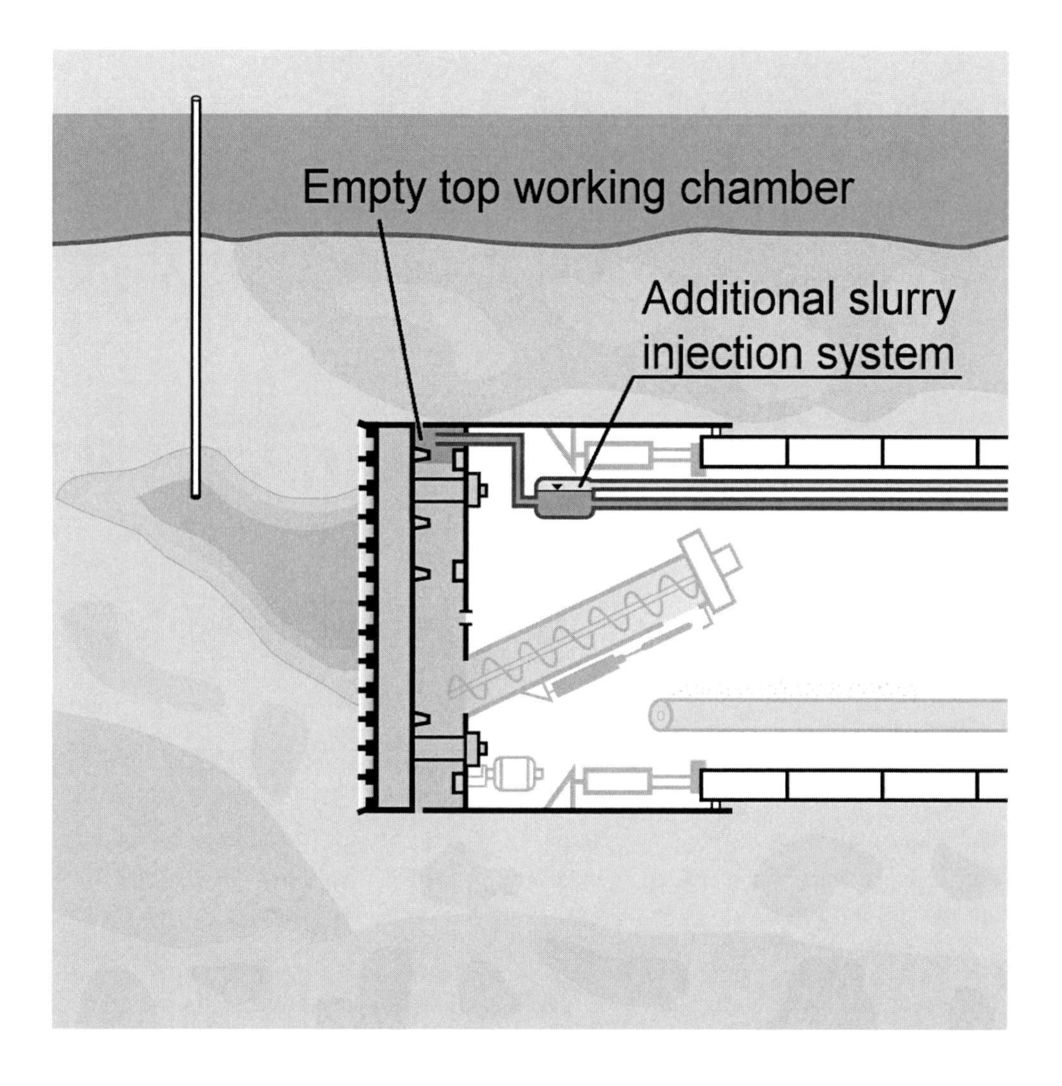

Figure 10: Active additional support system

4. Stones and boulders: a hindrance to pressurized TBM - Drives.

A tunnel drive in a glacial formed geology, as in Berlin, has to face stones and boulders embedded in a soft soil matrix. These ground conditions represents a challenging task for pressurized TBM drives.

At a 5.4 km sewer tunnel project in Berlin, three Slurry-TBMs have been used with a boring diameter of 3.8 m. Whereas the Berlin ground in general favours the application of slurry-TBMs, the frequently embedded stones and boulders often cause difficulties.

Figure 11: Glacial formed geology

For example, a collapse of the face created a sinkhole to the surface. The overburden was 15 m. The TBM had been operated with a penetration rate of 45–60 mm per rotation in sand and gravel with embedded stones. The sinkhole developed during attempts to free the cutter head which was apparently blocked by stones or boulders ripped from the face. The cutter head was freed by rotating in both directions without TBM advance and circulation of the slurry circuit.

An analysis of the machine and operation data shows that the TBM had been driven with a relatively high penetration rate of 45-60 mm per rotation. Encountered stones had been ripped of the face and crushed by the stone crusher. In this section where the cutter head was obviously blocked the number of stones had increased. The now larger number of stones banked up at the invert of the working chamber and blocked the cutter head.

Analysis of the data has demonstrated that the TBM had been operated in that section of the alignment with too high rate of advance which had caused the event.

The methodology to successfully excavate a tunnel in with a slurry-TBM has to follow some principles.

During excavation by the rotating cutter head, the excavation tool passes easily through the soft ground until it abruptly hits a stone or boulder. If the cutter head is fitted with an adequate tool arrangement and if the operation procedure is prepared for these conditions (mainly regarding the appropriate penetration rate) the tunnel drive will be not adversely affected. Otherwise the tools will be destroyed or the stones will be ripped out off the soil matrix.

At the beginning of the tunnel drive with slurry-TBMs, the cutter heads had been equipped with simple steel bars as excavation tools. Smaller stones were ripped from the face, whereas boulders or blocks were shattered at the face by hand-operated hydraulic splitters from the working chamber under compressed air.

Stones and pieces of boulders which could not be hydraulically converged because of their size were taken out of the empty working chamber by hand. Experienced crews did this during the period in which the segments were installed under the protection of the shield tail.

In 1986 cutter head were fitted for the first time with disc cutters to grind down stones and boulders in glacial deposits with a pressurized TBM. Today, cutter heads are fitted with a combination of disc cutters and scraper bits. The disc cutters are spaced up to 100 mm apart, and running 25-30 mm ahead of a series of scraper bits. The disc cutters should grind down the hard stones and boulders whereas the scraper bits should shave-off the weak materials such as sand/gravel, silt or clay.

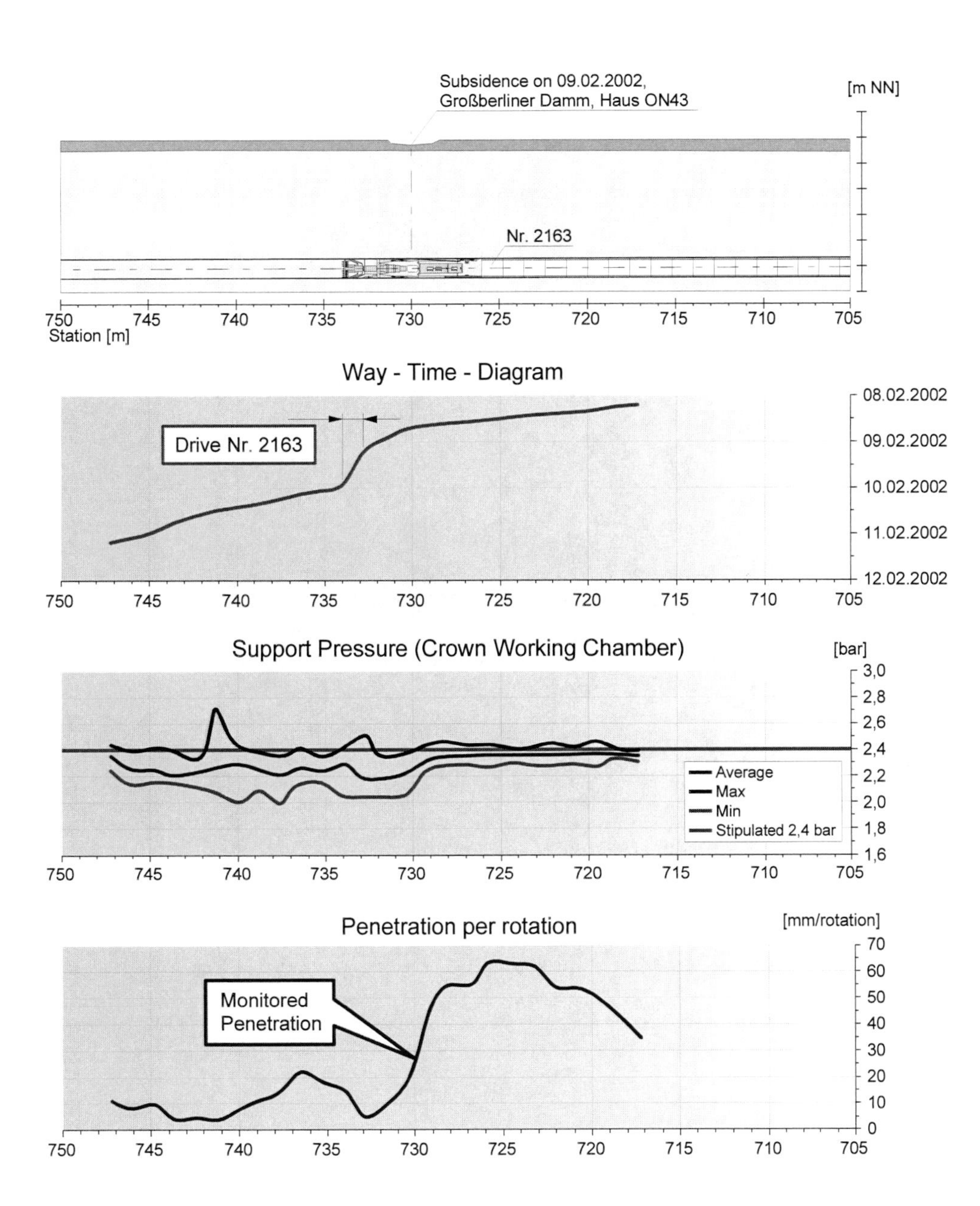

Figure 12: Data monitoring of TBM operation

Figure 13: Prototype Slurry TBM 1974

Because of the hardness of glacial boulders, the grinding procedure of the disc cutters has to follow the excavation principles of mechanized hard rock tunnelling. The load on a 17" disc should exceed 200 kN. The penetration rate per cutter head rotation is limited to 10 mm if granite boulders are encountered. The highly loaded cutting edge of the disc is only able to crush a 10 mm deep groove on a granite boulder and splits-off chips at the flanks of the groove.

Figure 14: Tool combination – discs and scrapers

It must be realized that the prevailing component of the forces induced by the rotating disc cutter in the stone or boulder is directed perpendicular to the face and consequently press the stone or boulder into the ground. The tangential component which may rip-off stones out of the matrix is rather small provided that the penetration rate does not exceed the above mentioned 10 mm and the disc cutter is not worn-out. This may happen easily when the disc passes through soft ground without rotating. To avoid stagnant disc cutters, the seals of the cutter bearings have to be de-stressed to the extent that the disc cutters can be rotated by hand.

If the penetration rate is higher than 10 mm or the disc cutters are not maintained, stones will be ripped-out of the matrix, endangering the stability of the face and damaging cutter head and tools.

There exist fundamentally different opinions on the removal of stones and boulders from the face at pressurized mechanized tunnel drive.

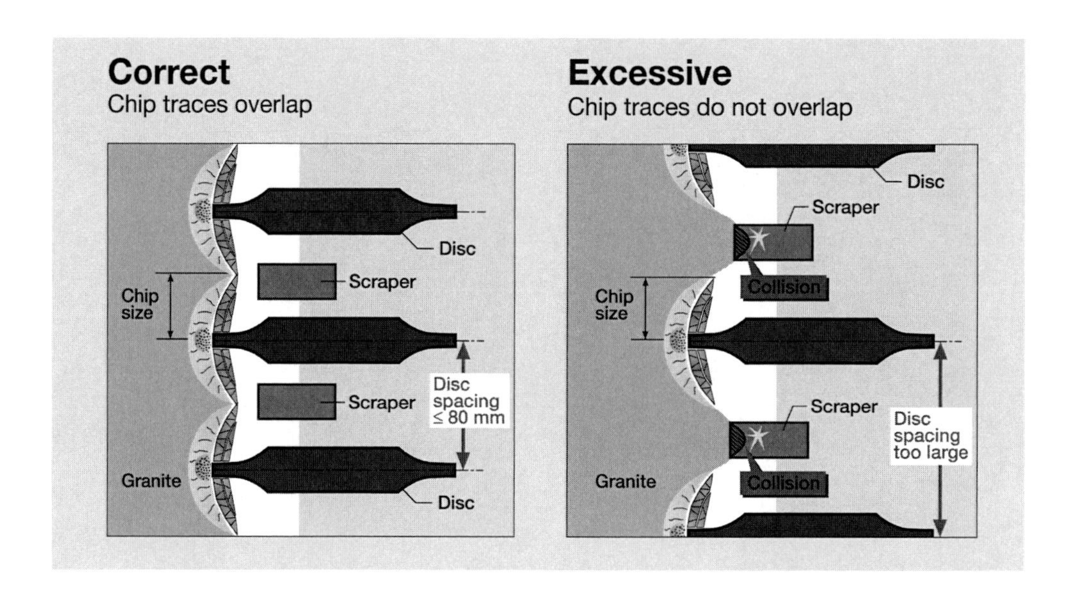

Figure 15: Disc spacing

Grinding down these obstacles at the face does not endanger the stability of the face and is effective independent of the size of stones or boulders. The rubble of disintegrated boulders can easily be pumped through transportation pipe lines. An additional stone crasher is not required for this procedure on larger TBMs with a sufficient diameter of transportation pipe line. However the rate of advance is limited by the restricted penetration rate.

Ripping stones and boulders out of the matrix facilitates high rate of advance. Single stones (smaller than 200 mm) may be ripped-off without local collapse of the face. However, boulders larger than 200 mm, cannot be pulled out without endangering the face stability.

Stones and boulders not ground down at the face fall down to the bottom of the working chamber and must be pushed through a stone crasher to the inlet of the transportation pipe line. The undefined forces for this transportation are generated by the inflow of excavated material and the suction effect of the circulation pump of the transportation pipe line.

However if the distance between the back the cutter head and the pipe inlet is too great, the ripped-off stones will be banked up and finally block the cutter head. This effect is adversely influenced by increasing numbers of stones.

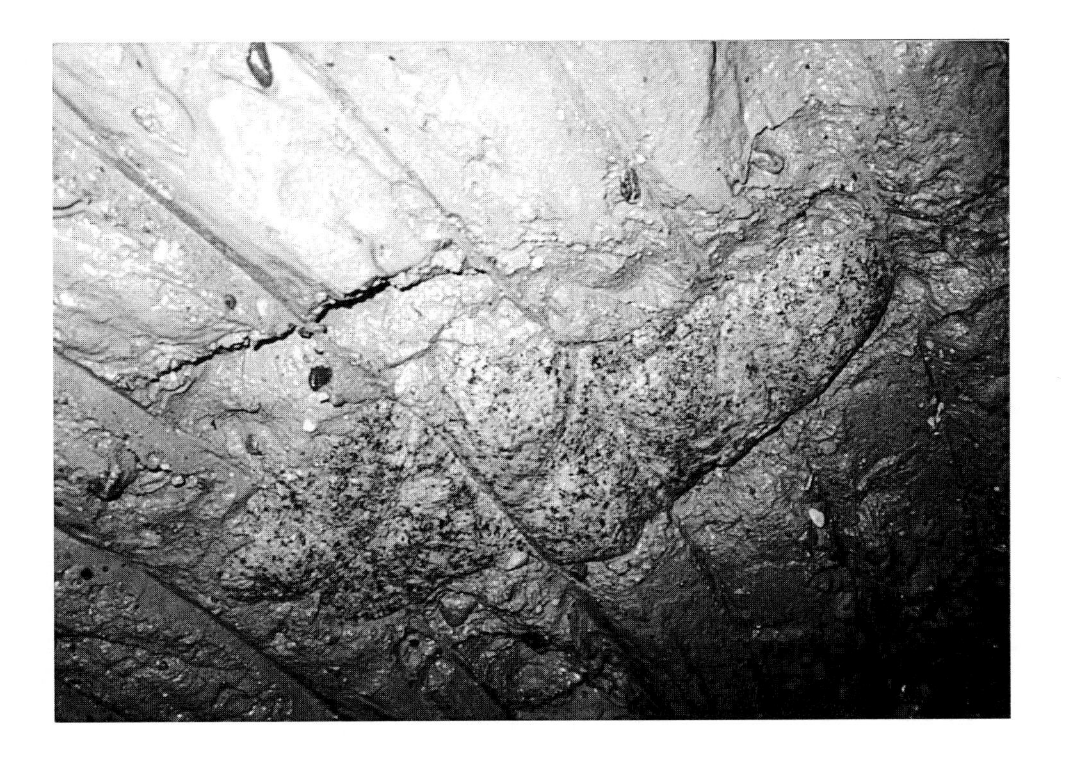

Figure 16: Boulder ground down

It is a frequently encountered opinion that a stone crasher installed in front of the inlet of the pipe can solve the transportation problems of ripped stones or boulders. There are stone crushers installed in some TBMs which may crush boulders with a diameter of 80 cm. But in ripping such boulders off the face, the cutter head will suffer enormous damage and the face will at least locally collapse. Additionally, the flow of excavated material is frequently interrupted during the closure sequence of the crasher. Also fine-meshed grill creates a hindrance to flow of loaded slurry and consequently a risk to a blockage of the cutter head.

The dimension of a stone crusher should be limited to a size which can crush boulders smaller than 30 cm. Larger boulders must be broken at the face to that diameter.

Figure 17: Inlet of transportation pipe line with grill and crusher

It must be realized that a stone crusher cannot be used in a EPB-TBM where the excavated material will be extracted from the working chamber by the protruding end of a screw conveyor. A stone crusher cannot be installed at the inlet of the screw conveyor. Therefore the diameter of the screw conveyor limits the size of ripped off stones or boulders from EPB-TBMs.

Estimating groundwater inflow into hard rock tunnels – the problem of permeability

Jack Raymer[1]

[1] Jordan Jones & Goulding, Inc., Atlanta, Georgia, USA
email: jraymer@jjg.com

Abstract:

In hard-rock tunnels, most of the inflow comes from a small portion of the tunnel length, some of the inflow comes from a large portion of the tunnel, and much of the tunnel is dry. The total inflow accumulates as the sum of all the inflows. This occurs because the rock-mass permeability along hard-rock tunnels typically has a practical range of over six orders of magnitude. These observations confound the traditional simplifying assumptions used in groundwater modeling, in which the permeability tends to be regarded as a given. This paper presents a practical method for estimating rock-mass permeability using the techniques commonly found on tunneling projects. It also looks at how some common simplifying assumptions can lead to unreasonable inflow estimates.

1 Introduction

Groundwater inflow into hard rock tunnels is very difficult to estimate accurately. In practice, estimates range from grossly low, which then results in large cost overruns and hazardous conditions in the workplace, to grossly high, which leads contactors to ignore them. This paper looks at some of the reasons why inflow estimates are difficult to make. It also suggests a few practical ways to get past some of these problems.

There are two main problems with inflow estimates. The first is a lack of simple, realistic equations or models that can be readily applied to hard-rock tunnels. This difficulty may turn out to be unavoidable, and this paper does not attempt to improve upon the situation. The second difficulty is that the *practical range* of permeability in fractured rock typically ranges over at least six orders of magnitude, and this range typically repeats again and again over the lengths of long tunnels. This range of permeability, combined with the length of the tunnels, makes hard-rock tunnels very different from well fields, major aquifers, and other applications of practical hydrology.

1.1 Definition of the problem

This paper is concerned with hard-rock tunnels in fractured rock that are excavated below the water table. These tunnels are commonly constructed under atmospheric pressure using either drill and blast methods or main-beam tunnel boring machines. Lining in these tunnels is typically installed only after excavation is completed, and only where needed. During construction, groundwater flows freely into these tunnels through fractures in the rock. Where the rock is tight and the potentiometric head above the tunnel is low, the inflow will be small. Where the rock contains large, open fractures or where the head is high, the inflow will be substantial. Where the rock contains both large fractures and high head, the inflows can be catastrophic. Kaneshiro and Schmidt (1995) give examples of some projects where ground water inflow has been problematic.

Tunnel designers must determine, and tell the contractor, how much water to expect over both the total length of the tunnel and in the heading area. The total flow is used to design appropriately sized pumping systems and water treatment plants. The inflows in the heading will affect the contractor's construction methods and schedule. Major delays can occur if either is underestimated. Excessive and unnecessary cost can result if either is grossly overestimated.

The problems, solutions, and costs of groundwater inflow vary over a practical linear range. Groundwater inflow accumulates over linear lengths of tunnel. Cost accumulates as inflow accumulates. In contrast, rock permeability, which turns out to be the most significant variable, is exponential in nature and ranges over many orders of magnitude. A significant part of the problem involves calculating linear effects from an exponential property.

1.2 Fundamental observation

In hard-rock tunnels, most of the inflow comes from a few places, some of the inflow comes from many places, and much of the tunnel is dry. The total inflow accumulates over the length of the tunnel and is the sum of all the inflows. This is the fundamental observation from hard rock tunnels, and the root of much of the trouble with estimating inflow using standard hydrologic methods.

This observation directly contradicts the standard assumptions of groundwater hydrology. It also points out a general oversimplification in the application of groundwater equations and models to real, natural systems. The main difference between hard rock tunnels and "standard" geologic settings is that the observed situation in hard-rock tunnels is so extreme as to force us to rethink our assumptions about how porosity and permeability are distributed over a rock mass.

1.3 Porosity and permeability

Porosity and permeability are the two properties of a rock that allow and limit flow. Both ideas are much more complex than a simple number and both deserve some discussion.

Total porosity is the space in a rock mass that can hold water. Effective porosity is the interconnected space through which water can move over a practical scale of time and distance. Porosity can be expressed as a number, usually as a ratio or percentage, and it can be described as a concept. Fracture porosity is porosity occurring in a network of interconnected fractures. (Interconnection is implied because the water has to be moveable at a practical scale.) Matrix (or intergranular) porosity is porosity occurring in the fine pores between grains, such as in sand or weathered rock. Limestones and volcanic rocks can have cavernous or vuggy porosity.

Pore systems in natural rock tend to have exponential, or fractal, relationships. Field geologists have long observed that fracture networks tend to follow scale-independent patterns, such that it is impossible to tell the scale of fractures in photographs without some independent reference, such as a coin, a person, or a trees on a mountainside. Scale independence is one of the main characteristics of fractal systems. As a general rule, systems that are can be expressed as dimensionless ratios, like porosity tend to be exponential in nature and to have fractal relationships.

Permeability is the ability of a rock mass to permit the passage of a viscous fluid through its pore space under a potentiometric gradient. Large pores and large fractures produce high permeability, which indicates that the rock is less resistant to fluid movement. Small pores and tight fractures have the opposite effect. Hydraulic conductivity is the same as permeability, except that it incorporates the fluid properties (water, in this paper) as well as the rock properties, whereas permeability is a function of only the rock characteristics. In this paper, the two terms tend to be used interchangeably, except that "hydraulic conductivity" is favored when a specific number is used and permeability is favored when talking about the general concept. In this paper, hydraulic conductivity is expressed in centimeters per second (cm/s). The term "average permeability," which is discussed later, is really average hydraulic conductivity.

Permeability is a product of the pore systems in the rock, and likewise is expected to have scale-independent, fractal relationships. Fractal properties tend to have log-normal distributions when they are sampled. This paper will show how the fracture permeability in hard-rock tunnels is log-normally distributed, and then discuss the consequences of the log-normal distribution when estimating inflow.

1.4 Log-Normal distributions

Log-normal distributions result whenever the logarithm of a property has a normal distribution when sampled. Log-normality can be suspected whenever a property involves ratios, ranges over several orders of magnitude, or has fractal relationships. Log-normality can be tested and demonstrated if the logarithm of a property is normally distributed when a large number of samples are collected.

Log-normal distributions have a geometric mean and do not have an average, or arithmetic mean. (The geometric mean is the exponential of the average of the logarithms of the sample.) A person could choose to calculate the average from a finite number of log-normally distributed samples, but the average has no real meaning with respect to the underlying distribution. The problem is that the average of log-normal samples does not tend toward the center of the distribution, but tends to increase as the number of samples increases. For example, if one were to perform 20 permeability tests and average the results, that average would tend to be lower than if one were to perform 10 more tests and take the average of 30 tests.

1.5 Gradient and inflow

Inflow to hard-rock tunnels is driven by a potentiometric gradient in the aquifer. The gradient develops in response to mining out the rock and groundwater, which creates a potentiometric sink in the groundwater surrounding the tunnel. The gradient, and therefore flow, is inward toward the tunnel. The rate of inflow is directly proportional to the gradient times the permeability. The permeability is typically a constant property of the rock and does not change in response to flow. The gradient is a transient property that changes in response to the loss of water into the tunnel and the addition of water from recharge and storage.

The potentiometric gradient is described in terms of the change in potentiometric head with respect to distance. Potentiometric head is the elevation to which a column of water will rise in a tightly cased well. Potentiometric head is expressed in units of length, and is used instead of pressure because it accounts for the weight of the water. Groundwater flows from higher head to lower head. In cases where the water is saline, and therefore denser, the potentiometric head is typically adjusted to a fresh-water equivalent head.

1.6 Project setting

The data in this paper are from several hard-rock tunnel projects in Atlanta, Georgia. Atlanta lies at an elevation of about 250 to 300 meters above sea level and is underlain by medium-grade metamorphic rocks and some granite. The rock is moderately fractured and has fracture porosity. The rock is overlain by about 10 to 30 meters of weathered metamorphic rock called "saprolite." The saprolite has matrix, or intergranular, porosity. The water table is typically high up in saprolite. Rainfall in Atlanta is around 140 centimeters per year. The area is well-drained with many small watersheds and no large lakes or rivers.

The tunnels from which these data were taken range in diameter from 2.5 meters to 9 meters, with 6 meters being most common. The tunnels range in depth from 10 meters to 120 meters, with 60 meters being typical. Inflow estimates to one of these tunnels using some of the methods described in this paper were made by Raymer (2001).

2 Inflow equations

Groundwater inflow equations are based on Darcy's Law and conservation of mass. Darcy's Law holds that inflow is proportional to the permeability times the gradient. Conservation of mass holds that the inflow equals the recharge plus the water released from storage. Most groundwater inflow equations hold that permeability is a given property of the rock mass and focus on the mathematically complicated relationship between inflow, recharge, storage, and gradient. These relationships are expressed as partial differential equations and can be solved with analytical or numerical models based on them.

This paper takes the opposite approach, because in hard rock tunnels permeability is anything but a given. If all of the factors that contribute to inflow are considered over a reasonable, practical range for a given hard-rock tunnel project, permeability will be found to be the least certain of all. Permeability is the most variable factor in the inflow equations, but the information on which to estimate it is typically the least reliable. This paper minimizes the roles of recharge, storage, and gradient, leaving those discussions to other workers. It emphasizes the role of rock-mass permeability – how to test it, how to estimate it, and how to use it.

This section discusses the simplest inflow equations in order to illustrate the role of permeability. It begins with the Thiem equation for steady-state radial flow to a well. It then imagines a horizontal Thiem equation and then proceeds to Goodman's equation for radial flow into a tunnel underlying a large body of water. These equations, while very simplistic, illustrate some major considerations that are applicable to much more complicated inflow equations.

2.1 Thiem equation

Darcy's Law is $Q = (KA)I$, where I is the gradient, KA is the permeability (K) across an area (A), and Q is the inflow. In this version of the equation (KA) are taken together in parentheses as a single term to emphasize the fact that actual water flows through a finite volume of rock mass, rather than a theoretical unity.

If Darcy's Law is configured for cylindrical coordinates around a vertical well, then the Thiem equation from well hydraulics results (Figure 1):

$$Q = 2\pi T\,(H_2 - H_1)\,/\,\ln(r_2/r_1) \tag{1}$$

where the gradient is incorporated in the term $(H_2-H_1)/ln(r_2/r_1)$. H_1 and H_2 are the potentiometric heads in the aquifer at two arbitrary points having radial distances r_1 and r_2 from the center of the well. The hydraulic conductivity (K) and unit area (A) are handled together in the term $2\pi T$, where T is the transmissivity. *Provided that the rock is uniformly permeable*, then $T=Kb$, where b is the vertical thickness of the aquifer.

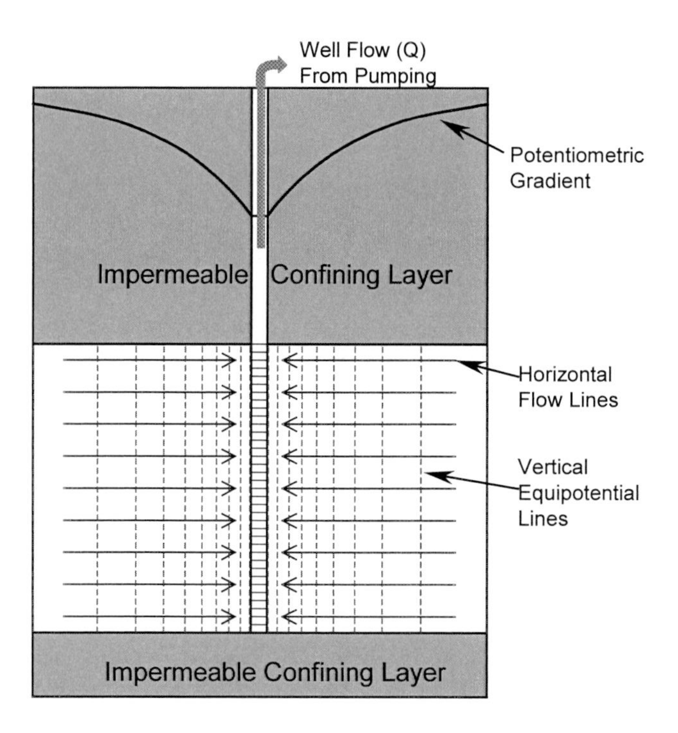

Figure 1: Thiem equation for radial flow to wells. Equipotential lines are vertical and concentric about the well. Flow lines are horizontal and radial toward the well. The aquifer extends infinitely, and at infinity recharge equals well flow.

The Thiem equation (as shown above) is based on the following assumptions:

- Flow is radial toward the well and non-turbulent. The well produces from the full thickness of the aquifer.

- The well has reached steady-state flow, meaning that the cone of depression around the well has encountered a supply of water sufficient to replenish the water produced from the well. At this point, the cone of depression stops expanding and water is no longer being released from storage.

2.2 Transmissivity and hydraulic conductivity

The Thiem equation, like most equations in well hydraulics, uses transmissivity (T) rather than hydraulic conductivity (K). (Transmissivity is also known as transmissibility.) The advantage of using transmissivity is that it does not require that hydraulic conductivity (permeability) be uniform along the length of the well, but only that it be uniform laterally around the well. Vertical variations in hydraulic conductivity, which are the rule in stratified aquifers, are averaged out in the overall transmissivity. Transmissivity is then a much truer coefficient of proportionality in natural aquifers than is hydraulic conductivity, because it accounts for spatial variability in permeability, at least in the vertical direction.

The dimensions of transmissivity are length squared per time (L^2t^{-1}). One of the lengths represents flow toward the well, as with hydraulic conductivity. The other length represents a perpendicular direction across which the hydraulic conductivity is distributed. This latter length dimension distributes the hydraulic conductivity over a specific amount of real space, rather than reducing it to an ideal unity.

2.3 Average permeability

Many textbooks tend to define transmissivity as the hydraulic conductivity multiplied by the thickness of the aquifer: $T=Kb$. This definition is not really correct, which leads to one of the key troubles with estimating inflow into hard rock tunnels. Transmissivity is the constant of proportionality in radial flow equations, such as the Thiem equation (Eq. 1). Lohman (1972) correctly observes that transmissivity equals the *average* hydraulic conductivity (K_{avg}) times the thickness. An even better definition would be to define the average hydraulic conductivity in terms of transmissivity: $K_{avg} = T/b$. This distinction does not matter if the rock is uniformly permeable: the permeability at any one point would equal the average. But natural rocks never have uniform porosity and permeability. This simplifying assumption is clearly untenable in hard-rock tunnels, where permeability fluctuates by many orders of magnitude over short distances and where permeability is log-normally distributed.

In fractured rock, the relationship between hydraulic conductivity and transmissivity needs to become a calculus problem rather than an algebra problem, where $K(x)$ is integrated over the thickness of the aquifer ($x = 0$ to $x = b$). $K(x)$ is the hydraulic conductivity at any point x along the axis of the well, which is typically vertical. The problem could be made even more complex if $K(\theta, r, x)$ is integrated in three dimensions rather than just one. In this paper, I am going to propose that integrating in only the axial dimension is sufficient to characterize the variability of fractured rock around a long tunnel. This proposition is based on my desire to keep the problem simple enough to solve with packer tests and pumping tests, and on my desire to avoid too much mathematical complexity.

2.4 Thiem for tunnels

The Thiem equation can be applied to tunnels by turning the well on its side. The axis of the well is now the centerline of the tunnel. As with a well, the cone of depression is radial around the tunnel. Unlike a well, the potentiometric surface is a bit more abstract, and will look like a trough along the axis of the tunnel if measured at springline (or any constant elevation).

For convenience, r_1 can be the tunnel radius and H_1 the potentiometric head at the tunnel radius. If the tunnel is under construction at atmospheric pressure, H_1 is the elevation of the tunnel wall. H_2 is the head of water some distance r_2 from the centerline of the tunnel. Transmissivity, which is the coefficient of proportionality, now has to be oriented horizontally along the axis of the tunnel, becoming the “horizontal transmissivity.”

Note that H_1 is not a specific point, but ranges between the crown and invert of the tunnel. This difference is trivial for deep tunnels. For large-diameter, shallow tunnels, however, this difference might become significant.

“Horizontal transmissivity” (T_h) is the coefficient of proportionality for the horizontal Thiem equation. It is placed in quotes because transmissivity is normally considered to represent the vertical thickness of the aquifer, rather than the horizontal length of the tunnel. The concept is important and worthy of indulgence to make a point. The fundamental observation shows that permeability varies over many orders of magnitude along the length of a hard-rock tunnel. This is similar to the horizontal stratification in a sedimentary aquifer penetrated by a vertical well. The “horizontal transmissivity” requires that the average permeability (K_{avg}) be considered when calculating inflow, rather than the idealized value of hydraulic conductivity. The average hydraulic conductivity is $K_{avg} = T_h/L$, where L is the length of the tunnel.

The assumptions of the Thiem equation still have to apply for tunnels. Flow has to be radial toward the tunnel. The tunnel has to be long, such that that non-radial flow around the ends of the tunnel is negligible compared to the total inflow. The water table has to be high above the tunnel and not draw down close to the tunnel. The cone of depression has to have room to expand radially to a point where the volume of water captured is large enough to provide for the inflow, even if an infinite supply of water is never encountered. These assumptions are reasonable for tunnels as long as the rock is nearly tight, the inflows very small, and the tunnel deep below the water table. In practice, tunnels that meet these conditions are less likely to have groundwater problems because the inflow rates will have to be very small.

2.5 Recharge and release from storage

The Thiem equation is based on the assumption that recharge to the flow system balances the production from the well. The safe yield of a well is considered to be the maximum production that will not exceed the recharge to the flow system. In contrast, a tunnel acts as a drain, taking all the water it can. Where fractures are large, the inflow will initially exceed the recharge, causing the overlying water table to drop. The ultimate long-term inflow in these areas will be limited by the recharge to the rock mass from the amount of water available in the watershed.

Rocks contain stored water that is released into the heading area of a tunnel shortly after a fracture is encountered. The release of stored water is responsible for "flush flows," or "heading inflows," which are rapid inrushes of water. Storage is directly proportional to the effective porosity of the rock mass, but is manifest in two different modes. In the first mode, stored water is released by the elastic expansion of the water and collapse of the rock mass when the hydrostatic pressure is released by tunneling. Expansion of the water and collapse of the pores forces water out into the tunnel. In the second mode, stored water is released by draining a rock mass or the overlying soil as the water table is lowered. In both cases, the rate of inflow is still limited by the permeability. The Thiem equation does not consider the release of water from storage.

2.6 Goodman's solution

Goodman (1965) considered the question of a tunnel lying beneath a lake or large river. He considered this lake or river to be an infinite source of water and applied the method of images (Lohman, 1972) to derive the following equation (Figure 2):

$$Q_L = 2\pi K H_0 / \ln(2z/r) \qquad (2)$$

where H_0 is the head of water above the tunnel, and z is the distance from the tunnel to the bottom of the lake. Goodman also divided T by the length of the tunnel to put the equation in terms of hydraulic conductivity (K) and inflow per unit length of tunnel

(Q_L). This equation only applies to steady-state inflow along the length of the tunnel. The inflow is steady state because the lake acts as an infinite recharge boundary, which causes the cone of depression to stop expanding.

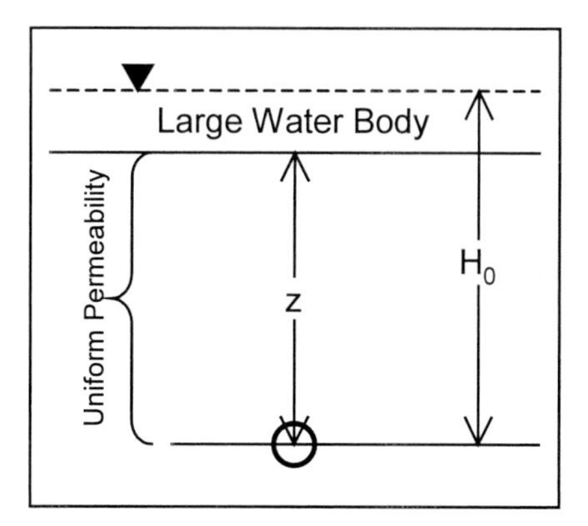

Figure 2: Goodman's (1965) model for tunnels beneath a large water body. *z* is distance from centerline to top of rock. H_0 is initial head between water surface and centerline.

Workers who have tried to apply this equation commonly report that actual inflows deviate greatly from the inflows predicted by this equation. Zhang and Franklin (1993) report that measured inflows range from 90 percent lower to 30 percent higher than predicted by Eq. 2. Heuer's (1995) work indicates that inflows from his projects have tended to be about one eighth (87.5 percent) lower than predicted by Eq. 2. My own experience from fairly shallow tunnels (typically less than 100 m) is in rough agreement with Heuer's. Other colleagues of mine report that inflows from deeper tunnels increase toward and then pass the inflows predicted by Eq. 2 as the tunnels become deeper.

Freeze and Cheery (1979) further modified Goodman's (1965) solution by replacing z with H_0. In this solution, the water table is modeled as an infinite recharge boundary. Freeze and Cherry's (1979) solution will give minimally lower estimates than Goodman's because H_0 is greater than z and both terms are in the logarithm. In a real tunnel, however, it is unlikely that recharge to the water table from precipitation could keep up with the yield from the larger fractures. This will serve to draw down the water table substantially around the tunnel and thus lower the head (H_0) to a point where recharge can keep up with the reduced inflow.

2.7 Nature of the variables

Goodman's equation, like the Thiem equation, contains some variables that are used linearly and some that are used in logarithms. The linear terms are K and H (or T and H_1–H_2). The logarithmic terms are z and r (or r_1 and r_2). When a term is used linearly, its linear distribution is needed for the equation. When a term is used in a logarithm, its exponential distribution is needed for the equation. How a term has to be used in an equation depends on the equation, not upon the natural characteristics of the term in question.

To be more specific, only the average K and average H are relevant to Goodman's equation or the Thiem equation, and only the geometric mean of z and r are relevant to these equations. This is a paradox, however, because z, r, and H are by nature linear properties whose central tendency is best described by an average, whereas K is by nature an exponential property whose central tendency is best described by a geometric mean.

If a term that is linear by nature is used in a logarithm, the effect becomes minimal because the range of the logarithm is small. But if a term that is by exponential by nature is used linearly, then the effect becomes large. This is the problem of permeability. Permeability is by nature exponential but is used in a linear manner. This makes flow, which is the linear result, highly sensitive to statistically insignificant variations in permeability.

2.8 Practical ranges of variables

The variables in Eq. 2 have practical ranges for tunnels. These practical ranges give insight into which are more important and which are less. In summary, K is the most important term and hardest to estimate, H_0 is less important and easy to estimate, and $\ln(2z/r)$ is of minor importance and easy to estimate.

H_0: In most situations, the maximum for the static head (H_0) is the elevation difference between the tunnel and highest water table around the tunnel. The minimum is the elevation of the lowest water table around the tunnel. These values can be estimated readily from topographic maps and piezometers. For the tunnels in Atlanta, H_0 ranges from about 120 meter to about 30 meters. The range in mountains will be much greater.

$\ln(2z/r)$: The rock cover above the tunnel should be known from borings. The tunnel diameter should be known from the design. Since both terms are in a logarithm, the practical range is quite small. For a 2 meter tunnel at 1000 meters deep, $\ln(2z/r) = 6.9$. For a 10 meter tunnel 30 meters deep, $\ln(2z/r) = 1.8$. This gives an extreme range between about 2 and 7. For the tunnels in Atlanta, $\ln(2z/r)$ ranges from about 3 to 5, which is small compared to H_0.

K: In fractured rock, permeability ranges over many orders of magnitude within a given rock mass. This variability is difficult to predict. The practical minimum for tunnels is around 10^{-6} cm/s, because even with large heads over long distances, the amount of inflow is very small. The practical maximum, however, can range up to 0.1 cm/s or higher, depending on the rock conditions. This gives a practical range of more than five orders of magnitude.

2.9 Possible sources of difficulty

The difficulties with Goodman's solution probably lie in three general areas. First, flow through fractured rock will not be radial toward the tunnel. Second, the bottoms of lakes and rivers are typically vertically stratified. Third, the hydraulic conductivity of the fractured rock may not be adequately characterized.

First difficulty: Individual fractures can be of significant size compared to the tunnel diameter. The larger (longer and more open) fractures will capture most of the flow and channelize it toward the tunnel. This will result in non-radial flow paths and possibly turbulence in fractures with large apertures. Non-radial flow and turbulence will reduce the inflow rate by a small percentage. This is probably the least important of the three difficulties.

Second difficulty: Lakes and large streams are typically underlain by horizontally stratified deposits. In many areas, the bedrock is overlain by stratified soils. These stratified deposits typically have low vertical permeability and will serve to impede recharge into the underlying bedrock. If inflow into the tunnel is high, then the leakage through the stratified soils may not keep up with the inflow, especially in the larger fractures. A steep gradient will develop in the stratified deposits and a flatter gradient will develop in the underlying fractured rock. The flatter gradient around the tunnel will cause a drop in the inflow rate.

Third difficulty: The permeability value used in the equation might not be the true average permeability of the rock mass. Because the range of permeability in fractured rocks is very high, even a small error in estimating average permeability will produce very different inflow estimates. These errors can come from not considering the limitations of the data and from taking the average of the test results, or worse yet, the geometric mean. Limitations of the test data will be discussed in the next section and estimating average permeability for the rock mass will be discussed in the section after that.

3 Testing permeability

Permeability can be tested in several ways. I favor packer tests for most situations in hard-rock tunnels. Packer tests can be performed economically in large numbers and can be designed to sample a volume of rock comparable to the diameter of the tunnel. Pumping tests can be a good supplement to packer tests in special situations where high inflows are expected and more information is needed than packer tests can provide. Pumping tests are not "better" than packer tests, however, because they tend to be very expensive and are less easy to control. Laboratory permeability testing of samples is not recommended because the test samples are too small to represent rock-mass conditions.

3.1 Packer tests

Packer tests are a method of measuring the permeability of the rock mass by injecting water into the rock at a given pressure and measuring the flow rate. For the tunnels in Atlanta, packer tests were performed by injecting water through a perforated pipe between an upper packer and a lower packer. The interval between the packers was 6 meters. The tests were performed at 6 meter intervals so that there was a continuous sequence of tests from the bottom of the borehole upward. The tests typically covered the interval from about 8 meters below the tunnel invert to between 20 and 55 meters above the tunnel crown. These tests were performed to the top of rock at shaft locations.

Permeability was measured as hydraulic conductivity and reported in units of centimeters per second (cm/s). An assumption of the testing procedure was that viscosity and density of the test water was very similar to the groundwater, such that no conversions would be necessary. If either water had been of very different temperature or salinity from the other, the test-water hydraulic conductivity would have been different from the groundwater hydraulic conductivity, and conversions would have been needed.

The packer tests were performed using a simple procedure described by the US Bureau of Reclamation (US Department of the Interior, 1980). The equipment consisted of two inflatable packers separated by 6 meters of perforated steel pipe (Figure 3). The pipe and packer assembly was attached to a string of steel pipe of at least 30 mm inside diameter. Pressure was measured by a pressure gauge at the ground surface and flow rate was measured using a mechanical totalizing flow meter.

The average permeability (K) for the test interval was calculated as the coefficient of proportionality between the applied gradient and the flow volume (Q), using the following formula:

$$Q = (2\pi KL)\,\Delta H / \ln(L/r) \tag{3}$$

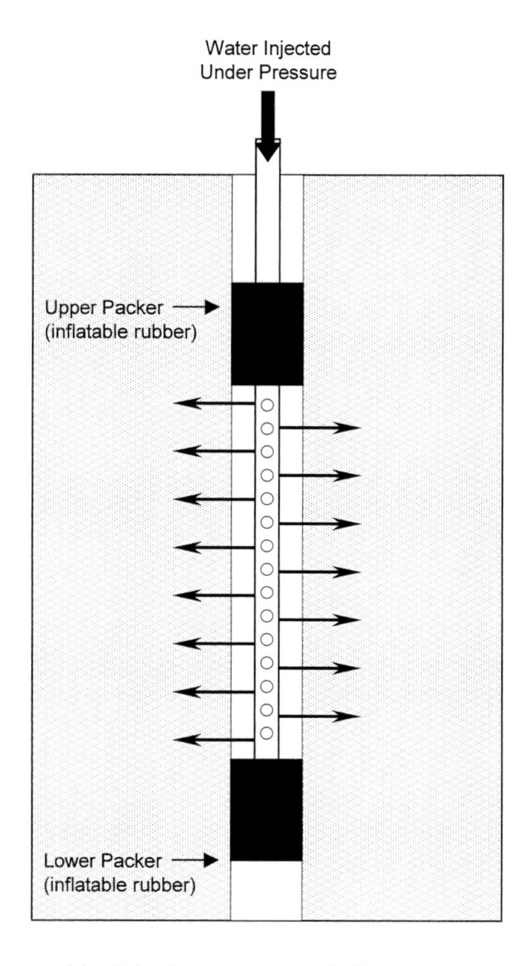

Figure 3: Packer test assembly. Injection pressure and flow rate measured with gauges at the surface. Water is forced out into the formation between the packers.

where ΔH is the applied excess head, L is the length of the tested interval (6 m), and r is the borehole diameter (96 mm). The applied head was adjusted for pipe loss, the static head in the borehole, and the height of the gauges above the ground surface.

The tests were run for approximately five minutes after the flow rate and pressure stabilized, which took from a few seconds to a few minutes. Each test was run at two pressures corresponding to 150 % and 160 % of the static head. No hydrofracturing was observed or expected at these pressures.

This method of testing is simple and inexpensive, both to run and to analyze. Because each test is simple, hundreds of tests can be performed on a project without requiring excessive costs or time. Ten to 15 tests can be run in a single borehole in one easy day.

3.2 Test results

Figure 4 is a histogram of 531 packer tests from tunnels in Atlanta. This figure shows that most of the results were in the range of 10^{-5} cm/s or less, and only a few of the results were greater than 10^{-3} cm/s.

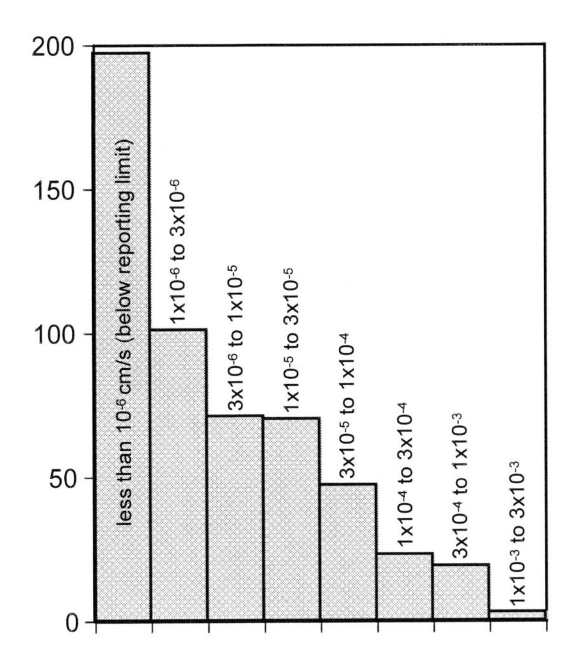

Figure 4: Histogram of 531 packer test results from several tunnel projects around Atlanta.

The packer test results were reported down to a minimum hydraulic conductivity of 10^{-6} cm/s. Below that limit, the results were reported only as $<10^{-6}$ cm/s. Between 10^{-6} and 10^{-7} cm/s, depending on the depth of the tests, the five-minute flow period became too short to accurately measure flow with the equipment being used. The reporting limit was set at 10^{-6} because it was the lowest round number at which flow could be consistently measured in all tests, regardless of depth.

For the tunnels in question, and probably for all but the deepest tunnels, 10^{-6} cm/s represents the lower limit of practical relevance. Below this limit, the amount of inflow becomes negligible, even under a high head. A lower quantitation limit down to around $3x10^{-7}$ cm/s could have been achieved in the deeper borings by running each test for about 30 minutes instead of five minutes. Much more expensive equipment and complicated procedures are needed to measure permeability consistently and accurately below about $3x10^{-7}$ cm/s. Even so, it will be shown later that some of the readings falsely high, even above the 10^{-6} cm/s reporting limit. This could have been caused by tiny leaks in the system, such as at pipe joints, or by round-off error from reading flow meters at the low end of their ranges.

3.3 Increments of transmissivity

Packer tests actually measure increments of transmissivity (T_i) corresponding to the length of the interval tested. As in the Thiem equation (Eq. 1), transmissivity is the coefficient of proportionality that relates the applied gradient [$H/\ln(L/r)$] to the inflow. Hydraulic conductivity (K) is an artificial concept calculated by dividing T_i by the test length L ($K=T_i/L$). The sum of all the packer tests in a borehole equals the transmissivity (T) of the formation around the borehole, assuming that the entire thickness of the formation were tested. This concept is important for two reasons. First, it illustrates the additive nature of flow. Second, it allows the packer test data to be compared to pumping test data.

Table 1 presents the packer test results from a particular borehole in Atlanta. The results are presented in terms of hydraulic conductivity (K) and increments of transmissivity (T_i). The total transmissivity for the borehole is the sum of the increments, which is $T = 4.06$ m^2/day.

Depth Range of Tests		**Test**	**Test Results**	
Top	**Bottom**	**Length (m)**	**K (cm/s)**	**T_i m^2/day**
15.4	21.6	6.2	1.0E-04	0.54
18.4	24.7	6.3	1.4E-04	0.73
24.5	30.8	6.3	4.4E-04	2.40
30.6	36.9	6.3	4.5E-05	0.24
36.7	43.0	6.3	1.7E-05	0.09
42.8	49.1	6.3	9.0E-07	0.00
48.9	55.2	6.3	4.3E-06	0.02
55.0	61.3	6.3	1.7E-06	0.01
61.1	67.4	6.3	9.0E-07	0.00
67.2	73.5	6.3	9.0E-07	0.00
			Total Transmissivity	*4.06*

Table 1: Increments of transmissivity from a continuous sequence of packer tests in a core boring in Atlanta. The total transmissivity of the formation is the sum of the increments.

A pumping test was also performed in the same borehole. Figure 5 shows the water-level recovery in the borehole following the pumping test. The horizontal axis is time since the end of pumping as a ratio of the total time since pumping began. The curve has three straight line segments, indicating that the gross transmissivity of the rock mass changed three times as the radius of recovery influence expanded (a situation that is typical of fractured rock). The earliest (leftmost) segment shows T = 8.2 m^2/day. The middle segment shows T = 4.1 m^2/day. The rightmost and latest segment shows T = 24.4 m^2/day. The middle segment closely corresponds to the packer test results and represents the general area around the borehole. The first segment probably represents an isolated large fracture very close to the borehole. The last segment may represent the broader formation some distance away from the borehole. Comparison of this curve with the packer test data shows that the packer tests are representative of the transmissivity in the general vicinity of the borehole.

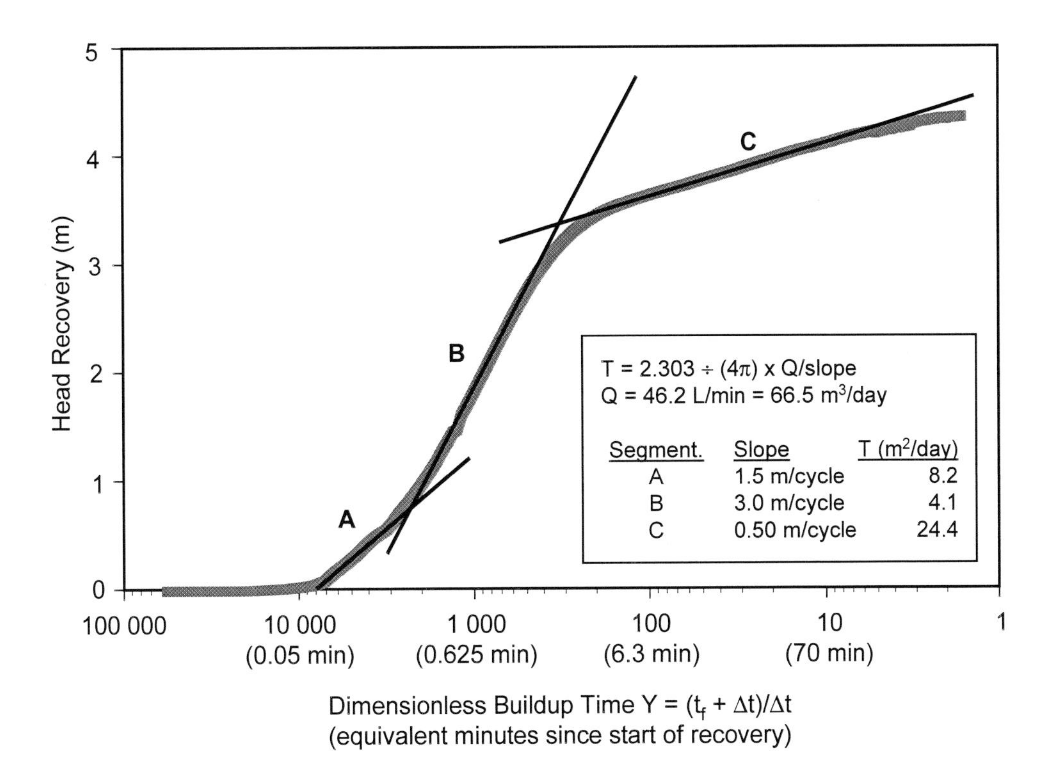

Figure 5: Recovery curve following pumping test from same borehole as packer test data in Table 1. t_f is time of production prior to beginning of recovery. Δt is time since recovery began. Straight-line segments correspond to transmissivity for various periods of time.

3.4 Application to "Horizontal Transmissivity"

The vertical transmissivity provided by packer tests can be applied to the "horizontal transmissivity" for a tunnel if one assumes that the vertical permeability around the tunnel is equivalent to the horizontal permeability around the tunnel. This assumption is built on two premises, which may or may not be appropriate for a particular tunnel. In summary, the principal variation in permeability is assumed to be along the length of the tunnel rather than around the circumference of the tunnel.

The first premise is that the rock is hydraulically isotropic. That is to say, the gross permeability in the vertical direction is roughly equal to the gross permeability in the horizontal direction. Such a premise is reasonable if the rock contains several fracture sets at different orientations. Figure 6 is a stereoplot for a tunnel in Atlanta, which shows four major fracture sets, with the southeastward dipping foliation being predominant. This large number of orientations allows the water to move vertically about as easily as horizontally. However, in horizontally stratified rocks, or in rocks where horizontal fractures predominate, the radial permeability measured from vertical packer tests will be substantially higher than the radial permeability around a tunnel (*i.e.,* the "horizontal transmissivity"). Freeze and Cherry (1979) report that horizontal permeability is typically three to ten times higher than vertical permeability in stratified rock masses. This could result in inflow estimates being overestimated by about three to ten times for tunnels in stratified rocks.

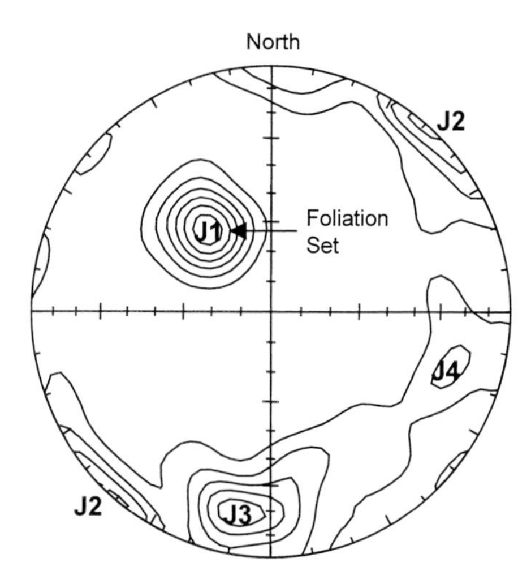

Figure 6: Stereonet of fractures in vicinity of Atlanta Tunnels based field mapping. Rock mass contains four major fracture sets (J1 through J4).

The second premise is that the vertical transmissivity measured in the packer tests is representative of conditions at tunnel depth. That is, the rock does not become more or less permeable with depth. Depth-related variability can be assessed and overcome by restricting the data domain to those increments of vertical transmissivity within some reasonable and representative depth range around the tunnel, such as within one diameter.

4 Permeability distribution

4.1 Cumulative curves

Figure 7 is a cumulative curve of the same data as shown in Figure 4. A cumulative curve is made by sorting the results in order and plotting them on a graph. The horizontal axis can be expressed either as the number of tests, or as a dimensionless percentile:

$$P_i = i/(n+1) \tag{4}$$

where i is an integer serial number assigned to each test, such that $i = 1$ represents the test with the lowest result, $i = n$ represents the test with the highest, and n is the total number of tests. Cumulative curves are useful for showing details in a distribution that might be lost when data are grouped into categories to make histograms.

The percentiles range from greater than zero to less than one ($0 < P_i < 1$). The percentile of a point i is the probability that any new point would have an hydraulic conductivity less than K_i. The median of the distribution occurs at P_i=50 percent.

The points form a curve which approximately follows a log-normal cumulative curve, as long as the points that are less than the reporting limit are allowed to have any value less than 10^{-6} cm/s. The roughness of the curve is attributed to random variability resulting from a limited number of tests. If the horizontal axis represents the percentage of tunnel length, then the area under the curve is the true average permeability (K_{avg}) of the tested rock mass. If the horizontal axis is multiplied by the length of the tunnel, then the area under the curve is the "horizontal transmissivity" along the tunnel axis. (Note that the vertical scale is logarithmic, such that the linear area under the curve fades away to nothing as permeability approaches the limit of $K = 0$).

The area under the curve is the integration of the log-normal cumulative curve over the percentage of a unit length. The limits of integration are $P_i = 0$ and $P_i = 1$. Note that permeability becomes infinitely high as it asymptotically approaches the limit of $P_i = 1$. At some point close to $P_i = 1$, the permeability, which is becoming infinitely large, will be overcome by the increment of unit length, which is becoming infinitely small, causing the integral to converge.

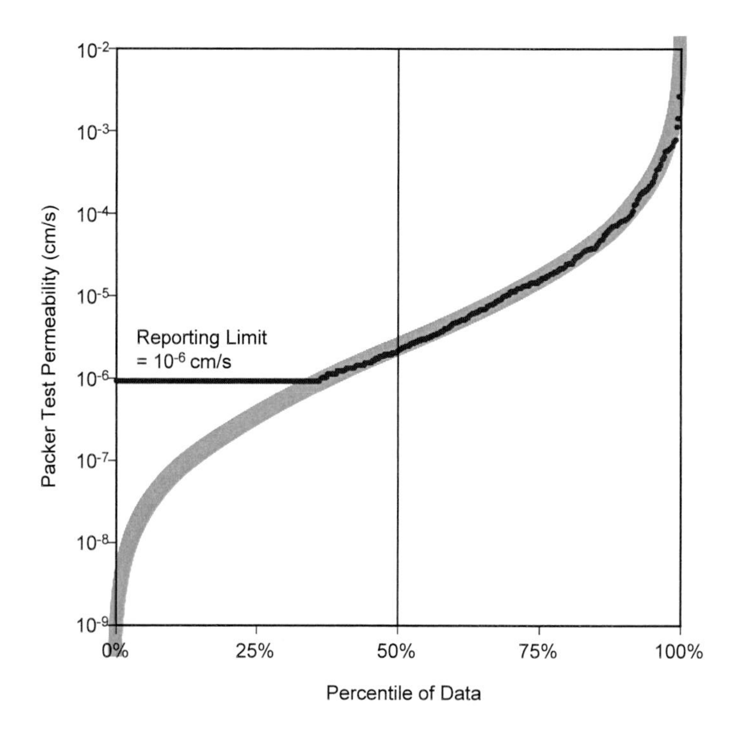

Figure 7: Log-scale cumulative curve of 531 packer test results from Atlanta tunnel projects. Shaded curve is log-normal distribution curve.

Figure 8 shows the same data but with K plotted on a linear axis. The points form a highly skewed distribution, indicating that the test results are not normally distributed. Because permeability is directly proportional to inflow for a constant gradient, the vertical axis could also become an inflow axis. The total inflow would be the area under the curve. This distribution illustrates the fundamental observation, that most of the inflow comes from a small percentage of the tunnel length.

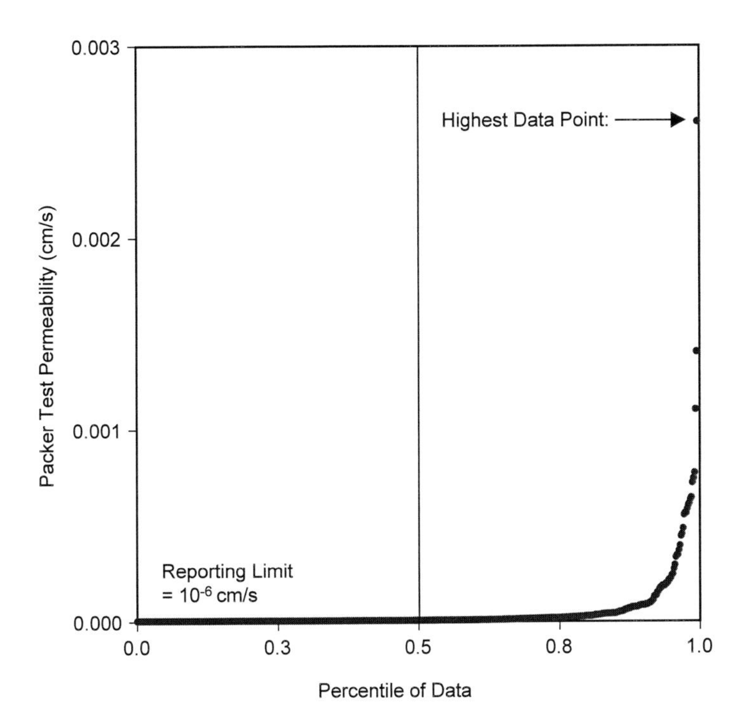

Figure 8: Linear-scale cumulative curve of 531 packer test results from Atlanta tunnel projects. If inflow is proportional to permeability, highest 5 % of data represents nearly all of the inflow.

4.2 Log normal plots

The cumulative curve becomes more powerful and easier to use if the percentile axis is transformed to a probability axis using the inverse normal distribution function ("=NORMSINV()" in Excel). This transformation makes the curve a straight line. The units of the horizontal axis are Z values. (This is capital "Z," which is different from the lowercase "z" used previously to represent vertical distance.) The permeability axis can also be linearized by taking the common logarithm of K_i. Figure 9 is a log-normal plot of the same data shown in Figures 4, 7 and 8. The data points now roughly follow a straight line, which is the log-normal model.

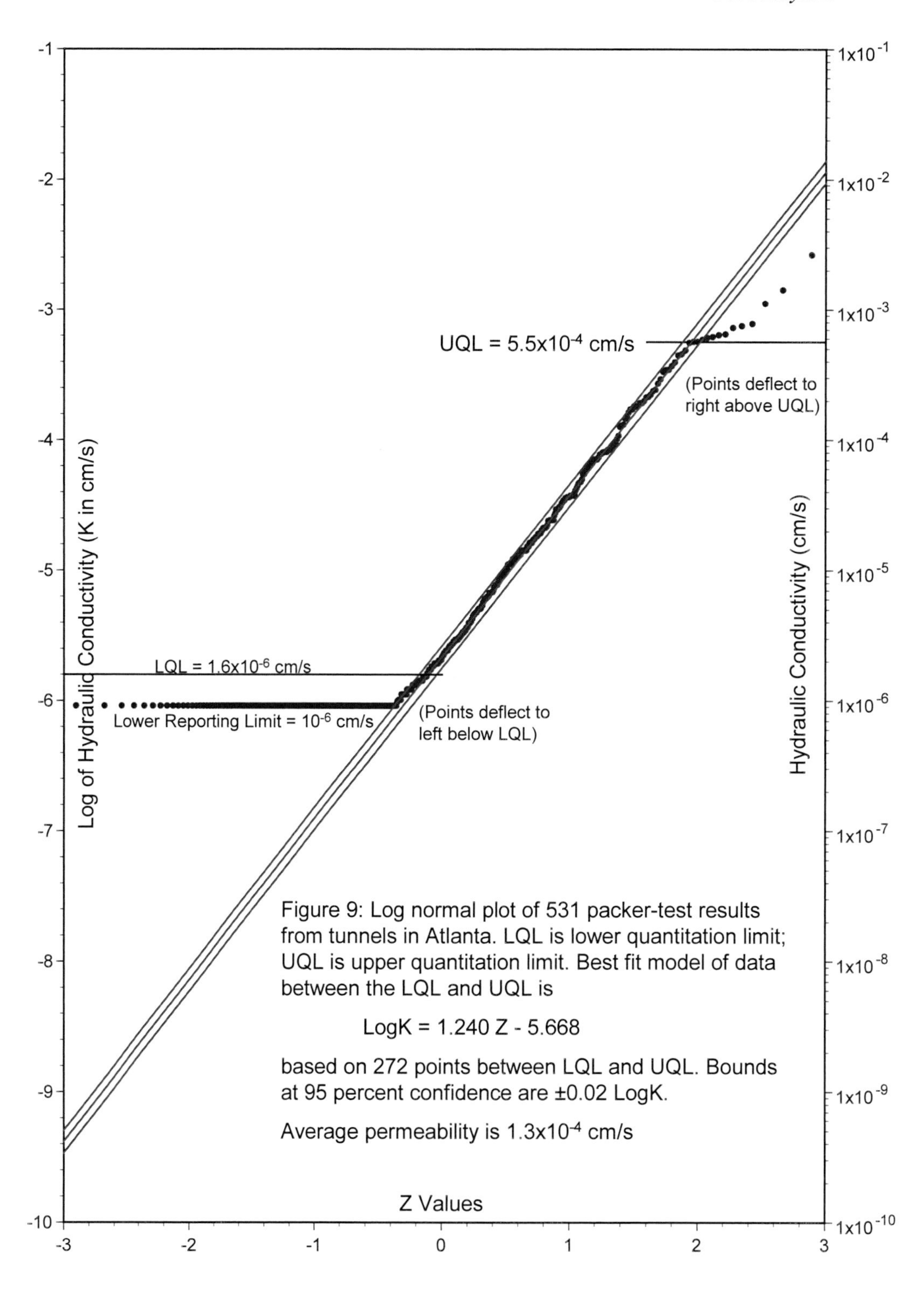

Figure 9: Log normal plot of 531 packer-test results from tunnels in Atlanta. LQL is lower quantitation limit; UQL is upper quantitation limit. Best fit model of data between the LQL and UQL is

$$\text{LogK} = 1.240\ Z - 5.668$$

based on 272 points between LQL and UQL. Bounds at 95 percent confidence are ±0.02 LogK.

Average permeability is 1.3×10^{-4} cm/s

Quantitation limits

The data in Figure 9 roughly follow a straight line between the lower quantitation limit (LQL) and the upper quantitation limit (UQL). The LQL and UQL bound the accurate range of the test method. Below the LQL, the line of data points deflects to the left because occasional tiny leaks in the system cause falsely high results. Above the UQL, the data deflects to the right because the pump and the pipe are not large enough to deliver enough water to adequately pressurize a highly permeable rock mass. For the tests shown in Figure 9, the LQL is 1.6×10^{-6} cm/s and the UQL is 5.5×10^{-4} cm/s.

The LQL is different than the reporting limit. The LQL is estimated based on the data. The reporting limit was determined during data calculation based on the accuracy of measurements. If no reporting limit had been used, the LQL would still show the lower limit of data accuracy.

The LQL and UQL will be different for any group of tests, depending on the equipment used, the pressures applied, and the depths of the tests. In general, there seems to be little advantage to trying to lower the LQL or raise the UQL, as long as there are a large number of points between them.

Log-normal model

A log-normal permeability model can be calculated from the log-normal distribution by using a least squares fit of all data between the UQL and LQL. This model, which has the form

$$\log K = m\,Z + \log K_0 \qquad (5)$$

where m is the slope of the line and K_0 is the intercept at $Z = 0$. (Statisticians might replace Z with σ in this usage, because Eq. 5 represents a model of the data, or expected value, rather than the real data. This distinction will not be maintained in this discussion.)

The log-normal model allows the high end of the permeability distribution, which represents nearly all of the inflow, to be estimated from a reasonable number of tests. The tests have to be accurate only between a reasonable UQL and LQL. This means that standard tests using simple, normal equipment and simple, normal procedures are adequate, which greatly reduces the cost of the investigation. Tests below the LQL and above the UQL are also important and valuable, however – not because their results are accurate in these extreme ranges, but merely because they exist to hold a place in the distribution.

Upper and lower bounds

The roughness of the data between the LQL and UQL is attributed to the random scatter of a limited number of tests. With a large number of test results, such as the 531 tests shown in Figure 9, the scatter about the best fit line is very small. As the number of tests between the LQL and UQL decreases, the scatter becomes greater. Greater scatter means that the best-fit line is less accurate and the area under the curve is less likely to be correct. In Figure 9, 95 percent of the tests between the LQL and UQL fit within bounds of ± 0.02 log$K$. With a smaller number of tests, this scatter tends to increase, which makes the bounds become wider.

In general, the bounds become further apart as the points becomes fewer. The maximum spread of the bounds should occur at around 30 points between the LQL and the UQL. With fewer than 30 points, however, the bounds tend to come back together because the number of points is too low. (For 2 points, best-fit line includes all the points.) In practice, the convergence of the bounds is slow between 20 and 30 points, so that with at least 20 points between the LQL and UQL, the best-fit line and the upper and lower bounds can be used with practical safety (Bulmer, 1979, p. 134).

The upper and lower bounds are important when estimating inflow to a tunnel. In most cases, the designer will want to state a baseline inflow rate that does not exceed a certain amount. Where "not to exceed" estimates are needed, the inflow should be based on the upper bound.

Area under the curve

The log-normal model can now be used in lieu of the data to calculate the average permeability (K_{avg}) or "horizontal transmissivity." Like the cumulative curve, the average permeability is the area under the curve, and the "horizontal transmissivity" is the area under the curve times the tunnel length. Unlike the cumulative curve, there are no natural limits of integration along horizontal axis. Instead the horizontal axis extends to positive and negative infinity. The lower bound of the area under the curve occurs at the limit of $K = 0$.

The area under the curve can be calculated in Excel using the method of Riemann Sums. The subscript "j" is used for each row of data in the spreadsheet (to avoid confusion with i, which was used already). The average permeability (K_{avg}) is the sum of column *G:* $\Sigma K_j \Delta P_j$. The formulas for each column of the spreadsheet are listed below.

Column A: Arbitrary values of Z_j in uniform small increments, such as 0.02, ranging from a high value, such as $Z = 8$ to a low value, such as $Z = -4$.

Column B: The negative of Z_j from Column A: $-Z_j$. The Z_j values should be sorted from highest at the top of the page to lowest at the bottom.

Column C: Values of the function $P_j(-Z_j)$: the percentile equivalent of $-Z_j$, calculated as =NORMSDIST(Column B). $P_j(-Z_j)$ values range from >0 to <1. Note that Excel cannot calculate values of NORMSDIST() for Z values less than –8.3, nor *Z* values greater than 7.86. These limits are well outside of anything I have needed in my work. If *Z* values beyond this range are needed, then a different method of calculating *Z* will have to be found.

Column D: Increments of $P_j(Z_j)$, calculated as $\Delta P_j = P_j(Z_j) - P_{j-1}(Z_{j-1})$.

Column E: Calculated values of Log K_j based on the model, where $K_j = m\ Z_j + K_0$, where *m* and K_0 were obtained from the equation of the model (Eq. 4).

Column F: Values of K_j, calculated as the antilog LogK_j from Column E: $K = 10^{logK}$

Column G: Values of $K_j \Delta P_j$, calculated as Column D * Column F. This calculates the rectangular area of each increment of *Z* up to the model, but omits the small triangular area that lies between the model and the top of the rectangle. The result can be made more accurate if this small triangle is added in: [Columns] $G_j = D_j * F_j + 0.5*(D_j-D_{j-1})*F_j$.

Limits of integration

Figure 10 shows the range in Z values that contribute to the average permeability of the rock mass for the data shown in Figure 7 and 9. The vertical axis is increments of

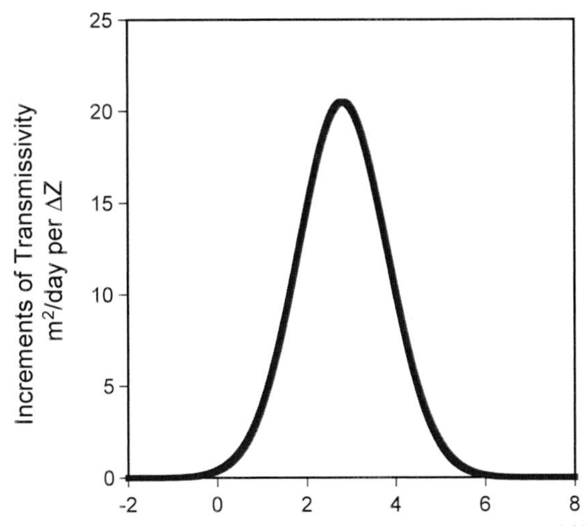

Figure 10: Range of *Z* values that contributes to average permeability calculation.

average permeability calculated by multiplying the permeability model by small increments of Z ($\Delta Z_i = 0.02$). The area under the curve is the average permeability of the rock mass. Half the average permeability lies above $Z = 2.32$ and half lies below. Below that 2.32, the permeability increases faster than the increment of length decreases. Above 2.32, the decline in the increment of unit length outpaces the increase in hydraulic conductivity, which causes the increment of average permeability to decline and then vanish.

About 90 percent of the average permeability is derived from the interval $0.68 < k < 3.96$. A Z value of 0.68 would correspond to a packer test result of 1.1×10^{-5} cm/s; $Z = 3.96$ would correspond to a packer test result of 2.2×10^{-2} cm/s. In normal sampling, it is expected that the 26,673 tests would be required to adequately cover this interval by using test data alone without the model. ($Z = 3.96$ corresponds to a probability of 0.0037% chance of reaching that value.)

4.3 Central limit theorem

The Central Limit Theorem from statistics applies to log-normal plots of packer test data. It "states that the sum of a large number of independent random variables will be approximately normally distributed almost regardless of their individual distributions," (Bulmer, 1979, p.115). This means that as the number of tests gets larger, a single, log-normally distributed population will tend to emerge. But as the number of tests gets smaller, the distribution will be more likely to become skewed, with multiple populations appearing.

Figure 11 is a plot of 31 tests from a particular section of an Atlanta tunnel. The rock in this area is massive granitic gneiss cross cut by large fracture zones. The fracture zones tend to be weathered such that they contain an appreciable amount of silt. The tests do not follow a good log-normal distribution. The tests from borings near the fracture zones follow one distribution and the tests from borings in the massive blocks appear to follow another. The fracture zone tests were used to estimate the average permeability; the tests from the massive areas were mostly below the reporting limit. When this data was included amongst the 531 tests, the distinction between the massive rocks and the faults was lost, but accounted for, in the overall statistical distribution.

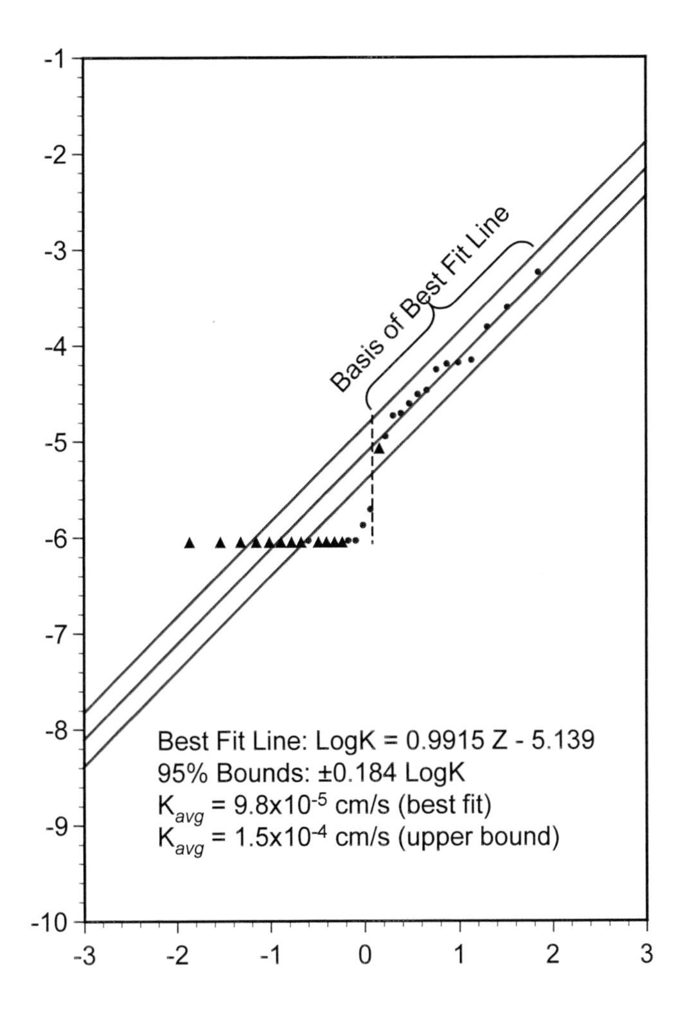

Figure 11: Plot of 31 test from a section of tunnel in Atlanta. Triangles represent tests from borings in blocks of massive gneiss. Circles represent tests from borings in fracture zones cross-cutting the massive gneiss.

4.4 Singular features

"Singular features" are isolated, abnormal features along a tunnel alignment, such as a large fault. Singular features are typically identified from field mapping or published literature, and then drilled and tested extensively. Singular features constitute a separate data domain and have a their own average permeability based on their own test results. The inflow from each singular feature is then added to the inflow from the rest of the tunnel.

Singular features present a particular problem because they all must be identified and tested individually. If a singular feature is missed, then its contribution to the inflow will not be accounted for. As a field geologist, I know that finding and testing all of the singular features on a job is much more difficult in practice than it sounds in theory. Tunnels are built in areas where such features are especially difficult to identify and test – under mountains, under cities, and under water. If the tunnel is deep, large features may not extend to the surface where they can be recognized. Published geologic maps are not specifically made for tunneling projects. These maps typically emphasize features of little importance to tunneling, such as fully welded faults with large stratigraphic offset, but omit features of great importance, such as fracture zones with intensely shattered rock but small stratigraphic offset.

Log-normal plots can overcome much of this problem in many situations. As it turns out many "singular features" are not really abnormal, they just occur at the high end of a log-normal distribution. This can be especially true for faults with intensely shattered rock. In many cases, such faults are simply zones of large fractures that occur in a repeating pattern throughout a region. If the tunnel is long, it may cross several of these large fracture zones. If a large number of borings are taken along the tunnel, including angle borings across the fracture zones, then the contribution of these zones can be considered statistically using log-normal plots.

4.5 Summary

Log normal distributions are a good method of estimating the average permeability of a rock mass. They are also good for estimating the percentage of a tunnel with a permeability above a given threshold. Log-normal distributions are easy to use and statistically stable. They make use of the kinds of data commonly seen on tunneling projects.

Log normal distributions overcome the problems of the UQL and LQL. They overcome the problem of not having the thousands of tests that would be needed to estimate the true average permeability of a mass of fractured rock. They use upper and lower bounds to address the random variability caused by having a limited number of tests.

Log normal distributions must be used with good geologic sense, such that the domain of test results represents the geological conditions around the tunnel. Log normal distributions can be used in many cases to bring "singular features" into the log-normal distribution, making them no longer abnormal and singular.

5 Conclusions

Inflow is controlled by the permeability, the gradient, recharge, and storage. In many common applications, such as well fields, the permeability can be tested and then taken as a given. The inflow equations then focus on the how the gradient develops in response to variable inflow, recharge, and storage. In hard rock tunnels, however, the permeability cannot be taken as a given, but is the most important single variable, ranging over a practical range of at least six orders of magnitude.

Large numbers of packer tests are a good way to characterize permeability in a rock mass. If the rock has several sets of fractures at different orientations, then the horizontal permeability is likely to equal the vertical permeability, at least over an interval of a few meters. If the rock is horizontally stratified, then the vertical permeability is likely to be substantially lower than the horizontal. In the latter case, packer tests will overestimate the vertical permeability.

The average permeability of a rock mass cannot be calculated simply by taking the average of the test results because permeability is log-normally distributed. Average permeability can be calculated by using the data to establish a log-normal distribution model, and then integrating that model over a unit length. Integrating the model over length incorporates the overwhelming influence of the highest test results, which end up controlling the average permeability and most of the inflow. Just taking the average of test results will give a result less than the true average permeability of the rock mass, sometimes by orders of magnitude.

References

[1] Bulmer, M.G., 1979, Principles of Statistics, New York: Dover Publications.

[2] Freeze, R.A. and J.A. Cherry, 1979, *Groundwater*, Prentice-Hall, Inc..

[3] Goodman, R., D. Moye, A. Schalkwyk, and I. Javendel. 1965. Ground-water inflow during tunnel driving, *Engineering Geology* 2:39.

[4] Heuer, R.E. 1995. Estimating rock-tunnel water inflow. *Proceeding of the Rapid Excavation and Tunneling Conference*, June 18-21, 1995. 41.

[5] Kaneshiro, J. Y. and B. Schmidt, 1995, Fracture and shatter zone inflow into hard-rock tunnels – Case Histories. Rock Mechanics, Daemen & Schutz (eds.) Rotterdam: Balkema.

[6] Lohman, S.W. 1972, Ground-water hydraulics. U.S. Geological Survey Professional Paper 708.

[7] Raymer, J.H., "Predicting Groundwater Inflow into Hard-Rock Tunnels: Estimating the High-End of the Permeability Distribution," 2001 Proceedings of the Rapid Excavation and Tunneling Conference, Society for Mining, Metallurgy and Exploration, Inc..

[8] U.S. Department of the Interior, 1980. Earth Manual: A Water Resources Technical Publication.

[9] Zhang, L. and J.A. Franklin, 1993, Prediction of Water Flow into Rock Tunnels: an Analytical Solution Assuming an Hydraulic Conductivity Gradient. *Int. J. Rock Mech, Min. Sci. & Geomech.* Abstr., vol. **30**, pp 37-46.

Part III

Measurements

Geotechnical instrumentation of tunnels
Part 1: Performance monitoring for tunnel design verification

Helmut Bock

Geotechnical Consultant, Past-President of Interfels GmbH, Stoltenkampstr. 1, D - 48455 Bad Bentheim, Germany, Tel. + 49 - 5922 - 2700, Fax: +49 - 5922 - 2799
QS-Consult@t-online.de,
http://home.t-online.de/home/QS-Consult

Preface

It is common practice to monitor the performance of tunnels during and after construction. Monitoring is carried out by combined geodetic and geotechnical instrumentation and monitoring methods.

Traditionally, the objective of monitoring is verification of the tunnel design. Key physical parameters are measured and compared with predicted values. In case of significant deviations an adjustment of the tunnel design may be indicated. Beyond this, geotechnical instrumentation may also be used in the quality assessment of certain tunnel construction procedures. In critical situations (e.g. tunnelling beneath settlement-sensitive structures in inner-city areas) it can yield construction control signals for the entire tunnelling operation.

Part 1 of this contribution gives an account of recent developments and practices in performance monitoring for tunnel design purposes. A separate Part 2 focuses on geotechnical instrumentation to assist with construction control.

1 Introduction

In underground construction, geotechnical instrumentation is used for purposes which are summarised in Table 1 below.

For evaluation of the tunnel design, a set of elementary tunnelling instrumentation is commonly employed which is set out in Figure 1.

The discussion of the instrumentation shown in Figure 1 (together with some selected additional instrumentation) will be in accordance with the common engineering purpose which can be classified as follows:

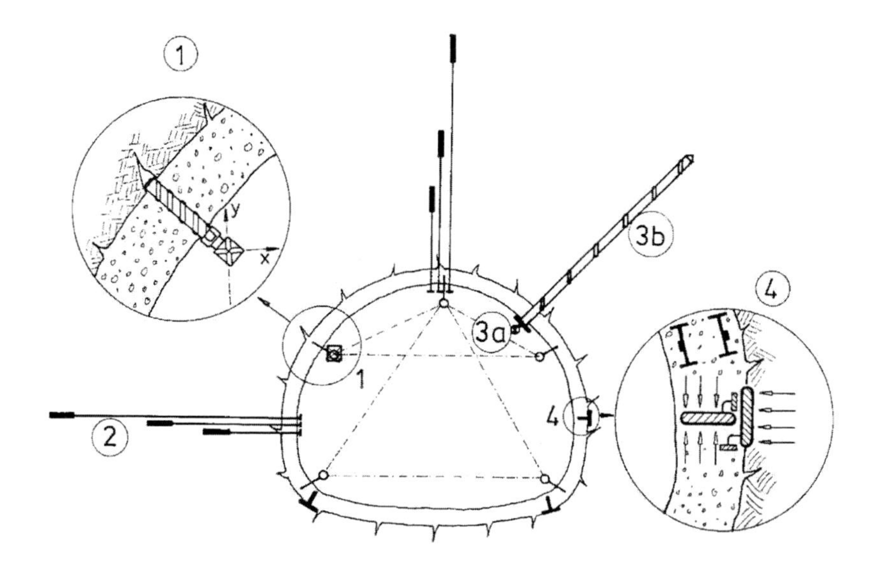

No.	Measuring Objective	Instrument
①	Deformation of the excavated tunnel surface	Convergence tape Surveying marks
②	Deformation of the ground surrounding the tunnel	Extensometer Inclinometer
③	Monitoring of the ground support element "anchor"	a. Anchor load at collar b. Distribution of anchor forces
④	Monitoring of the ground support element "shotcrete / concrete shell"	Total pressure cells Concrete embedment gauges

Figure 1: Elementary tunnelling instrumentation (out of Interfels catalogue)

No.	Purpose	Example	Suitable instruments	Part
1	**Empirical proof** of *new equilibrium* after excavation	- Convergence - Surveying methods	Tape extensometer. Total station.	1
2	**Comparison** measurement vs prediction → for improved tunnel design	- Displacement monitoring in the surrounding ground - Stress monitoring of shotcrete lining	Borehole extensometer and inclinometer. Total pressure cell (TPC). Flat jack compensation.	1
3	**Quality control** of construction procedures	- Excavation profile - Drillhole direction control in freezing tunnel	Tunnel scanner. Deflectometer.	2
4	**Control** of entire tunnel construction operations	Soilfrac ® and real-time settlement monitoring in inner-city tunnelling	Multi-point liquid levelling system. Electrolevel gages. Motorised digital level.	2

Table 1: Application of geotechnical instrumentation in underground construction

- Instrumentation for an *empirical proof of a new equilibrium* after tunnel excavation: Convergence and/or geodetic deformation monitoring (Section 2).
- Instrumentation for the monitoring of displacements and stresses to compare measured and predicted values for *evaluation and improvement of the tunnel design* (Section 3).

2 Instrumentation for an empirical proof of a new equilibrium after tunnel excavation: convergence tape measurements and surveying methods

Convergence measurements and monitoring of the displacements of the excavated tunnel surface by surveying methods are part of the routine operations in today's tunnel construction. In essence, the change of the displacements is monitored and correlated with tunnel construction procedures such as excavation, installation of the ground support and closure of the invert. *"Stabilisation of the entire system and its safety, the necessity of additional support and, in reverse, the permissibility of reducing the support system is judged on the basis of convergence or displacement measurements at or near the excavated tunnel surface" (Leopold Müller, 1978; p. 607).*

2.1 Convergence tapes (tape extensometers)

For convergence measurements, the distance between two points on the excavated tunnel surface is measured by specially manufactured tapes (so-called "convergence tapes" or "tape extensometers"). A common technical feature of such tapes is the high reproducibility of their tensioning force in a measuring position. The measuring points are defined by convergence bolts which are attached to the linings, e.g. placed in the shotcrete in a manner that their measuring heads are pointing towards the excavation and remain accessible throughout the lifetime of the monitoring project. After installation of the bolts, a first round of reference (or "zero") convergence measurements is carried out. By way of follow-up measurements (and subtraction of the measured values from the respective values of the zero measurement) the *change of the distance* between the bolts can be determined.

The accuracy of convergence measurements is in the range of ± 0.003 to ± 0.1 mm and depends on a number of factors, amongst them the type of instrument, the material of the tape and the type of coupling of the tape to the bolts (ref. to Table 2). Note that convergence measurements are generally more accurate than geodetic displacement measurements (ref. to Table 2 and Section 2.2).

In today's European tunnelling practice, convergence tapes are hardly used anymore. Instead, geodetic monitoring is quasi-standard and has substituted tape measurements almost entirely. This is despite the lower system accuracy of the geodetic method. Only in special or sensitive projects (e.g. underground research laboratories; nuclear repositories) are convergence tapes still in use. Generally in these applications, tapes of the Accuracy Class I or II (ref. Table 2) are employed.

Accuracy Class	Measuring Principle Essential Construction Features	(Manufacturer)	Resolution [mm]	System Accuracy [mm]
I	Convergence tape with invar wire Coupling: Universal joint. Bolts with measuring stop.	(SolExperts AG).	0,001	± 0,003
II	Convergence tap with steel tape Coupling: Universal joint. Bolts with measuring stop.	(Interfels GmbH).	0,01	± 0,05
III	Convergence tap with steel tape Coupling: Eyebolt-hook connection.	(North American manufacturers)	0,05	± 0,5
IV	Geodetic tunnel surveying Total station (tachymeter) with integrated coaxial distance measurement	(e.g. Leica)	1	± 2 to 3
V	Tunnel Scanner	(ref. to Part 2)	appr. 1	appr. ± 10

Table 2: Instruments and methods for convergence measurements in tunnelling

The main reason for the substitution of convergence tape measurements by geodetic deformation monitoring is associated with the following principal disadvantages of convergence tape measurements:

1 high degree of interference with tunnelling operations

2 no realistic perspectives for automation

3 restricted to *relative* displacement measurements.

2.2 Geodetic deformation monitoring

In today's tunnelling practice, *geodetic monitoring represents the real backbone of tunnel performance monitoring* in terms of volume of work and turnover figures.

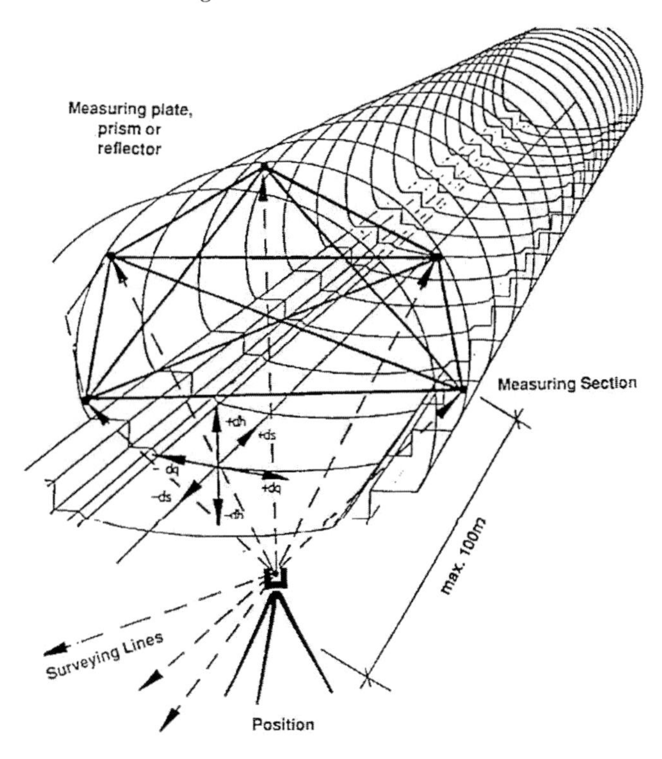

Figure 2: Principal sketch of a geodetic deformation monitoring set-up in tunnelling (Intermetric catalogue)

As indicated in Figure 2, displacement measuring sections are routinely installed every 10 to 25 m along the tunnel axis. Commonly, each section consists of five reflector targets (Figure 3) which are ± equally spaced along the tunnel periphery. This set-up is modified in partial tunnel excavations (Figure 4). The targets are surveyed, up to a distance of maximally 100 m, by precision total stations with automatic data acquisition. Each set of readings consists of two angles and one distance measurement. Total stations can be freely positioned to minimise interference with the tunnelling operations. Recent developments include motorised instruments with automatic target recognition. Such instruments can be temporarily or permanently positioned at the tunnel sidewall and can be operated by non-specialists.

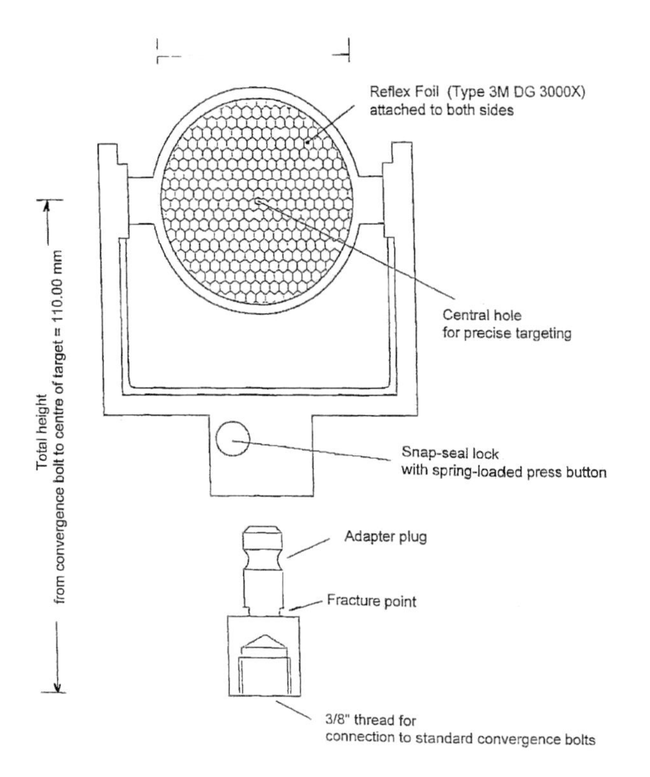

Figure 3: Reflector targets (left) for geodetic deformation measurements in tunnels (top) (Interfels catalogue)

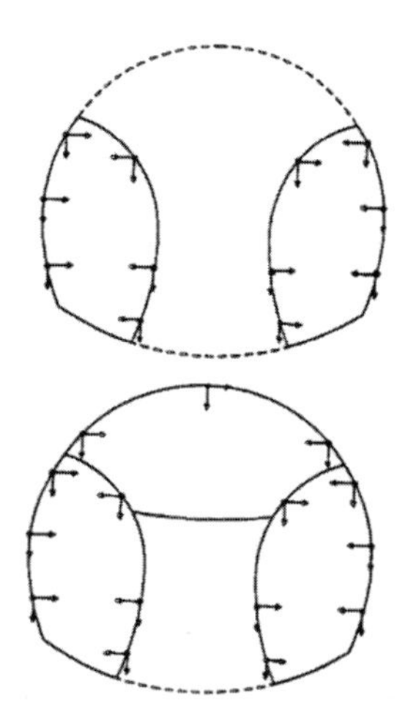

Figure 4: Common layout of targets for geodetic deformation measurement in partial tunnel excavations - © IUB

The results of geodetic tunnel deformation measurements are usually presented graphically. One standard graph is that of the displacement of the 5 measuring points in a cross section as depicted in Figure 5. Note that *absolute displacements* were determined and graphed. The relative displacements between any two measuring points (= "convergence" in the narrow sense) is then simply calculated by subtraction between the two displacement values.

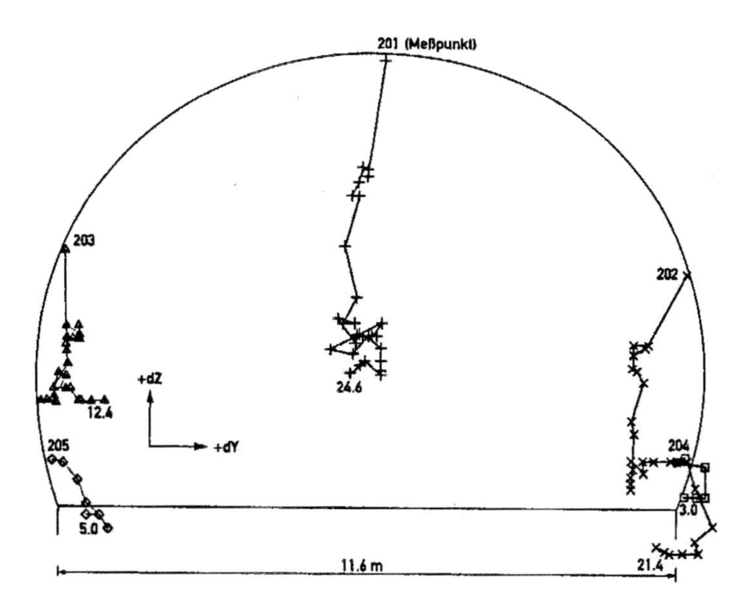

Figure 5: Example of a geodetic deformation measurement in tunnelling. Presentation of the absolute displacements of the 5 targets (in millimetres) for various time steps.

Much emphasis is currently given to the *integration of geodetic tunnelling surveying with traditional geotechnical monitoring*. The key to this approach is a suitable, integrated acquisition and evaluation software. Geotechnical instrumentation companies, which have the leading edge in this regard, are GeoData of Austria with its ITMS (Integrated Tunnelling Measuring System), SolExperts of Switzerland with its GeoMonitor System and Sol Data of France (for company addresses ref. to Appendix 1). Similarly, some traditional surveying companies are also moving towards such an integrated approach by establishing special geotechnical instrumentation units. An example is Intermetric GmbH, a Stuttgart-based surveying company, which has specialised in integrated geotechnical / geodetic monitoring services for tunnelling. Recently, Intermetric has released its Software System "iGM" (intermetric Geotechnique Monitor). iGM is an automatic data acquisition and evaluation program for up to 500 sensors in a measuring network. iGM can manage data from motorised digital levels, motorised total stations (tachymeters) and all types of geotechnical instruments including extensometers, inclinometers, level gauges, temperature gauges, vibrating wire sensors, total pressure cells and strain gauges.

2.3 Engineering assessment of convergence and geodetic deformation measurements

Apart from the graph in Figure 5, the measuring results of the convergence tape or geodetic deformation measurements are typically graphed as time-displacement curves, presented in Figure 6. Essential for the assessment of such measurements is a synoptic presentation of the tunnel construction work with the time-displacement curve.

The main criterion for the engineering assessment of the measuring results is simply the question of whether or not the convergence movements have come to a halt. If positive, the time-displacement curve will tend to converge towards a horizontal line. This is taken as proof that the tunnel system, consisting of the surrounding ground and the various support materials, has found a new equilibrium. If negative, the actual time-displacement curve would be inclined, perhaps even weeks or months after completion of the last tunnel construction procedure. The base of this engineering approach therefore is the consideration of the ***deformation velocity*** (and of the deformation acceleration). The *absolute magnitude of the deformation* is *not* considered in this regard.

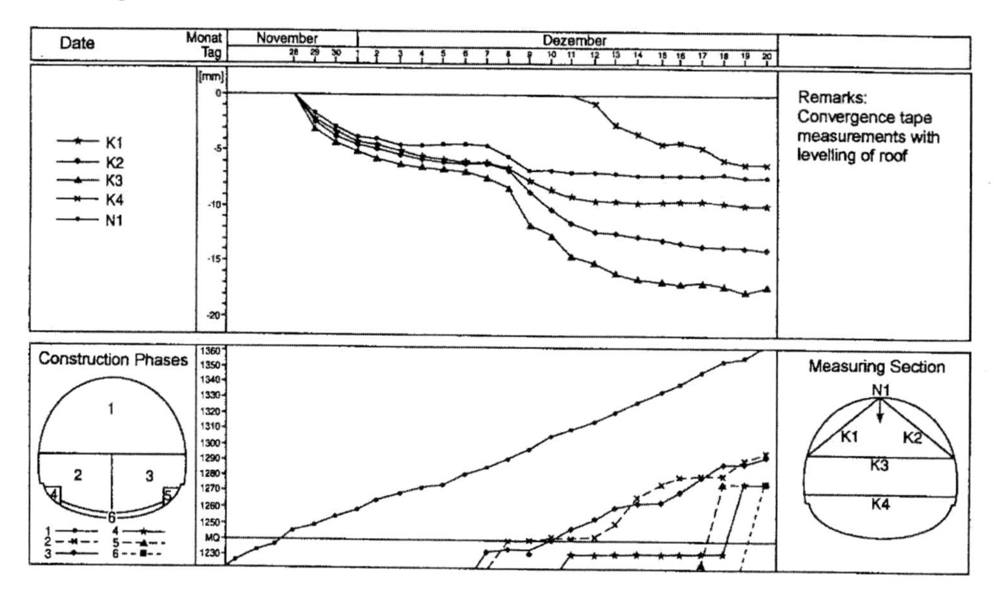

Figure 6: Graphical presentation of conventional convergence tape measurements (top) in a time versus relative displacement diagram. Note: Influence of tunnelling operations (bottom) on convergence.

The common interpretation of convergence and geodetic deformation measurements therefore remains at a rather ***empirical level***. It does not yield any in-depth insight into the mechanics of a tunnel system. In particular, it does not provide any information on the actual safety margin of the tunnel and its linings.

2.4 Load bearing capacity reserves and safety factors of shotcrete linings as deduced from geodetic deformation measurements

Efforts to correct the above-mentioned situation are currently being undertaken by Rokahr and co-workers of the University of Hanover. These include extensive on-site testing at a number of Austrian and German tunnelling projects. Rokahr also points to the fact that the common empirical interpretation of convergence and geodetic deformation measurements does not yield any clear evidence on the actual load bearing capacity of a shotcrete tunnel. On the basis of such measurements alone, it would not be possible to specify any factor of safety against failure of the tunnel construction and its lining.

In realising that the stress-strain relationship of shotcrete is highly time-dependent, Rokahr claims that science has advanced to the stage where this relationship can be specified with a sufficient degree of accuracy and confidence. Employing numerical modelling procedures he converts the measured strain of the shotcrete lining (as deduced from the displacement measurements) into stresses. On the basis of this information, it is then possible to specify the actual degree of loading, the load bearing capacity and the actual factor of safety of the shotcrete lining.

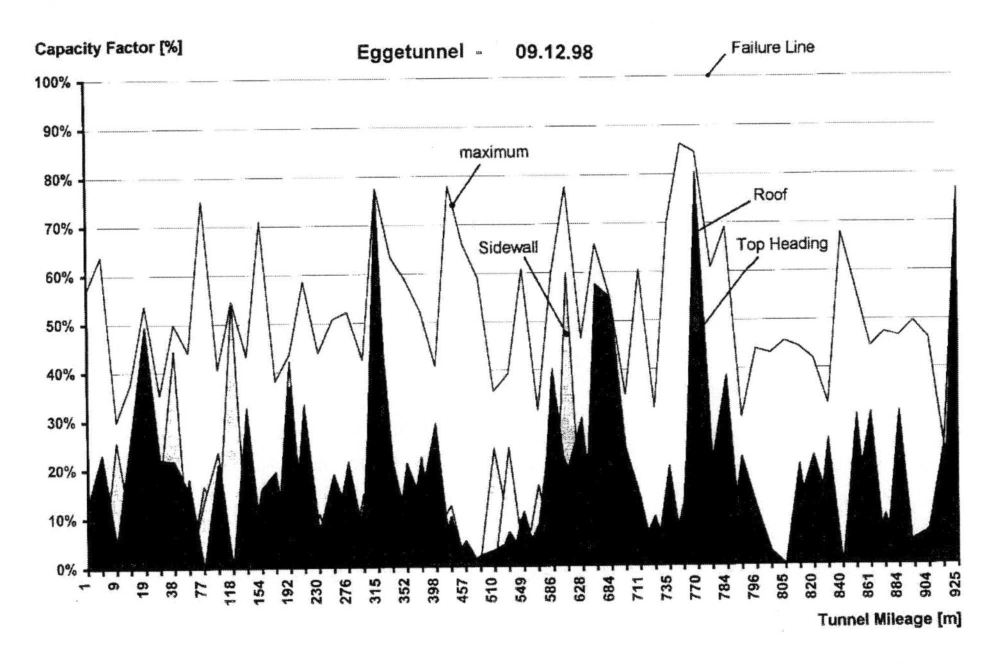

Figure 7: Capacity factor of a shotcrete lining of a 925 m long tunnel section (Egge railway tunnel - NW section , currently under construction; after Rokahr, 1999)

Figure 7 shows an example for a 925 m long shotcrete tunnel section. It indicates that, whilst subject to significant local fluctuations, the capacity of the shotcrete lining (i.e. the actual load divided by the load bearing capacity) was maximally occupied to about a level of 50 to 80%. This is equivalent to a factor of safety against failure of the shotcrete lining of between 1.25 and 2.0.

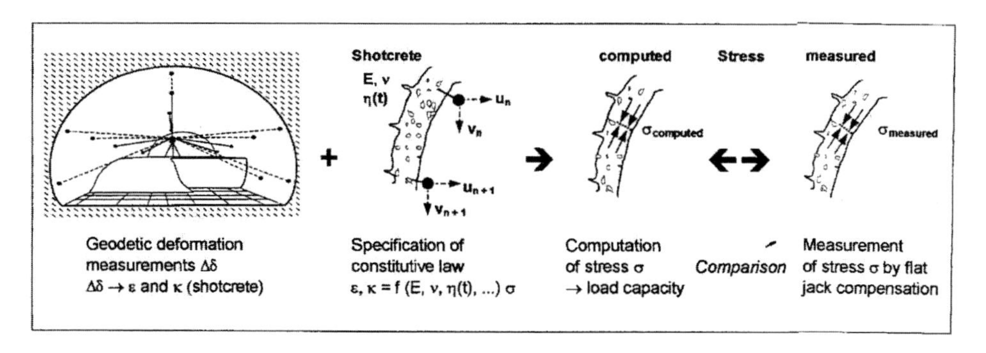

Figure 8: Evaluation and checking procedures of the load bearing capacity determination of shotcrete linings based on geodetic deformation monitoring and stress measurements by the slot cutting and flat jack compensation method.

The derivation of the load bearing capacity of the shotcrete lining is based on a number of steps as indicated in Figure 8. Each step is intrinsically associated with assumptions and errors. It is therefore highly desirable to check the end result (i.e. the computed stresses acting in the lining) by direct measurement. Recently this has been done by employing the *Slot Relief & Flat Jack Compensation Method.* This method has a proven record of reliability and, in the author's opinion, is the best method for stress determination of concrete and similar materials at accessible surfaces. The slot relief and flat jack compensation method is carried out in a sequence of three steps, as indicated in Figure 9. Firstly, measuring marks are set next to the intended cut (Figure 9a). Then the slot is cut by a diamond saw, typically ϕ 450 mm, and measurements taken on the convergence of the measuring marks due to slotting (Figure 9b). Finally, a flat jack, which conforms to the shape of the slot, is inserted into the slot and hydraulically inflated until the point of full reversal of the measured convergence (Figure 9c). The point is termed "compensation point". Accordingly, the pressure acting in the jack at the compensation point is termed "compensation pressure". This pressure is (nearly) equivalent with the stress in the shotcrete to be determined. Note that no knowledge of material parameters (such as the Young's modulus E) is required for this test.

In the example of the tunnel project shown in Figure 7, the result of the comparison between computed stresses (as deduced from geodetic displacement measurements) and measured stresses (employing the flat jack compensation method) is shown in Figure 10. There is excellent agreement between the two methods.

It is the expectation of the author that within the near future the evaluation of the capacity of shotcrete linings will become a widely applied standard in tunnel construc-

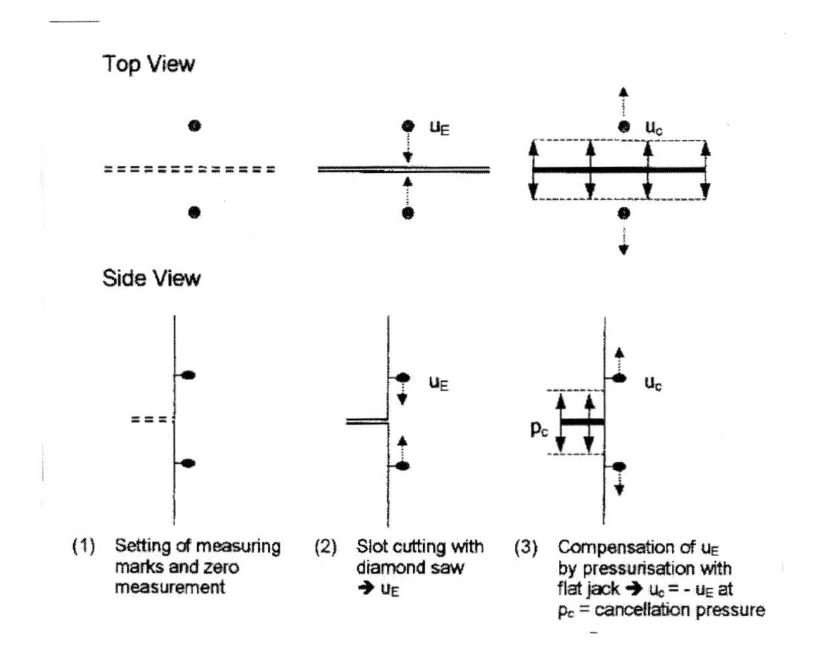

Figure 9: Measuring principle of the Slot Cutting and Flat Jack Compensation Method

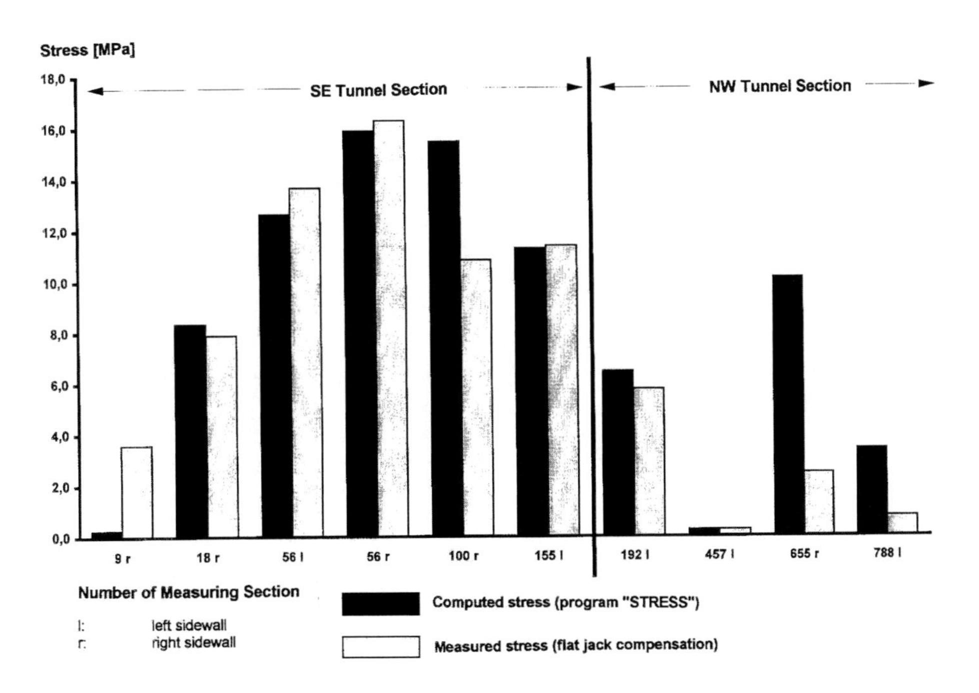

Figure 10: Comparison between computed and measured tangential stresses of the shotcrete lining of the Egge Tunnel (Rokahr, 1999)

tion. The base requirement for this will be *extensive geodetic deformation monitoring*, accompanied with regular *checking of the computed stresses by the slot cutting and flat jack compensation method.*

3 Instruments for monitoring of displacements and stresses for better design

In the previous Section 2 it was already indicated that convergence and / or geodetic deformation measurements alone are insufficient for a full judgement as to the mechanical behaviour of the tunnel system. Two reasons can be identified in this regard:

- The ground surrounding the tunnel is not directly monitored. This, however, is necessary as the ground has a definite load bearing capability and is one of the contributing factors in the overall stability of the tunnel system.
- A mechanical description of a tunnel system remains incomplete if it is solely based on displacements and its derivatives. Knowledge of the forces (and stresses) are definitely required for completeness.

In a tunnel monitoring program which is specifically set up for control of the tunnel design, at least all of the standard instrumentation, as indicated in Figure 1, should be installed to provide a sufficiently broad data base for comparisons to be made between measurements and predictions. This monitoring program consists of the following:

- Deformation measurements of the excavated tunnel surface, as already described in Section 2.
  Instrumentation: Total stations and reflector targets (in special cases: convergence tape)
- Deformation measurements of the ground surrounding the tunnel.
  Instrumentation: Borehole extensometer (especially 3-point extensometer)
  Comment: Hardly used anymore in Europe as installation interferes with tunnelling operations. However, the situation is completely different in near-surface tunnelling (ref. to below).
- Control of the ground support element "anchor" or "rock bolt".
  Instrumentation:
  - *Anchor load cell → monitoring of the forces at the head of the anchor*

   - *Measuring anchor → strain monitoring over the length of the anchor. This yields information on the required length of the anchor.*

- Control of the ground support element "shotcrete".
  Instrumentation: Total pressure cells (TPC) → passive hydraulic flat jacks for monitoring of the radial and tangential stresses.
  Comment: In Europe, many tunnel engineers are somehow disenchanted with TPCs for shotcrete stress monitoring. Clearly, the performance of TPC is critically dependent on a number of factors, amongst them the TPC design, local conditions and, in particular, the quality of the installation. Some engineers prefer concrete embedment strain gauges instead of TPCs. This requires an in-depth knowledge of the time-dependent stress-strain relationship of the shotcrete (ref. also to the approach of Rokahr → Section 2.4).

- Control of the ground support element "steel arches" (not indicated in Figure 1).
  Instrumentation: Strain gauges and total pressure cells
  Comment: Not commonly used in Europe.

3.1 Instruments for measurement of the displacements of the ground

In near-surface tunnelling, the above standard instrumentation will be modified accordingly. In particular, this applies to the measurement of the ground movements by borehole extensometers which are now installed from the ground surface and not, as previously, from within the tunnel excavation.

Extensometers which are installed from the ground surface offer the following advantages:

- No interference with tunnel construction operations (this is seen as *the* major advantage)

- Installation and measurements are not restricted to the post-excavation phase. All ground deformations can be monitored including those ahead of tunnelling which are of particular concern in inner-city tunnelling.

- Problem-free installation and convenient measuring operations with high-definition probe extensometers.

The measuring example of Figure 11 gives evidence of these advantages.

Borehole extensometer measure the particular component of ground displacements which is directed *along the axis* of the borehole. For monitoring the complete deformation state of the ground instruments must be employed, in addition to extensometers, which measure the displacement components acting *across* the borehole

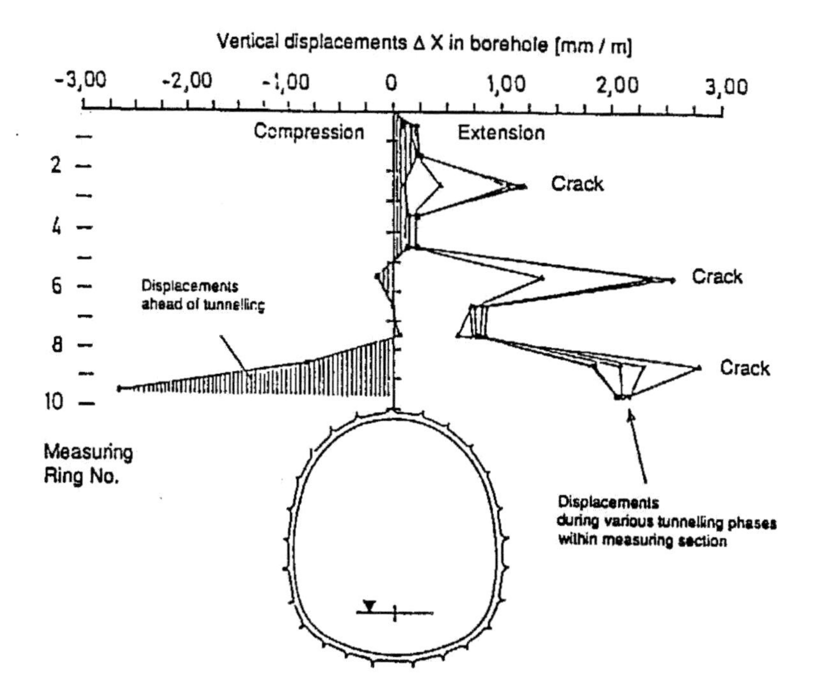

Measuring task: Evaluation and control of loosening of the tunnel roof strata

Main results:

(1) *Ahead of tunnelling* (hatched in the graph) there is a *state of compression* at the tunnel face and in the roof strata

(2) As soon as tunnelling has passed the measuring section, *the characteristics* of the ground deformation of the roof strata *changes from compressive to extensional.*

(3) Clearly, the degree of extension is *discontinuous.* Local peaks were recorded at depths of 9.0 m; 7.0 m and 4.0 m *(→ cracks).* This is indicative of an onion-style loosening structure in the tunnel roof strata.

(4) The four follow-on measurements, carried out during the various tunnel excavation phases, do not differ significantly. This is indicative that the tunnel system has found a new equilibrium after excavation.

Figure 11: Monitoring of the loosening of the roof strata of a near-surface railway tunnel by the probe extensometer "INCREX" (after Estermann, 1991)

axis. This is achieved by standard inclinometers (in vertical boreholes) or by horizontal inclinometers or deflectometers (in inclined or horizontal boreholes), either stationary as fixed borehole chains or as mobile borehole probes.

Figure 12 shows a measuring example of a deep-seated tunnel in which the ground displacements were measured by the conjunctive use of mobile extensometer and inclinometer.

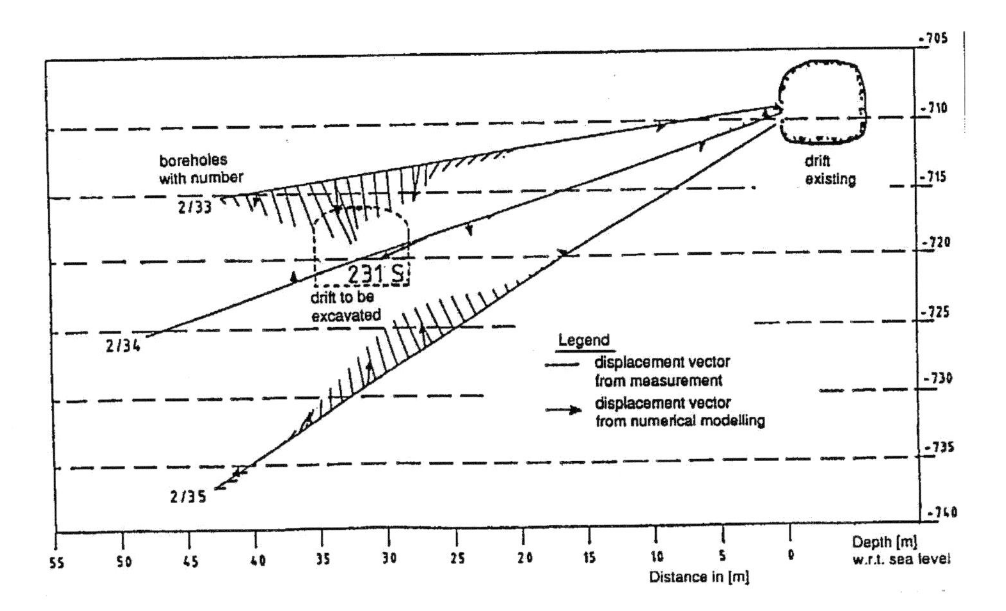

Measuring task: High-definition deformation measurements in the ground for comparison with predictions of Finite Element computations.

Main results:

(1) Most detailed evaluation of the ground displacement state in the vicinity of the newly excavated drift as shown by displacement vectors. Note that each vector has an axial (determined by the probe extensometer INCREX) and a lateral component (determined by horizontal inclinometer) with regard to the borehole axis.

(2) All displacement vectors are oriented towards the new excavation (even in the invert strata).

(3) There is a reasonable degree of agreement between measured and predicted displacement values both with regard to direction and magnitude (➔ Indication that the geomechanical model is acceptable).

Figure 12: Ground displacements due to the excavation of a deep-seated tunnel as measured by the conjunctive use of probe extensometer and probe inclinometer (after Diekmann and Kern, 1991)

The measuring example of Figure12 also indicates the intrinsic purpose of this type of monitoring which is to provide the basis for a comparison between measurements and predictions. The displacements, as predicted from numerical modelling studies, are indicated in Figure 12 by arrows. It is up to the Geotechnical Engineer to judge the degree of agreement achieved and to decide whether or not the geotechnical model will be acceptable or has to be refined.

Mobile borehole probes for high-definition ground displacement measurements around underground openings are in common use. The most popular extensometer probe is the "Sliding micrometer", manufactured by SolExperts AG with a wide distribution in countries such as Switzerland, Germany and Eastern European countries. The "INCREX" probe, manufactured by Interfels GmbH, is popular in Italy, Austria and parts of Germany. The market for mobile inclinometers is highly contested between Glötzl GmbH, SisGeo s.r.l. in Italy and Slope Indicator (e.g. in Switzerland). The least common instrument is the mobile deflectometer, however, in recent years Interfels GmbH with its newly designed deflectometer probe has won a sizeable market share. Combined probes are manufactured by SolExperts AG, namely the extensometer / inclinometer probe "TRIVEC" and the extensometer / deflectometer probe "LADEX".

3.2 Instruments for measurement of ground stresses

Whilst a complete, well-proven and widely used set of instrumentation exists for monitoring of deformations in the ground, this is not necessarily the case for stress measurements and stress monitoring. From both a conceptual and technical point of view, the measurement of ground stresses (and the change of stresses) is generally much more difficult than that of the displacements. Some engineers make a virtue of this situation when arguing that they can do without any stress measurement and stress monitoring. However as mentioned before, stresses are an intrinsic part of any geomechanical system and cannot be ignored if our considerations are to be complete.

Continuing problems with measuring and monitoring of ground stresses have lead a number of instrumentation manufacturers towards the development of improved or innovative stress measuring methods. Table 3 gives an overview of testing methods which are currently in the market. The table also gives an indication on the author's evaluation as to the various methods' future development potential.

Figure 13 shows a stress measurement example. It indicates the distribution of the circumferential stresses in the sidewall rock of a tunnel as determined by borehole slotting. The high-definition measurements clearly delineate an approx. 2.0 m deep plastic zone. This information is important for the tunnel design, e.g. indication on the post-failure characteristics of the ground, confirmation of the rock loads and selection of proper lengths of the rock bolts.

Detailed and systematic investigations with objective comparisons between the various stress measuring and monitoring methods as well as comparisons between measured and predicted values are yet to be carried out in tunnel construction. Such investigations, however, are absolutely essential for improving our knowledge of tunnel systems, thereby permitting better design, safer and more efficient construction procedures.

Ground Stress Measuring Method (Manufacturer)	**Absolute** $\sigma_1\ \sigma_2\ \sigma_3$	**Change** $\Delta\sigma$	**Remarks Limitations**	**Development Potential**
Overcoring of a strain cell CSIRO - Australia: "Hollow Inclusion Strain Cell" (Mindata P/L, Seaford, Vic.)	$\sigma_1\ \sigma_2\ \sigma_3$	(no)	World-wide in use Expensive; Limited number of tests	Mature method. No major further potential
CSIR- South Africa: 3-ax. strain cell	$\sigma_1\ \sigma_2\ \sigma_3$	(no)	as above	as above
Hydraulic Fracturing (numerous small companies)	$\sigma_1\ \sigma_2$	no	World-wide in use limited for 2-D	medium
Borehole Slotting Stressmeter (Interfels)	$\sigma_1\ \sigma_2$	limited	Numerous measurements. Momentarily restricted to 2-D and t(max) = 40m.	high
Hydraulic Total Pressure Cells (TPCs) installed in boreholes (Glötzl; Interfels)	no	yes	Suitable for soil, soft rock and rock salt	medium
Hard Inclusion Cells (Vibrating Wire) (Geokon; Irad Gage)	yes	yes	For hard rock only. Requires overcoring	low to medium

Table 3: Methods for measuring and monitoring of ground stresses

4 Conclusion

The following trends can be identified with regard to performance monitoring of tunnels and other underground structures for design purposes:

- In terms of volume of work and turnover figures, *geodetic deformation monitoring represents the real backbone* of today's tunnel performance monitoring work.

- Much emphasis is currently given to the *integration of geodetic tunnelling surveying with traditional geotechnical monitoring methods*. The key for this approach is the availability of a suitable integrated acquisition and evaluation software.

- The interpretation of the common convergence and geodetic deformation measurements remains at a rather empirical level. This is seen as a major deficiency in current practice. It can be expected that in the near future a more

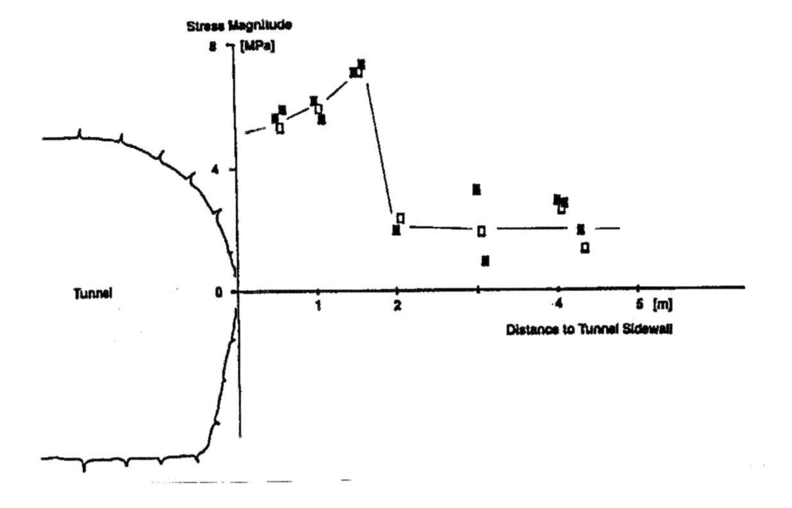

Measuring task: High-definition stress measurements in the sidewall rocks of a tunnel.

Main results:

(1) Detailed evaluation of a stress profile with delineation of
➔ the primary ("geologic") stress state amounting to appr. 2.0 MPa
➔ the secondary stress state with an amount of maximally 8.0 MPa

(2) Delineation of a 1.5 to 2.0 m wide "plastic zone"

(3) The stress profile is indicative of a brittle failure behaviour of the ground.

Figure 13: Stress measuring example: Distribution of the circumferential stresses in the sidewall rocks of a tunnel (Hagerbach / Switzerland)

rigorous *evaluation of the load bearing capacity of shotcrete linings* will become a widespread standard. The base requirements for such evaluation will be extensive geodetic deformation monitoring, in-depth knowledge of the material law of shotcrete and regular checking of the computed stresses by the slot cutting and flat jack compensation method.

- Measurements and monitoring of the *ground stresses,* principally desirable from an engineering point of view, *are still not in widespread use*. New high-definition stressmeters can delineate primary and secondary stresses around a tunnel at reasonable costs.

- Overall, it appears that, amongst tunnel engineers, there is no longer a strong interest in detailed instrumentation programs for checking and improving of the tunnel design as was the case some decades ago. This is in marked contrast to booming instrumentation demands for control of the construction procedures, as will be discussed in Part 2 of this contribution.

Bibliography

[1] Diekmann, N. and Kern, P., 1991. Investigations into the stability of drives in an underground mine. INCREX in combination with mobile inclinometer. - Interfels News, **4:** 15 - 18.

[2] Estermann, U., 1991. Application and data processing of INCREX measurements for near-surface tunnelling.- Interfels News, **4:** 3 - 9.

[3] Müller, L., 1978. Rock Construction, Vol. 3: Tunnelling (in German). - 945 p., Stuttgart (Enke).

[4] Rokahr, R. B. and Zachow, R., 1999. A new method for the daily control of the load bearing capacity of a shotcrete lining (in German). - Unpubl. Report, 7 p., Hanover (Univ. Inst. for Subsurface Construction)

Appendix: List of Companies

Geodata	Hans-Kudlich-Str. 28	A - 8700 Leoben	Austria
	Tel. +43 - 3842 - 26 555 -0	Fax: +43 -3842 - 26 55 55	www.geodata.at
Glötzl	Forlenweg 11	D - 76287 Rheinstetten	Germany
	Tel. +49 - 721 - 5166 -0	Fax: +49 - 721 - 5166 - 30	www.gloetzl.com
Interfels	Deilmannstr. 5	D - 48455 Bad Bentheim	Germany
	Tel. +49 - 5922 - 98 98 - 0	Fax: +49 - 5922 - 98 98 98	www.interfels.com
Intermetric	Industriestr. 24	D - 70565 Stuttgart	Germany
	Tel. +49 - 711 - 780 0392	Fax: +49 - 711 - 780 0397	www.intermetric.de
SisGeo	Via Serpero S. P. 179	I - 20060 Masate / Milano	Italy
	Tel. +39 - 02 - 957 64130	Fax: +39 - 02 – 9576 2011	www.sisgeo.com
Sol Data	6, rue de Watford	F – 92000	Nanterre France
	Tel. +33 - 1 - 4776 5790	Fax: +33 - 1 - 4692 0365	www.soldatagroup.com
SolExperts	AG Schulstrasse 5	CH - 8603 Schwerzenbach	Switzerland
	Tel. + 41 - 1 - 806 29 29	Fax: +41 -1 - 825 0063	www.solexperts.com

For list of references ref. to Part 2

Helmut Bock, Ph.D., MSc.

Chartered Engineer in Geomechanics and Geomonitoring Systems, Bad Bentheim, Germany. President of the German Society of Engineering Geologists. From 1989 to 1997 he was President of Interfels GmbH, Bad Bentheim, providing geotechnical instrumentation products and field services. Previously he was Professor of Geomechanics at James Cook University in Australia. He has been involved in in-situ rock testing methods and geotechnical instrumentation for more than 25 years, and has numerous publications and patents.

Geotechnical instrumentation of tunnels Part 2: Instrumentation to assist with tunnel construction control

Helmut Bock

Geotechnical Consultant, Past-President of Interfels GmbH, Stoltenkampstr. 1, D - 48455 Bad Bentheim, Germany, Tel. + 49 - 5922 - 2700, Fax: +49 - 5922 - 2799
QS-Consult@t-online.de
http://home.t-online.de/home/QS-Consult

Preface

In Part 1 of this contribution an account was given of recent developments and practices in performance monitoring for the verification of the tunnel design.

In this Part 2, geotechnical instrumentation for the control of the tunnel construction will be discussed.

1 Introduction

In line with the ISO 9000 ff standards, industry is facing new demands for improved quality management of its operations. The geotechnical industry is no exception in this regard. On tunnelling sites, such development is reflected in increased and more rigorous controls of the quality of construction work and procedures. Geodetic and geotechnical instrumentation and services are amongst the most important tools in this regard. Currently, this market sector of geotechnical instrumentation performs with significantly more dynamism than the traditional sector of tunnel performance monitoring for design verification.

With reference to common geotechnical design and construction procedures (Figure 1), the quality control of the construction by geotechnical instrumentation can be understood as a feed-back loop. In the diagram of Figure 1, this loop shows up as the smallest loop possible within the geotechnical design procedure. Note that the geotechnical instrumentation for verification of the design, as discussed in Part 1, can be visualised as an element within a greater feed-back loop.

The discussion of geotechnical instruments, applied to assist with the quality control of tunnel construction procedures, will be dealt with in the following two sections: The discussion will be supported by way of a number of monitoring examples.

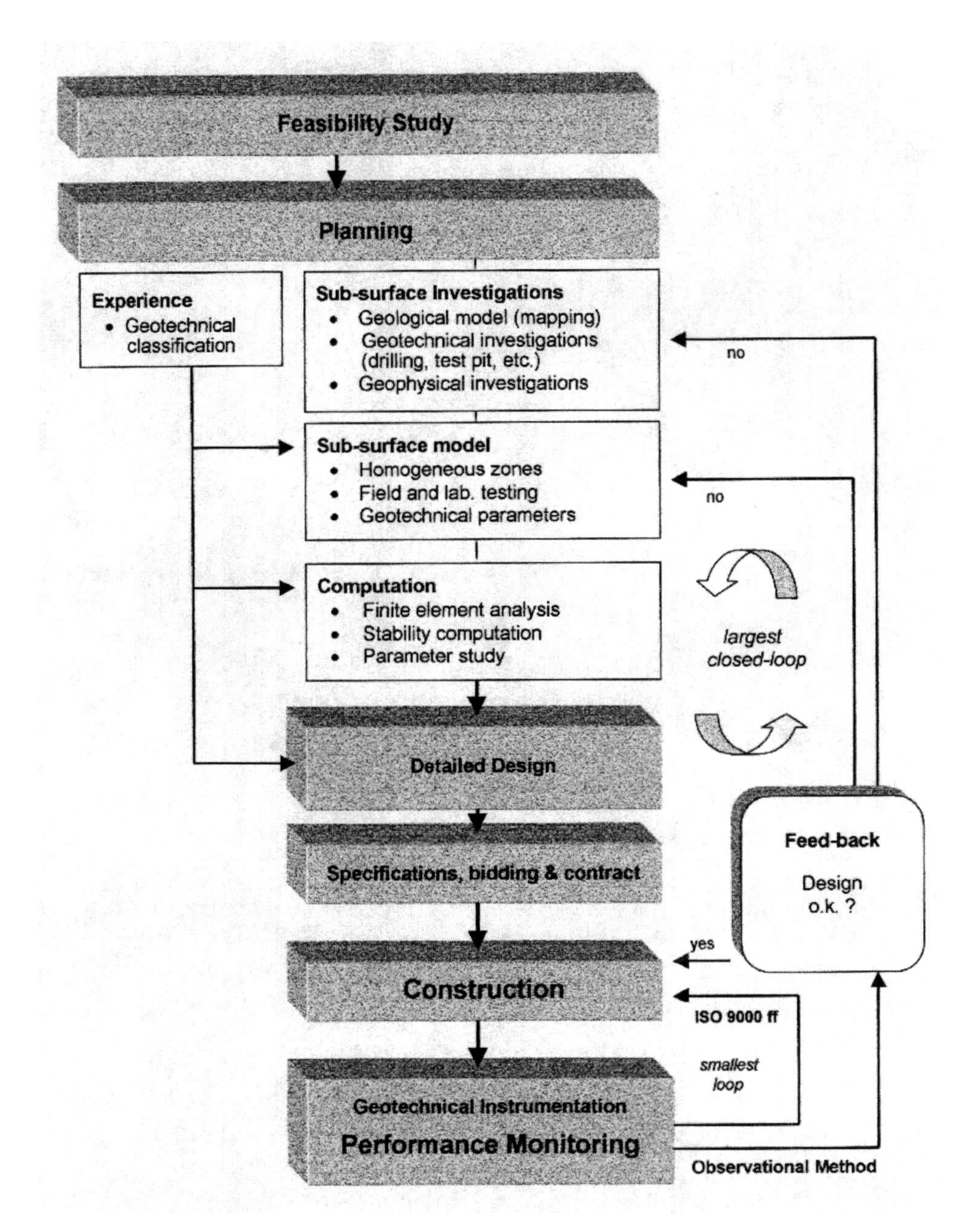

Figure 1: Performance monitoring by instrumentation as part of the geotechnical design procedure. Note the various feed-back loops for verification of the design (ref. Part 1 of this contribution) and the control of the construction procedures (ref. this Part 2).

2 Instrumentation for the control of selected tunnel construction procedures

One of the standard controls used in tunnelling construction is acoustic emission and ground vibration monitoring. Such monitoring is routinely carried out for all types of tunnelling operations, not only for drill-and-blast excavations but also for partial and full-face tunnel boring machines.

In the following, two more recent developments in the control of specific tunnel construction measures will be discussed. These developments are quite noticeable within the European market. They are:

(1) Innovative tunnel scanner: Control of the excavation (over- and underprofiles) and of the thickness of the shotcrete lining (Section 2.1).

(2) New probe deflectometer: Control of the quality of drilling by deviation measurement of horizontal and inclined boreholes (Section 2.2)

2.1 Tunnel scanner for control of the excavation profile and of the concrete thickness.

Tunnel scanners are widely employed for a variety of purposes, amongst them profile scanning and clearance control of a tunnel prior to its commissioning.

A new tunnel scanner has recently been released under the trade name ”DIBIT” and has already made a significant impact on the European market. Developed by a number of Austrian companies, "DIBIT" is a fully digitised photogrammetric measuring system for the documentation of tunnel advances. It enables the control of specific underground construction procedures such as the excavation of the tunnel, the contouring of the tunnel profile and the application of shotcrete lining as a primary support.

As indicated in Figure 2, the recording system consists of two CCD cameras which are mounted on a portable frame. The system produces digital stereoscopic images of the tunnel surface. The position of the camera frame is automatically determined by a total station with automatic target recognition which is positioned up to a distance of maximally 100 m. For this purpose three reflector targets are permanently mounted on the frame. Positioning can be carried out in conjunction with routine geodetic deformation measurements which were already discussed in Part 1 of this contribution. On-site recording of the scanner takes only a few minutes and can be carried out by non-specialists. Digital images are automatically stored in a field-PC which is integrated within the system. By means of the stereoscopic images, the 3-D

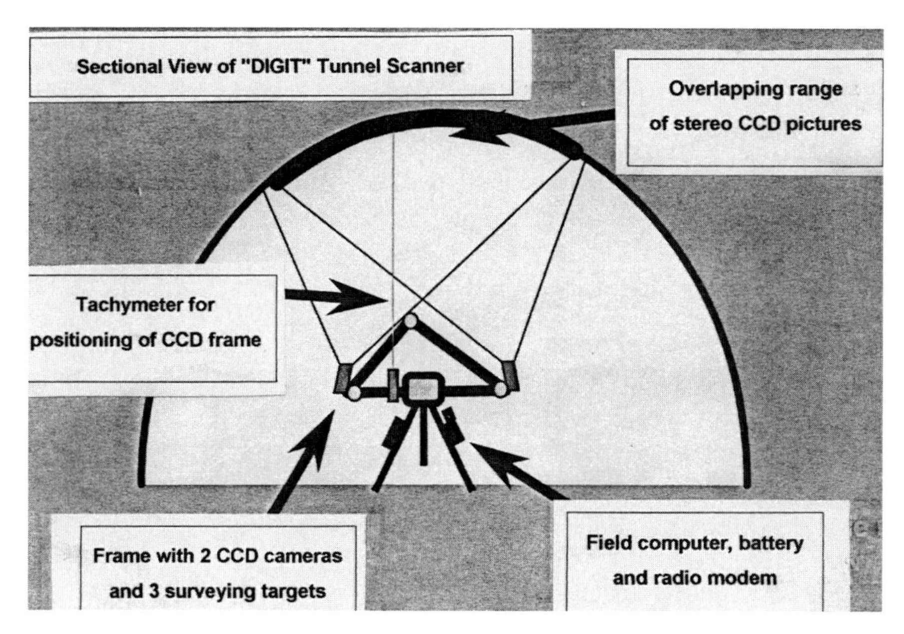

Figure 2: Principal features of the "DIBIT" tunnel scanning system.

co-ordinates of the tunnel surface surveyed can be automatically calculated. At the current stage of development, the accuracy of the system is in the range of ±5 mm for each co-ordinate.

Integrated PC software permits numerous evaluation options, amongst them:

- Comparison of actual and nominal excavation surfaces:
  Determination of over- and underprofiles not only in selected cross sections (Figure 3) but also over the full length of the tunnel (Figure 4).

- Comparison of the tunnel surfaces prior to and after shotcreting:
  Determination of the thickness of the shotcrete lining with automatic specification of the minimum, maximum and average thickness (Figure 5).

- Comparison of the tunnel surface at different instances in time:
  Determination of the deformations of the tunnel surface (convergence) (Figure 6). Note, however, the comparatively low system accuracy of tunnel scanners (ref. to Table 2 of Part 1).

Besides the above-mentioned possibilities, the particular value of the "DIBIT" tunnel scanner lies in the production of digital image data for an objective documentation of various tunnelling stages.

The "DIBIT" system is manufactured and sold by the Austrian company "DIBIT Messysteme" and "DIBIT" Testing Services are offered by "GeoConcept" in Germany (for addresses ref. to Appendix 1).

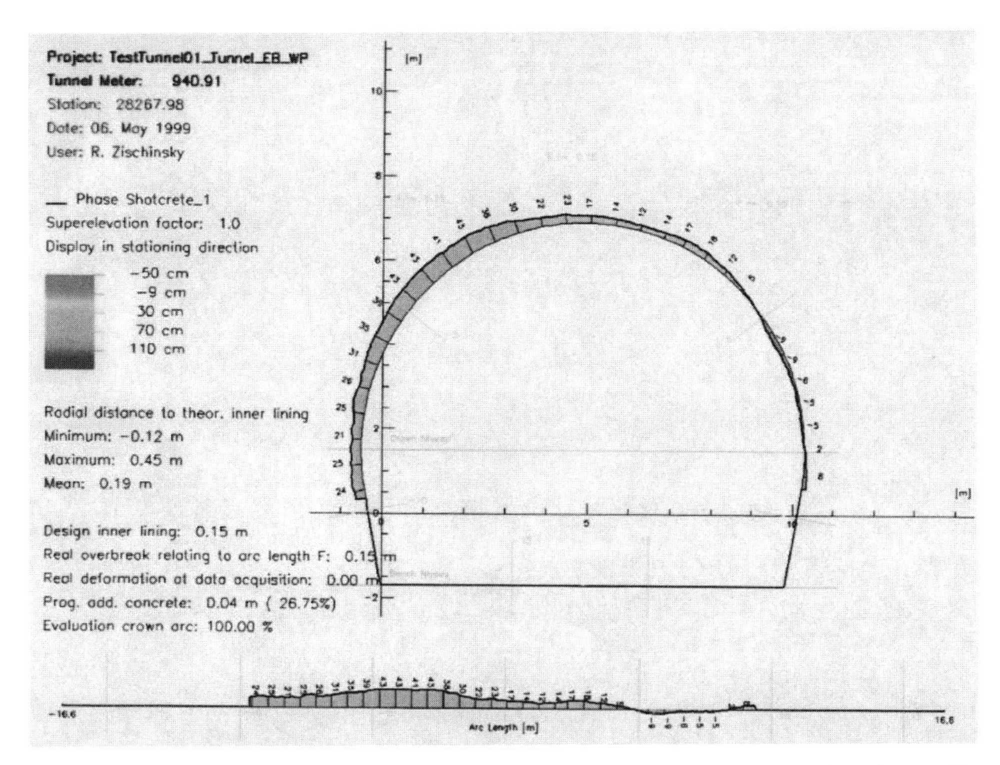

Figure 3: Tunnel scanner "DIBIT": Comparison of actual and nominal excavation surfaces in a profile.

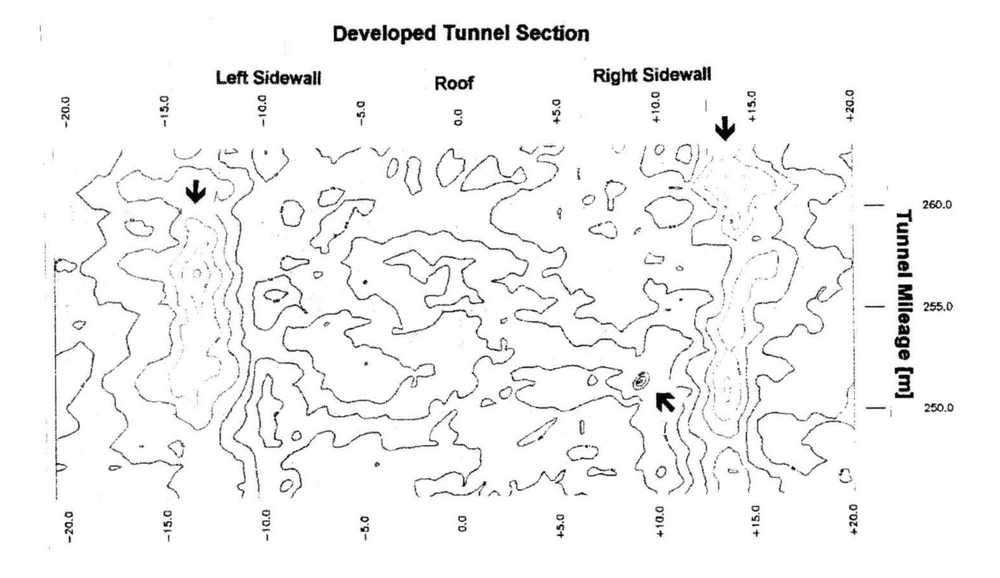

Figure 4: Tunnel scanner "DIBIT": Comparison of actual and nominal excavation surfaces in a plan view. Arrows indicate systematic over-profile in the lower sidewalls and local under-profile at rock bolt heads.

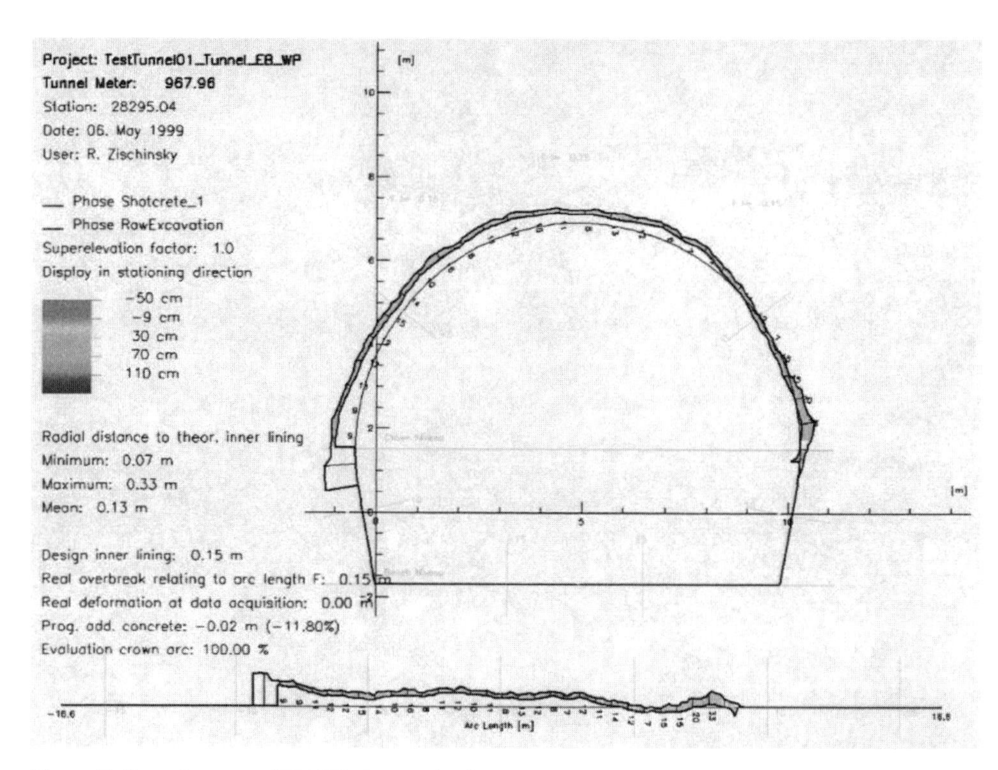

Figure 5: Tunnel scanner "DIBIT": Determination of the thickness of the shotcrete lining. Note: Relatively homogeneous thickness of the lining (average of 130 mm) in this example.

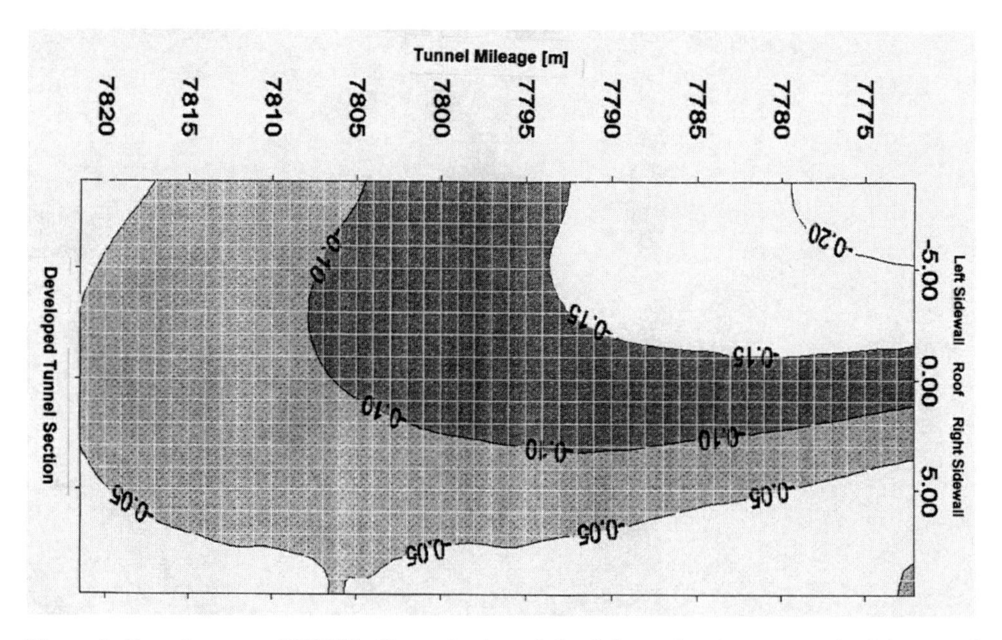

Figure 6: Tunnel scanner "DIBIT": Determination of the deformation (convergence) of the tunnel surface.

2.2 Deflectometer measurements for control of the drilling work in tunnelling

Both horizontal and inclined drilling is regularly carried out as part of the tunnel construction. Examples include drilling of anchor boreholes, of exploration boreholes in the face of the advancing tunnel and of sets of horizontal drillholes in ground freezing tunnelling. Part of the construction control procedure is the measurement of the drillhole deviation. Experience has shown that horizontal and inclined drillholes are particularly prone to deviation, whereby the degree of deviation depends on the quality of the drilling equipment, the experience of the crew, the length of the drillhole and, last but not least, the geologic conditions.

There are various systems on the market which measure deviations in horizontal and inclined boreholes, such as gyro probes (DMT), earth magnetic field probes (Reflex Instrument AB), electro-optical ("Maxibor" of Refflex Instrument AB) and photographic probes (e. g. "Multi-Shot"). In typical geotechnical applications with comparatively shallow boreholes of depths in the 10 to 100 m range, portable deflectometer probes are most commonly used. As described in Dunnicliff (1988, 1993: p. 273) a deflectometer probe consists of two beams of equal length connected by an articulated joint, with two angle transducers arranged to sense the two independent angular rotations between the two beams (Figure 7).

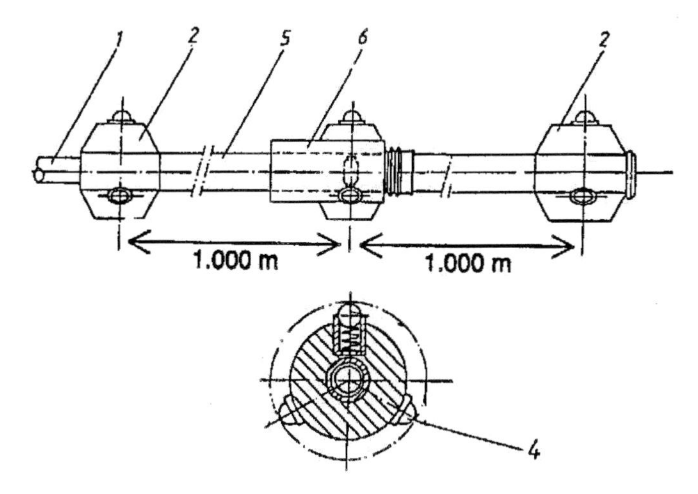

Figure 7: Schematic of a probe deflectometer (Interfels)
1 Insertion rod
2 Centring housing
4 Contact to casing
5 Tube
6 Water-proof housing of articulated joint with two built-in potentiometer transducers

Early angle measuring configurations were based on full bridge bonded resistance strain gage transducers (Slope Indicator), on tensioned wire passing over knife edges with induction transducers (Interfels) or on electro-optical transducers (Glötzl). Recently, Interfels released a new deflectometer with built-in potentiometer rotation transducers which, in the opinion of the author, surpasses all previous deflectometer versions in regard to system accuracy and robustness of construction. Bock et al., 1997 reported a system accuracy for a 40 m long borehole of ± 25 mm for both inclination and azimuth components. Note that standard steel casings with $\phi_i \geq 82$ mm were used in the survey.

The following measuring examples are from surveys undertaken with the new Interfels deflectometer.

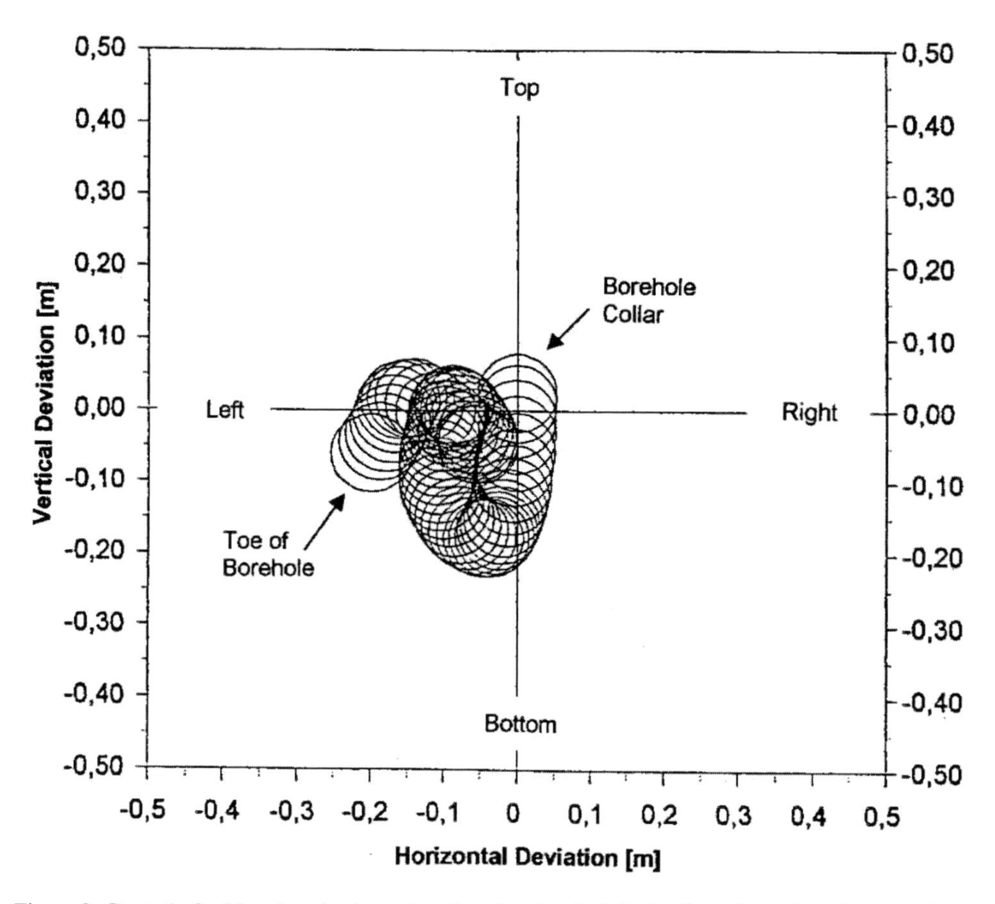

Figure 8: Control of a 56 m deep horizontal exploration borehole in the face of an advancing tunnel by a deflectometer survey. Shown is the top view into the borehole with deviation from the nominal axis in steps of 1.0 m.

Figure 8 shows the deviation of a 56 m deep exploration borehole which was drilled in the face of an advancing tunnel. The borehole was surveyed in steps of 1.0 m and was carried out by one trained technician. Setting up of the equipment and carrying out normal and reverse measurements took approximately 70 minutes.

As can be seen clearly in Figure 8, the borehole runs like a corkscrew. The deviation of the components at the toe of the borehole amounts to about 0.18 m (downward) and 0.20 m (to the left), respectively. This is equivalent to a deviation of approximately 0.33 % of the total depth and well within the specified limits.

Figure 9 refers to a ground freezing tunnelling project in highly permeable Quaternary sediments consisting of layers of sand and gravel. From a vertical shaft a set of horizontal holes were driven to form a frozen soil cylinder around the contour of the tunnel to be constructed. It is commonly known that significant drillhole deviations can lead to gaps in the frozen ground cylinder with the potential of a disastrous water and soil ingress during tunnel excavation. In the case of Figure 9, a total of 63 horizontal drillholes was surveyed each to a depth of 36 m.

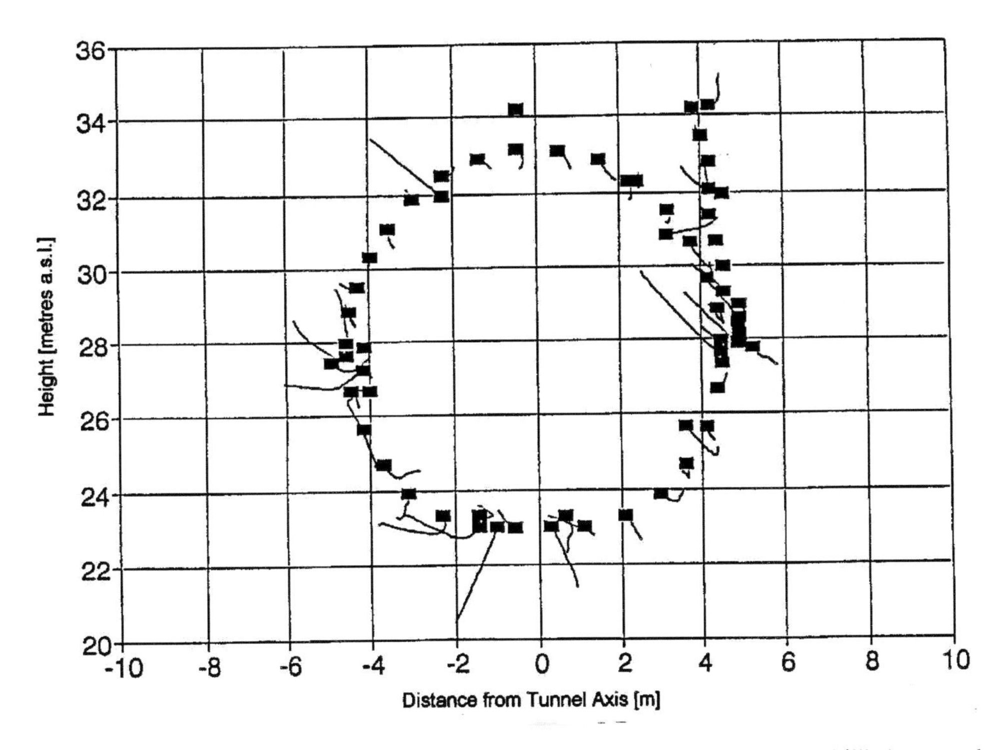

Figure 9: Front view of the wall of the starting shaft with the location of 63 horizontal drillholes around the contour of the ground freezing tunnel. All drillholes were surveyed by a probe deflectometer. The measured borehole deviations are indicated.

In the project a substantial number of drillholes were found to deviate more than the specified value of 0.5 % of the end depth. Additional drillholes were required to correct the situation.

3 Real-time monitoring for the control of entire tunnelling operations

It is obvious to carry out construction controls not only *after completion* of specific work procedures but continuously *in parallel with the ongoing construction.* Monitoring data can then be used not only as a base for the quality assessment of the construction work, but also for the direct control of the operations. For example, in the previously discussed case of drillhole deviation measurements, it is obvious that modern directional drilling rigs are to be employed in the first instance. With these rigs, the inclination and azimuth of the drilling head are continuously monitored and adjusted as required. However, it should be realised that, up until now, such rigs have been significantly more costly than conventional drilling rigs and are not always suitable in geotechnical applications.

A precondition for the direct control of any construction procedure is on-site *real-time monitoring*. Such monitoring is the actual "hit" in geotechnical instrumentation. Key geotechnical parameters are continuously monitored and immediately processed by automatic data acquisition and evaluation procedures. Real-time monitoring contributes to lowering the risk of unforeseen events to an absolute minimum. This is often achieved with a surprisingly high degree of success. Beyond this it opens up the possibility for innovative construction procedures which would not be possible otherwise.

With regard to this, reference is made of the Soilfrac®compensation grouting method developed by the Keller Company. The Soilfrac method is increasingly being employed where tunnelling is undertaken beneath settlement-sensitive structures such as buildings, railroads, freeways or pipelines.

The measuring example shown in Figures 10 and 11 refers to a particular project which is widely considered to be one of the best documented early European real-time monitoring projects (Otterbein and Raabe, 1990). The construction of a four-lane freeway right through the City of Bielefeld / Germany included tunnelling in highly weathered shales beneath a number of 6-storey buildings. The roof of the 25.0 m wide twin tunnel was a mere 4.5 m distant from the strip footings of the buildings. The cross sectional area of the tunnel amounted to 220 m^2. The maximally allowable settlement difference was specified by the Structural Engineer as being $\Delta s_{max} \leq$ 1mm/m($\leq 0.1\%$). Conformance to the specifications had to be documented for all tunnelling phases.

Tunnel construction with such stringent settlement requirements is only possible by means of special construction measures. In the example, compensation grouting was carried out during various tunnelling phases targeting the zone between the tunnel roof and the foundation of the buildings. Fine-tuning of the compensation grouting

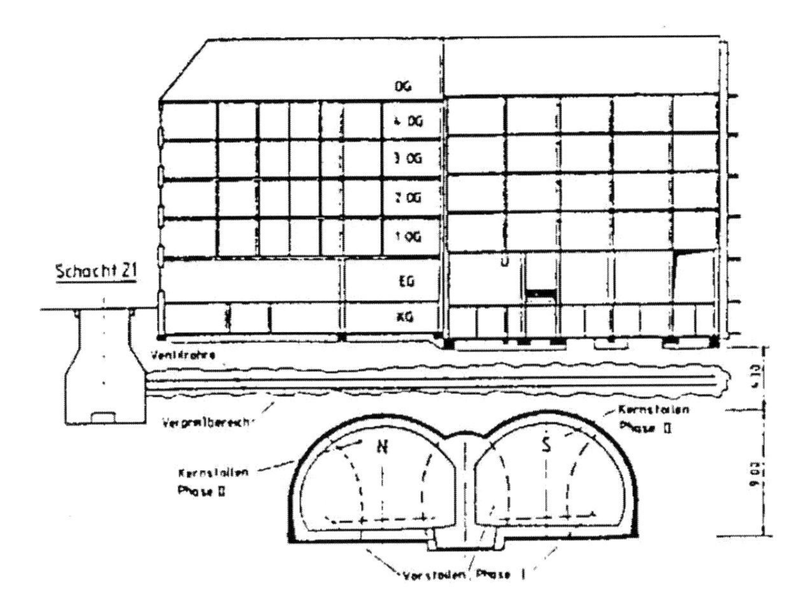

Figure 10: Tunnelling beneath settlement-sensitive buildings in Bielefeld / Germany

procedures was carried out in such way that, on one hand, no excessive settlement differences occurred during the tunnel excavation and, on the other hand, no undue heave occurred due to excessive grouting pressure (Figure 11).

The key to the successful application of this method was real-time monitoring of settlement and heave. In total, 76 electronic liquid level gauges were mounted in the cellars of the buildings, in a set-up similar to that depicted in Figure 12. Each gauge was connected through tubing to an automatic level controller, which held the elevation of the liquid constant by means of a mini-pump, reservoir and an overflow unit. LVDT float sensors monitored the height of the liquid within each gauge. When settlement or heave occurred, the sensor detected an apparent change in the height of the liquid. In fact, the gauge and sensor had moved relative to the elevation of the liquid surface, which had remained constant. The system was connected to a data logger and a PC for continuous monitoring and automatic, real-time updates of graphs and tables. The monitoring system was thus part of a closed loop feed-back circle on construction operations as indicated in Figure 13.

The system accuracy of the electronic liquid level system is about ±0.3 mm. This is somehow better than the system accuracy of alternative real-time settlement monitoring systems such as motorised digital levels and chains of 10 interconnected electrolevels. Table 1 summarises the advantages/disadvantages and limitations of the various real-time monitoring methods.

One of the most complex controls of tunnelling procedures through geotechnical instrumentation is the "Integrated TBM Contol System" as proposed by Kaalberg and

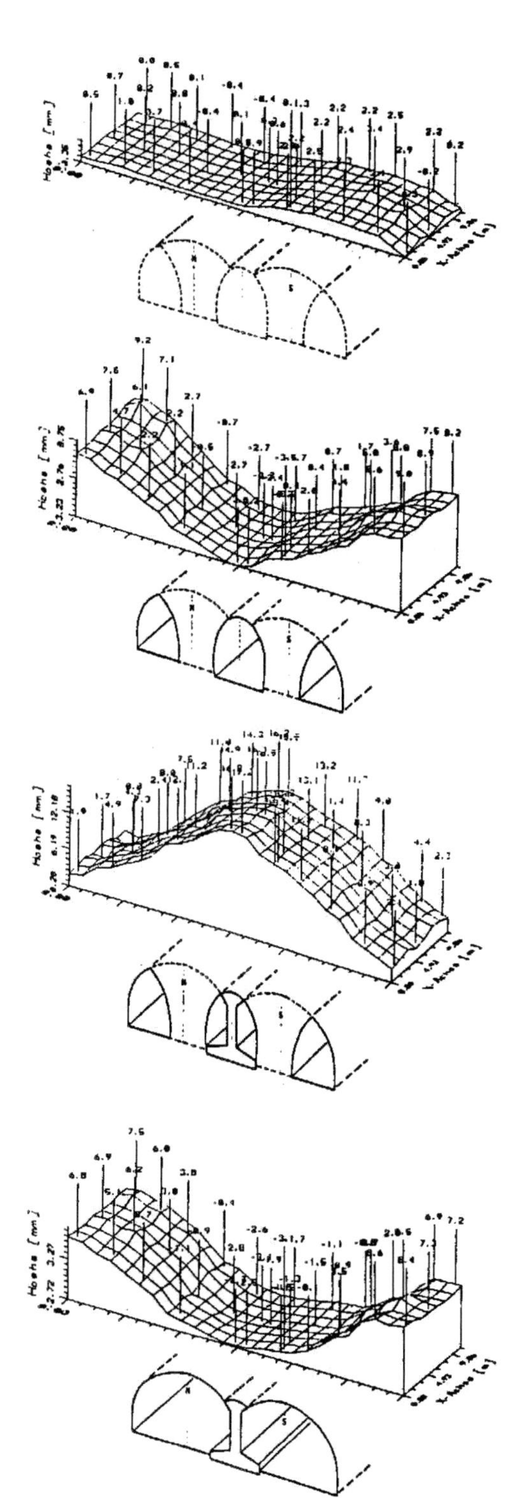

Figure 11:

Compensation grouting in a tunnel project beneath settlement-sensitive buildings in the City of Bielefeld with graph of the settlement / heave of the foundations in four construction phases as recorded by an electronic liquid level system.

(a) Top: Very light consolidation grouting prior to start of the excavation

(b) Second from top: Settlement trough after partial tunnel excavation

(c) Second from bottom: Overcompensation of the settlements which have occurred in (b) through grouting.

(d) Bottom: Settlement trough after completion of the tunnel construction.

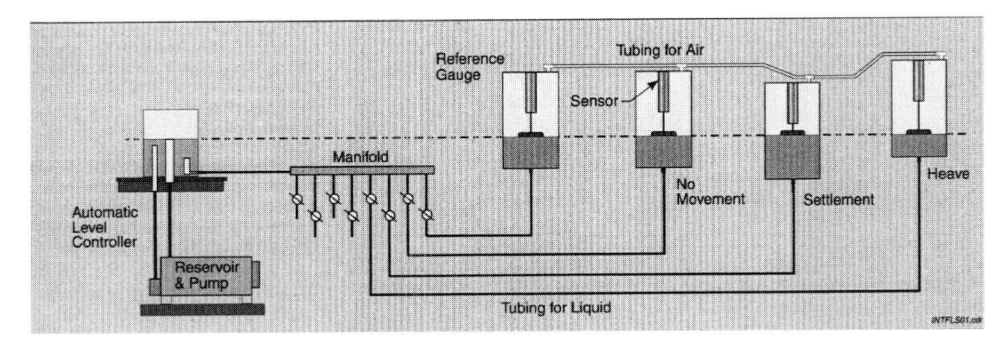

Figure 12: Multi-point Liquid Level System for real-time monitoring of settlement and heave (Interfels system adopted by Slope Indicator, 1998).

No.	Monitoring Method	System Accuracy [mm]	Typical Distance Covered	Advantage	Disadvantage
1	Motorised Digital Level	± 0.3 to ± 1.0	5 - 100 m	Almost no restrictions in number of monitoring points. Low cost. Now well established.	Requires unobstructed line of sight. Restricted inside of buildings and In foggy conditions.
2	Electronic Liquid Level System	± 0.3	5 - 100 m	For use *inside and* outside of buildings. High reliability by level controller and temperature compensation.	Expensive installation. No possibility to accommodate greater height differences in one system.
3	Electrolevel (Chain of 10 elements)	± 0.5	1 - 30 m	For use at accessible surfaces *and in boreholes*. For monitoring lines in any direction. No moving parts.	Difficult and expensive to cover larger distances (say > 30 m). No information in case of a failure of a single element.

Table 1: Comparison of alternative real-time settlement monitoring systems.

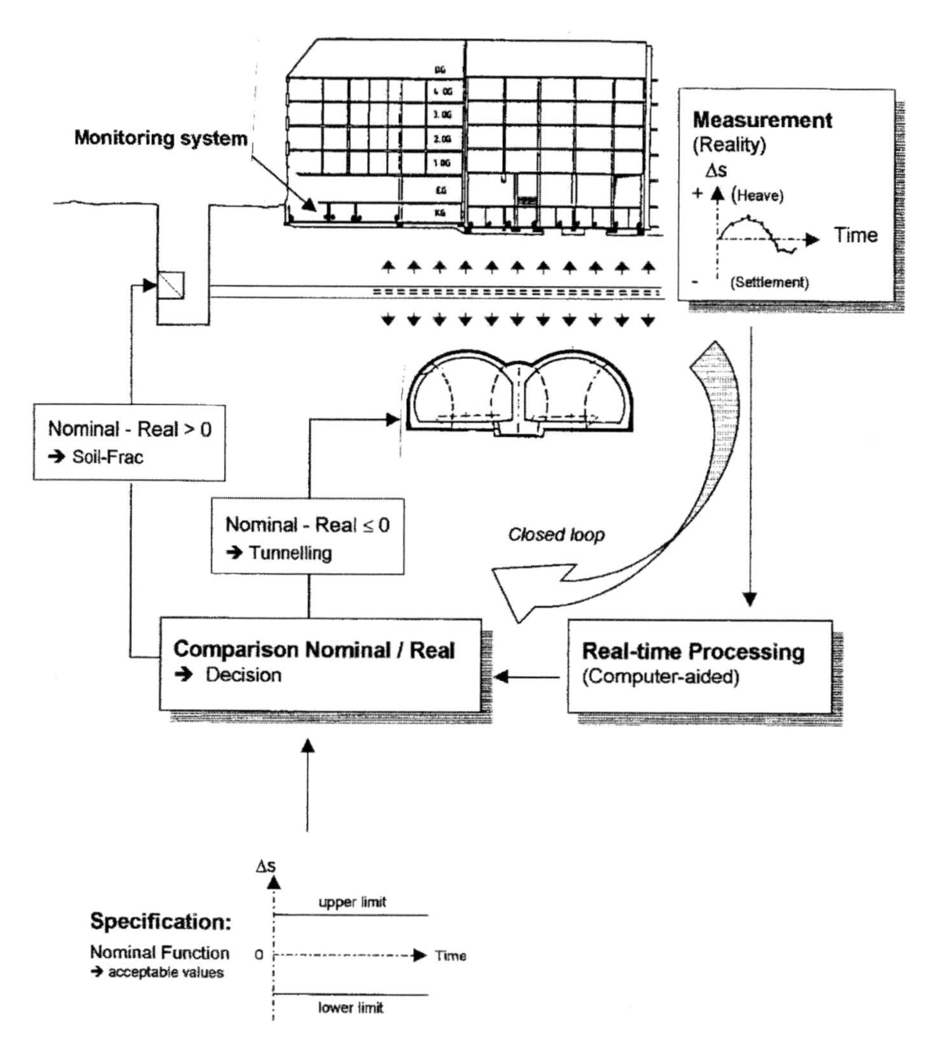

Figure 13: Real-time monitoring system as part of the closed-loop construction control in inner-city tunnelling.

Hentschel (1997) and Doom et al. (1999). This system is conceived to control of the TBM operations in near-surface, inner-city tunnelling in soft ground. Specific reference is made to conditions in the City of Amsterdam, characterised by an inhomogeneous ground structure, by historic buildings founded on wooden piles and by numerous cases of settlement damage occurring over past centuries. The main characteristics of the "Integrated TBM Control System" is the integration of *geotechnical parameters* into the control of the TBM. Until now, machine parameters were exclusively used for this purpose, however this is insufficient for TBM operations in settlement-sensitive environments such as in Amsterdam.

As indicated in Figure 14, numerous settlement measuring points are placed at the buildings (targets for motorised tachymeter and/or electrolevels), in the ground and on or near the piles (multiple-point borehole extensometers). Geotechnical monitoring data, together with the TBM machine data, constitute the information base of the "reality" which is continuously updated in parallel to the TBM operations. This base provides the input for a complex close-loop control mechanism as depicted in Figure 15. The control signal acts on the TBM actuator for adjustment of TBM shield forces and, in particular, adjustment of the contact grouting pressure within the shield's specially designed tail in order to avoid detrimental settlements.

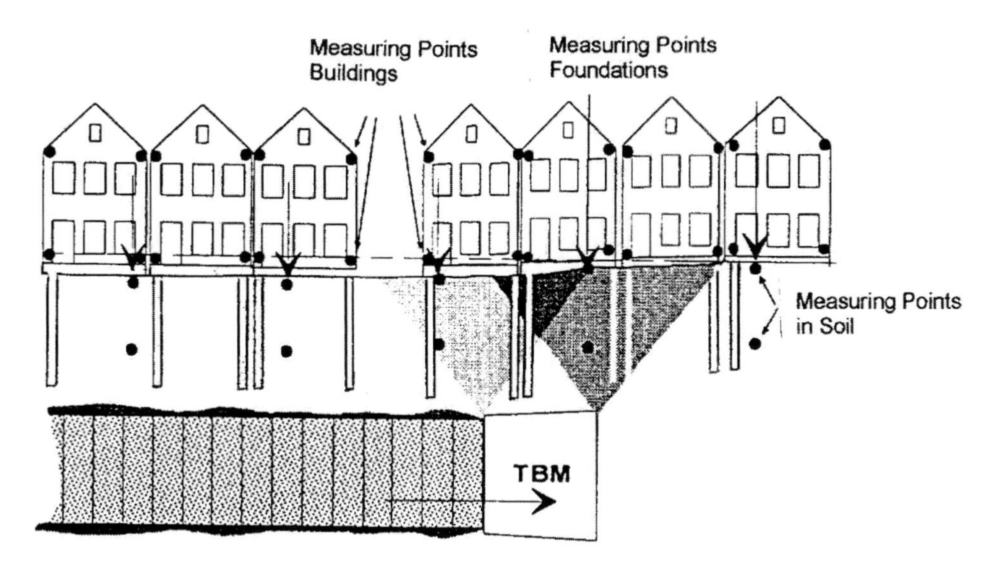

Figure 14: Geotechnical and geodetic monitoring points with signals to be used as feed-back for the control of a TBM in the settlement-sensitive environment of the City of Amsterdam.

Critical to the success of the "Integrated TBM Contol System" is the definition of a suitable *control function*. Obviously, this function must incorporate not only monitoring signals of the buildings, the ground and the piles, but also relevant soil and building parameters in order to process all information in a realistic model. Note

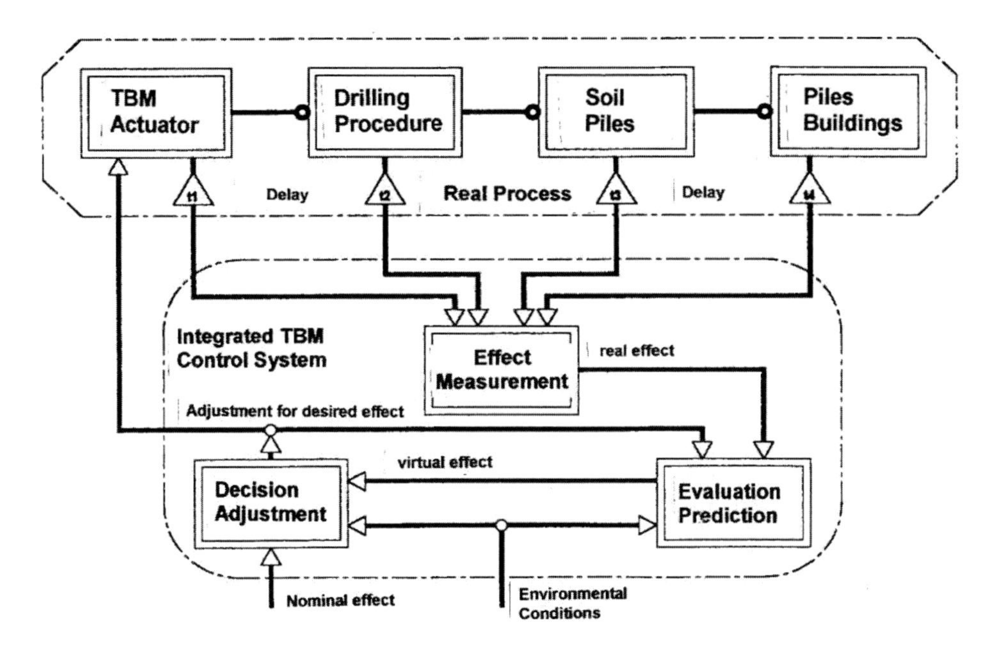

Figure 15: Function diagram of the "Integrated TBM Control System". Note that both TBM machine data and geotechnical field data are used as feed-back parameters (from Doorn et al., 1999).

that soil parameters are highly time-dependent putting an additional degree of complexity into the model and into the control procedures.

It remains to be seen whether the proposed "Integrated TBM Contol System" will ultimately be successful. Nevertheless, for the time being, it marks a definite peak in continuous efforts to achieve better control of construction procedures through geotechnical monitoring.

4 Conclusion

The following trends can be identified with regard to the application of geotechnical instrumentation in the control of tunnel construction procedures:

- Generally, the geotechnical instrumentation market sector applied in the control of construction procedures is significantly more dynamic than the traditional sector of performance monitoring for verification of the design.

- "DIBIT", a new tunnel scanner based on a fully digitised photogrammetric measuring system, has made a significant impact in the European market. It enables the control of specific underground construction procedures such as the excavation of the tunnel, the contouring of the tunnel profile and the application of shotcrete lining as a primary support. It also allows a rough determination of the tunnel convergence.

- A new deflectometer probe has been developed by Interfels for deviation measurements of horizontal and inclined boreholes. In tunnelling, this new probe has been successfully employed in the surveying of anchor boreholes, exploration boreholes in the face of advancing tunnels and in boreholes for ground freezing tunnels.

- Real-time monitoring is being increasingly employed in the control of settlement-sensitive tunnelling operations. Three types of automatic settlement monitoring instrumentation are currently in use. These are (in the order of current preference): (1) Motorised digital level, (2) Multi-point liquid level system and (3) Electrolevel.

- The Soilfrac® grout compensation method has found widespread application in inner-city tunnelling beneath settlement-sensitive buildings. A precondition for the employment of this method is real-time monitoring of settlement and heave.

- The "Integrated TBM Contol System", proposed for the metro construction in Amsterdam, marks a definite peak in continuous efforts to achieve better control of tunnel construction procedures by way of geotechnical instrumentation and monitoring.

Acknowledgement

Thanks to Patrick Hartkorn for geodetic terminology and to Ellie for reviewing.

Bibliography

[1] Bock, H.; König, E. and Estermann, U., 1997. Deviation measurements in horizontal drillholes - State-of-the-art. (in German). - Taschenbuch f. d. Tunnelbau, **1997:** 169-183, Essen (Glückauf)

[2] Diekmann, N. and Kern, P., 1991. Investigations into the stability of drives in an underground mine. INCREX in combination with mobile inclinometer. - Interfels News, **4:** 15 - 18.

[3] Doorn, Cor. Th. v.; Vlijm, H. M.; Kaalberg, F. J. and Hentschel, V., 1999. Particular requirements onto the TBM for the construction of the North-South metro line in Amsterdam - Minimising of the settlements by a settlement-oriented control of the TBM (in German). - Proceed. 10. Kolloquium for Construction Techniques, Bochum 25th March 1999, p. 47 - 58, Rotterdam (Balkema).

[4] Dunnicliff, J., 1988, 1993. Geotechnical instrumentation for monitoring field performance. - 577 p., New York (Wiley).

[5] Estermann, U., 1991. Application and data processing of INCREX measurements for near-surface tunnelling.- Interfels News, **4:** 3 - 9.

[6] Kaalberg, F. and Hentschel, V., 1997. Tunnelling in soft ground with high water table and wooden pile foundations - Development of a settlement-oriented and settlement-minimising TBM procedure (in German).- Research + Practice - New accents in Tunnel Construction, STUVA Congr. '97, **37:** 72 - 79, Düsseldorf (Alba).

[7] Müller, L., 1978. Rock Construction, Vol. 3: Tunnelling (in German). - 945 p., Stuttgart (Enke).

[8] Otterbein, R. and Raabe, E.W., 1990. Tunnel Bielefeld B 61n - Real-time settlement monitoring as the key for safe tunnelling under buildings. - Interfels News, **2:** 1-6.

[9] Rokahr, R. B. and Zachow, R., 1999. A new method for the daily control of the load bearing capacity of a shotcrete lining (in German). - Unpubl. Report, 7 p., Hanover (Univ. Inst. for Subsurface Construction).

Appendix: List of Companies

DIBIT Messysteme Ges.m.b.H.	Att.: Dr. Helge Grafinger	
Gewerbepark 3	A - 6068 Mils	Austria
Tel. + 43 - 5223 - 466 463 18	Fax: + 43 - 5223 - 466 463 20	www.dibit-scanner.at/
DMT Welldone Drilling Services GmbH	Attn.: Dr.-Ing. Werner Vorhoff	
Am Technologiepark 1	D - 45307 Essen	Germany
Tel. +49 - 201 - 172 1454	Fax: +49 - 201 - 172 1447	www.dmt-gmbh.de/
GeoConcept Messtechnik GmbH	Attn.: Gerhard Weithe	
Wilhelm-Bläser-Str. 8	D - 59174 Kamen	Germany
Tel. +49 - 2307 - 995 110	Fax: +49 - 2307 - 995 112	www.geoconcept.de/
Glötzl GmbH	Attn.: Rainer Glötzl	
Forlenweg 11	D - 76287 Rheinstetten	Germany
Tel. +49 - 721 - 5166 -0	Fax: +49 - 721 - 5166 - 30	www.gloetzl.com/
Interfels GmbH	Attn.: Jan Evers	
Deilmannstr. 5	D - 48455 Bad Bentheim	Germany
Tel. +49 - 5922 - 98 98 - 0	Fax: +49 - 5922 - 98 98 98	www.Interfels.com/
Refelex Instrument AB	Attn.: Yngve Lennerstrand	
P.O. Box 118	S - 18622 Vallentuna	Sweden
Tel. +46 - 8511 - 80 610	+46 - 8511 - 80 620	www.reflex.se/
Slope Indicator Company		
3450 Monte Villa Parkway	Bothell, WA 98041-3015	USA
Tel. +1 - 425 - 806 2200	Fax: +1 - 425 - 806 2250	www.slopeindicator.com/

Geophysical investigations: Integrated seismic imaging system for geological prediction during tunnel construction

G. Borm and R. Giese

GeoForschungsZentrum Potsdam, Germany
e-mail: gborm@gfz-potsdam.de, rudi@gfz-potsdam.de

Abstract

The Integrated Seismic Imaging System ISIS provides high resolution seismic images of the rock mass ahead of a tunnel face for online geological prognosis. Specific features of ISIS are the newly developed repetitive pneumatic impact hammer with laser triggering and an array of glass fiber reinforced rock anchors containing 3D-geophones to be be readily installed during the excavation process. An interactive software package enables online data acquisition, processing, 3D-visualization, and engineering geological interpretation.

1. Introduction

A major problem in driving a tunnel is the prior knowledge of the geological environment and the geotechnical parameters to be expected. An increasing extent of underground construction works needs to be carried out under difficult soil and rock conditions. Besides exploratory drilling, non-destructive geophysical methods may be used to indicate lithological heterogeneities within distances of up to several hundred meters. Seismic imaging, with its large prediction range and high resolution capability, is the most effective method.

Since the early nineties, the Tunnel Seismic Prediction System TSP of Amberg Measuring Technique AG, AMT Switzerland, has been applied successfully in tunneling projects worldwide [3, 5]. The concept of the Integrated Seismic Imaging System ISIS [1] was developed by the GeoForschungsZentrum Potsdam GFZ in cooperation with AMT as a follow-up of the TSP.

2. System components

The components of ISIS are the receiver rods, a pneumatic impact source, and a user-friendly software package.

2.1 Geophone anchors

Mounting the receiver system can be readily integrated in the tunnel working routine. Standard glass-fiber-reinforced polymer resin anchors are used as housing for 3D-geophones. The anchor rods act as seismic antennas and simultaneously as stabilizing elements [2]. The geophones are mounted in 3 orthogonal directions at the tip of the rock anchor (fig. 1). Complete seismic signals of up to several kHz are detectable with these.

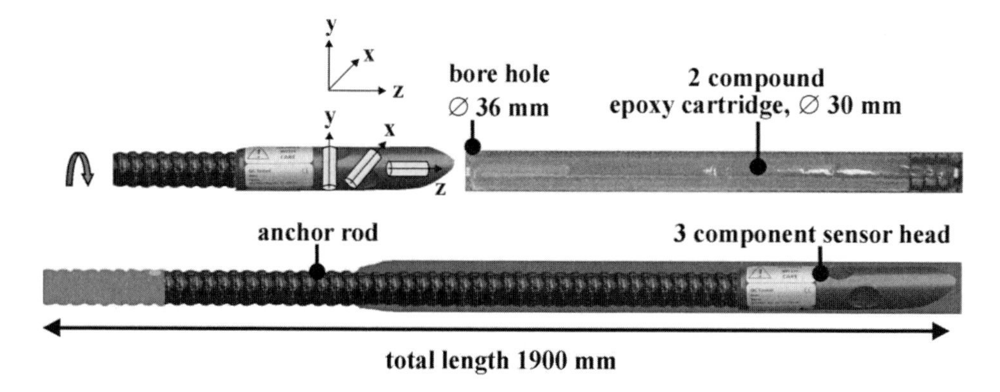

Fig. 1: Rock anchor with 3-component geophones.

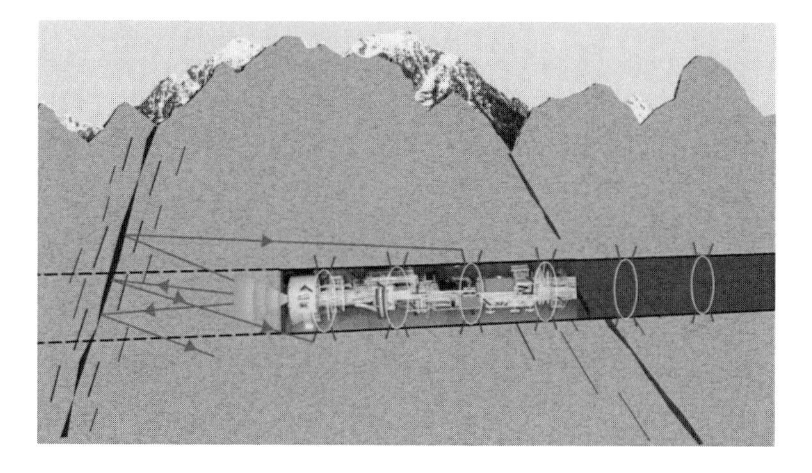

Fig. 2: Scheme of the Integrated Seismic Imaging System for geological exploration during excavation of a tunnel.

The geophone anchors are cemented in boreholes with a two-component epoxy resin to esure optimum coupling of the geophones to the rock. Besides using hand-held drilling machines, installation of the geophone anchors can also be performed by using drill rigs, similar to the standard installation of rock bolts. Properly orientated, the reciever rods can span a complex radial and axial geophone array close to the tunnel face (fig. 2).

2.2 Seismic impact source

The seismic impact source of ISIS is a mechanical hammer using a pneumatical cylinder driven by compressed air [4]. Power for the impact is generated with help of a mass (5 kg) moving on roller bearings inside the housing (fig. 3). By a driving pressure of 8 bar an acceleration of 80 g and an impact force of 4 kN with a duration of 1 msec is achieved.

The impact hammer can be applied in all directions with an appropriate carriage (fig. 4). The hammer is prestressed against the tunnel wall with a force of 2 kN to ensure sufficient coupling to the rock.

The impact source is able to generate mechanical pulses with frequencies of up to 2 kHz repeated every 5 seconds. The triggering of the signal is achieved by a laser light-barrier inside the housing. The maximum error in triggering time is less than 0.1 ms. This small time lag, together with the excellent repeatability of the transmitted signals at each source point, leads to a significant improvement of the signal-to-noise-ratio by using vertical stacking, which is a statistical procedure to amplify correlated signals, such as reflections from discontinuities, and to reduce noncorrelated signals such as the noise from construction machinery.

2.3 Interactive software

The software package is integrated in a Windows-based architecture. Data acquisition and processing is done on a standard field notebook PC. An interactive process control guides the user straigh to the results. All parameters for the given data processing modules are set automatically, but manual processing of the data is also possible by choosing optional parameters in dialogue boxes.

The program is able to extract the complete seismic wave field and separate its compressional and shear components. Both are simultaneously processed by means of sophisticated velocity analysis and depth migration [6]. The resulting velocity field indicates lithological changes and discontinuities in the rock mass ahead of the tunnel face and in the tunnel wall.

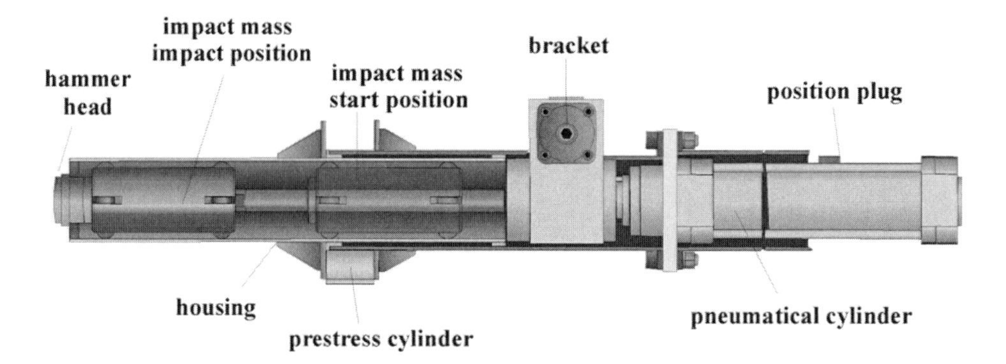

Fig. 3: Mechanical design of the pneumatic impact hammer.

Fig. 4: Impact hammer in operation at the tunnel wall.

The seismic data is fed into a numerical procedure for depth migration where energy reflected by geological heterogeneities is analysed. The most prominent reflections are automatically detected and visualized by the program. The interactive 3D-display is able to image geological structures and faults in every perspective. The 2D-display can present the geological targets in horizontal and vertical cross-sections and even depict the distribution of rock mechanical parameters derived from the velocity field.

3. Field tests

Underground seismic measurements with ISIS were made in the Faido access tunnel of the Gotthard Base Tunnel (Central Switzerland) during construction. The Faido adit was excavated with drill and blast. On its total length of about 2700 m, the tunnel is driven in metamorphic rock mainly consisting of folded gneiss (fig. 5). At the deepest part, the overburden exceeds 1300 m.

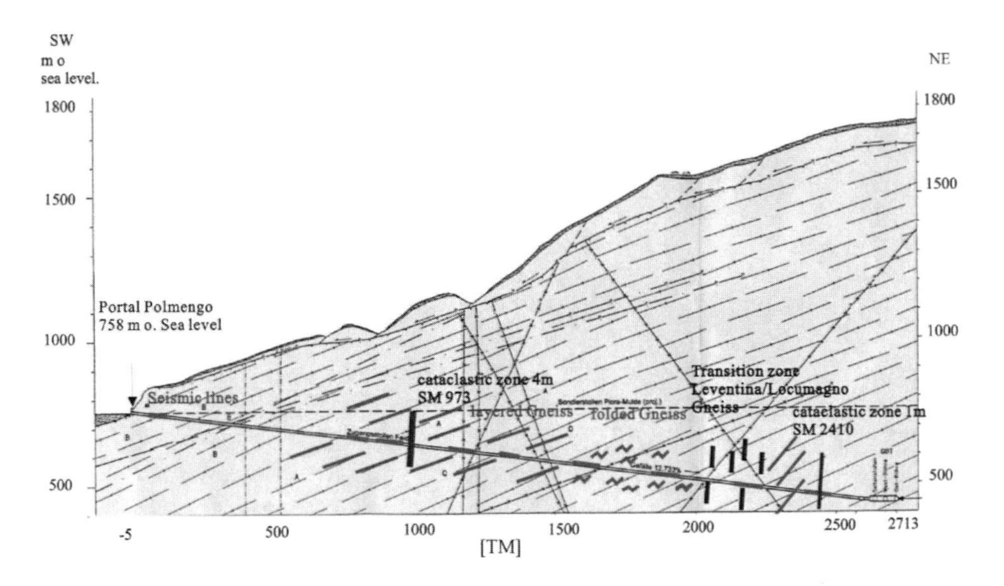

Fig. 5: Geological profile of the Faido access tunnel with seismic lines.

3.1 Seismic layout

Figure 5 shows the arrangement of the seismic lines along the Faido adit. The measurements were repeated every 200 m to get a continuous seismic information along the whole tunnel. During each campaign, seismic energy was recorded from

30 to 60 source points distributed at intervals of 1.0 to 1.5 m along the tunnel wall. The seismic signals were recorded with 8 to 10 three-component geophone anchors installed in 2 m deep drillholes at intervals of 9 m. Reflected energy was recorded over travel distances of up to 250 m.

3.2 Seismograms

Figure 6 depicts the seismogram traces for a compressional wave velocity component parallel to the tunnel axis recorded at one of the geophone anchors where the distance (*offset*) to the source points were successively increased from 16 to 72 m. The direct waves are compressional, shear and surface waves. Irregularities of the tunnel wall cause a marked scatter of the surface waves. The data are low-pass filtered to eliminate high-frequency noise. An automatic gain was applied to raise the amplitudes to a common level *(trace normalizing)*. Phases belonging to the direct wave field are visible in the first 10 ms at 15 m offset and up to 30 ms at 72 m offset.

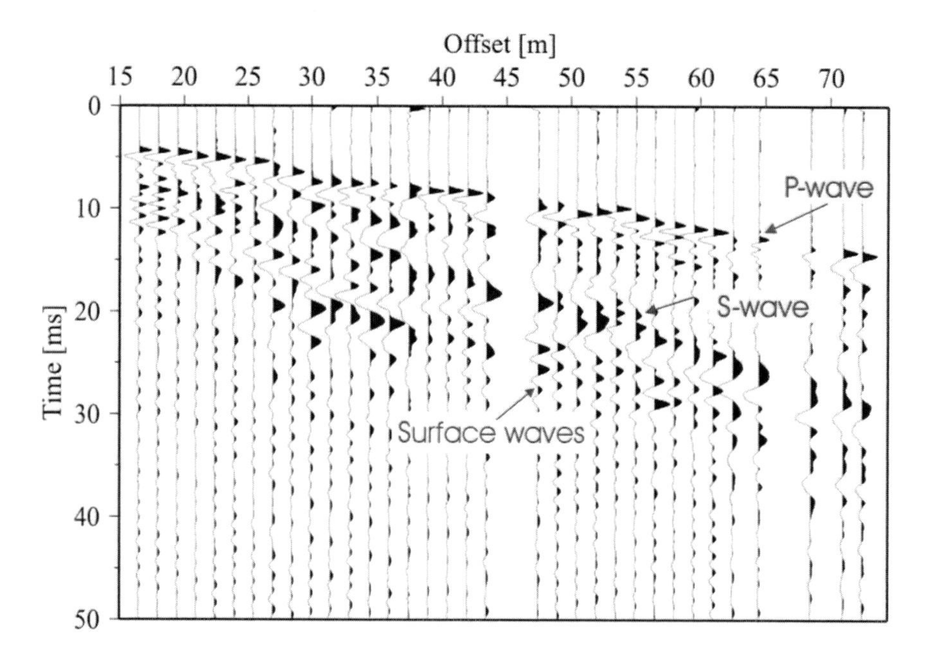

Fig. 6: Seismic data of a velocity component parallel to the tunnel-axis of the profile between tunnel-meter 890 and 955. The first breaks of P- and S- waves at receiver-point offsets from 16 to 72 m are indicated by arrows.

3.3 Tomographic inversion

Arrival times of the direct compressional and shear waves were used for tomographic inversion (fig. 7). This is a numerical procedure, where the seismic wave field is first calculated by forward modelling and then compared to the measured one. By minimizing the differences in travel time between the theoretical approach and the actual field data, the velocity model is iteratively improved until a suitable match to the observed response is obtained.

The 2D velocity model for the seismic wave velocities reveals a disturbed zone extending 2 to 3 m deep into the rock mass. This zone is characterized by strong variations of the compressional wave velocity from 3500 to 5800 m/s and of the shear wave velocity from 2000 to 3000 m/s. High velocity zones correspond to

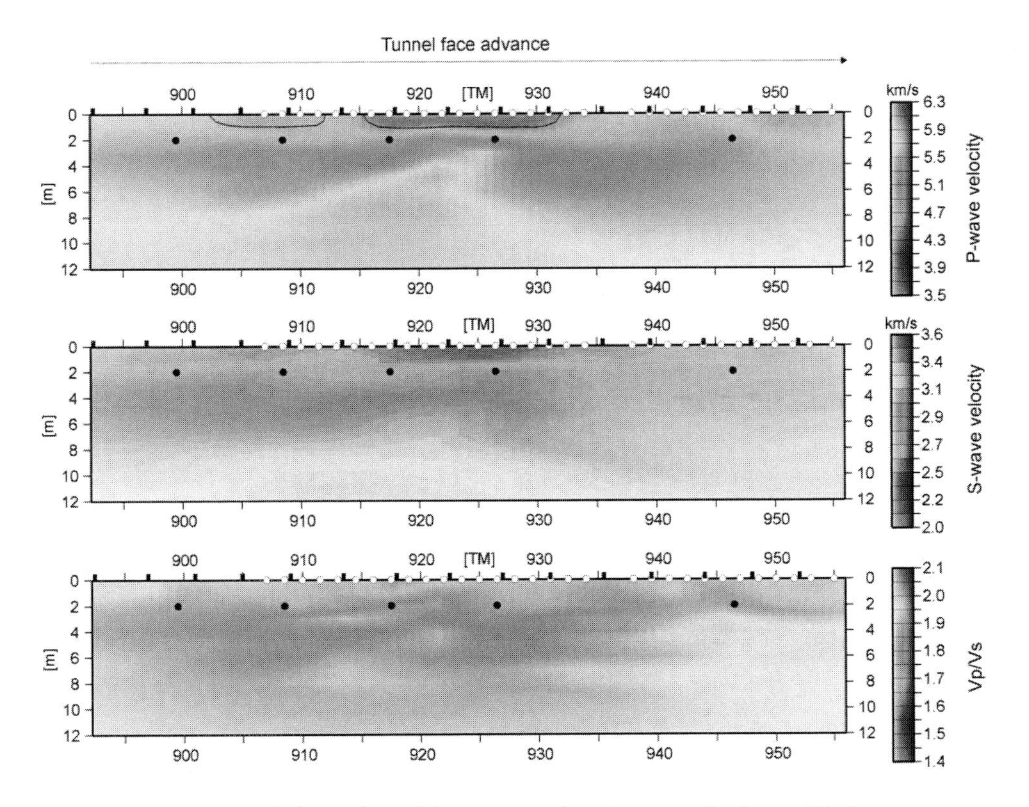

Fig. 7: Tomographic inversion of (a) compressional wave velocity v_p, (b) shear wave velocity v_s and (c) ratio v_p/v_s near the cataclastic zone between tunnel-meters 916 and 932. Source and receiver points are marked by circles and dots, respectively, and solid bars depict tunnel excavation advances per shift. Dashed lines delimit the outcrop of cross-cutting cataclastic faults at the tunnel wall.

hard inclusions such as quartz, whereas low velocities indicate mainly faults and fracture zones. The excavation disturbance zone of the tunnel is most specifically characterized by steep gradients in the seismic velocities, by attenuation of seismic waves as a result of intrinsic damping and/or scattering, by fractures, and by an irregular shape of the tunnel wall.

3.4 Geological interpretation

The analysis of the tomographic inversion exhibits regions of reduced seismic velocities close to the tunnel wall and a major one cross-cutting the tunnel between tunnel-meter TM 916 and 932. The tomographic inversion reveals a significant decrease of v_p and v_s some decameters before reaching this zone and an increase of the ratio v_p/v_s. Direct geological inspection of the site after excavation revealed a 4 m thick water bearing cataclastic fault surrounded by zones of disturbed rock.

4. Perspectives

The components of ISIS for geological prediction ahead of underground construction works are sustainably improved through practical testing. In addition to this, wireless communication of the data between the registration site and a remote station such as a control center is to be introduced. Further development of ISIS will be focussed on an automatized geological expert program for interpretation, characterisation and evaluation of the measured seismic attributes and their variations in space and time based on neural networks.

Ackowledgement

We gratefully acknowledge the excellent cooperation and most efficient support of Amberg Engineering AG Zürich - namely to Felix Amberg, Thomas Dickmann, and Bruno Röthlisberger - and for the kind offer of the opportunity to carry out the several ISIS testing campaigns at the AlpTransit construction site of the Gotthard Base Tunnel.

Bibliography

[1] Borm, G., Giese, R., Otto, P., Dickmann, Th., and Amberg, F.: Integrated Seismic Imaging System for geological prediction ahead of underground construction. Rapid Excavation and Tunneling Conference (RETC), June 11.-13., San Diego USA, 2001

[2] Borm, G., Giese R., Schmidt-Hattenberger, C., and Bribach, J.: Verankerungseinrichtung mit seismischem Sensor (Anchoring system with seismic sensor). German Patent Appl. 198 52 455.2; European Patent Appl. 99120626.9-2316; Japanese Patent Appl. HEI 11-322268, 1999

[3] Dickmann, Th. & Sander, B.K.: Drivage concurrent Tunnel Seismic Prediction, Felsbau, Vol. 14, pp. 406-411, 1996

[4] Otto, P. & Borm, G.: Pneumatischer Impulsgenerator für Untertageseismik. Deutsche Patentanmeldung: 199 44 032.8, 2000

[5] Sattel, G., Frey, P., and Amberg, R.: Prediction ahead of the tunnel face by seismic methods pilot project in Centovalli Tunnel, Locarno, Switzerland. First Break, Vol. 10, pp. 19-25, 1992

[6] Yilmaz, Oe.: Seismic data processing. Investigations in Geophysics, Vol. 2, Society of Exploration Geophysicists SEG, Tulsa/Oklahoma, 1987

Part IV

Management aspects

What tends to go wrong in tunnelling

Sir Alan Muir Wood

Consultant, Halcrow Group, London

Abstract: The process of development of a tunnel project should proceed continuously in such a manner as to reduce the prospect of uncertainties developing to hazards and then to risk. At each phase in development, particular forms of hazard threaten to intervene. This account, based on the Author's experience, selects examples of factors leading to failures in optimisation or even to physical failure. The nature of the contractual structure is often a material factor. Emphasis for success is placed upon the fundamental need to address the project as a system, with the several processes functionally inter-related

1 Introduction.

Murphy's Law (with doubtless equivalent in many other languages): *If something can go wrong, it will go wrong.*

My title may suggest a certain inevitability of tunnel misfortunes. In fact, my intention is quite the opposite. The view of the media, at least in my country on selective evidence, is that tunnel projects obey a natural law of exceeding targets of time and cost, i.e. that the risk is high. My personal experience has been to the contrary.

Tunnelling is an activity which depends on overall holistic surveillance. Where the project has been approached as a system, so that each contributory cause to significant uncertainty can be adequately investigated and where there is informed continuity in managing the system overall, the risk is controllable and capable of being contained within the budget. While there are many ways in which the several aspects of a system may be handled, the essential feature is that the project must be treated as a system of inter-related processes. There is no other way of undertaking effective economic tunnel projects — or any other project!

My object is to consider the evolving processes that contribute to the tunnel system, emphasising, with examples, common pitfalls that may be avoided by anticipation of events. Present day notions of Quality Assurance (QA) lead to a commendable recording of decisions taken and actions confirmed. The down-side of QA is that possible hazards for attention tend to be confined to those identified for QA. The belief grows that no other issues matter. The two most important characteristics of the successful tunneller are the constant alertness to the unexpected (often first apparent in a trivial manner) and a reliance upon observation (whether or not this is adopted as a formal tool of control – see Section 9 below).

In my experience, there is much engineering competence available in tunnelling around the world. By far the greatest barrier to success is the interference by administrative protocol, causing sub-division of responsibility in the mistaken belief that this controls cost, coupled with the excessive numbers of Parties looking over each other's shoulders and 'signing off'. This is a wretched designation which implies that one particular item has been accepted without qualification, with no regard for any future change in view, if and when other factors may change – as they often do, on completion of a specific duty. All these practices lead to atomisation of the project, increasing uncertainty as to who is responsible for what, preventing an overall strategy. Diffuse responsibility leads to a reliance on others who may each have limited knowledge of the scene overall. Tunnelling is an activity which depends for success on an overall holistic surveillance of all the specialised contributory activities. In consequence, the seeds for problems are sown before any activity identifiable as actual tunnelling has even started. How can there be continuity for a World Bank project for which the engineering responsibility for initial planning, investigation and design is not allowed to continue into the subsequent phase of construction?

Where a tunnel project for a public activity has the support of private finance, at a chosen moment in time transfer of responsibility will occur, from the public initiator to the private executor. There are ways of achieving continuity, by for example bringing the private interests into discussion of project strategy before formal takeover or by the process of 'novation' whereby the engineering team responsible for the initial design is absorbed into the private organisation. In either instance, much attention to detail is required to ensure the possibility for success, taking account of the issues raised in the following text.

2 Initial planning

An account of this nature would best be presented in an interactive manner between the activities recounted in the several sections. In print, this is impractical so the reader needs to use the mind as a spread-sheet to provide the linkages. We immediately stumble into one such problem. Optimisation of the potential of the project starts in the first stage of planning, when the project is no more than 'the gleam in the eye'. How do we start the optimisation process when so little is known about the options and when it is likely that many decisions on allied issues are being taken, so that options may be closed off by those with no appreciation of the potential loss from this premature action? In many instances, a little knowledge of topography and geological structure will allow tunnelling costs to be presented in broad terms but sufficient to indicate which options merit retention to the next stage, and which other options may be ignored. While there will be many aspects that enter into the project optimisation overall, the tunnelling activity will probably represent both a major fraction of the total and also one with the greatest degree of variability, without causing major increase in the cost of other features. Optimisation must reflect total life costs so estimates of construction and operation need to be combined, however rudimentary, at this stage.

It may not be apparent initially that a tunnel is a necessary part of the project. In such circumstances it is only too likely that initial planning will not receive any input from tunnelling expertise. The late introduction of a tunnelled option is yet less likely to be capable of optimisation, other features having by this time been settled irrevocably. One feature that is too readily overlooked concerns the occupation of the ground surface. It may be objected that a tunnel frees the land surface. So it does, but linkages to the surface, entailing portals and approaches for transport tunnels, are inflexible in their needs. Their disposition, minimising lengths and avoiding conflict with others services, may make a major contribution to cost.

Around 30 years ago, the south half of a ring road to the South of London was being proposed. Late in the stage of initial planning when the chosen line permitted little deviation, the thought occurred that a below-ground solution would have merit. A study was undertaken in two sections. One half did indeed indicate that, taking account of land costs and other benefits, tunnelling would be a favoured option, making use of disused railway marshalling yards and similar sites to provide access from the surface. It was also evident that a relatively small shift in the line of the road would have allowed better use of the topography for tunnel access at a further saving in differential cost. These benefits remain unrealised since the entire project was abandoned in preference to an outer ring road. The lessons learned, however, remain to be applied elsewhere. Operational research used to teach us to approach optimisation by identifying the source of the greatest potential improvements; this remains appropriate to project definition, particularly where tunnels may be involved.

3 Procurement 1

Procurement as a process is here divided into two parts, Procurement 1 and Procurement 2, so that the process of initial design, which intervenes, does not arrive too early or too late. The essence of procurement of people to engage in the project is that it is conceived as building a team, in stages as the project requires; only thus can the essential purposeful continuity in the evolution of the project be achieved. The tunnel project is 'fuzzy' at the time of Procurement 1, to be brought into sharper focus in a relatively short time. The need is for an intermediary, who may be the project designer. Where, however, the shape of the project is most uncertain, it may be more appropriate to appoint in the first instance what has been termed the 'surrogate operator' [7]. This appointment is for one familiar with tunnel construction and operation processes, also capable of communicating with the Client on the other aspects of the project: planning; finance; legal; contractual. However achieved, this is a key appointment — of course it may be provided through the office of the Client where an enlightened Client is in the business of commissioning comparable projects, understanding the benefits of a team approach.

A vital feature of the initial appointment is that the fee to be paid should not be based in any way upon competition against the cost of the work to be undertaken. The good engineer will recognise the likely scope of work and will foresee the nature of problems to be encountered. The lesser engineer, on the other hand, will attempt to transfer the greatest degree of responsibility and risk to others, minimise his own costs and, when the problems are realised, find cause to claim for extra costs. This is one of the major causes of cost and time over-runs. Quality and competence in the members of the team represent positive qualities of the project's resource, not costs to be reckoned negatively by the accountant.

Development Banks used to operate what was known as the 'two-envelope' scheme of bidding for the appointment of engineering consultants, with the understanding that the first envelope, concerning resources, competence and comprehension of the needs of the assignment provided guidance for the choice of the preferred engineer. Only then was the second envelope opened, providing the basis for negotiation with this preferred engineer. Recently, the two-envelope scheme has become more widely adopted, but the essential feature abused. The two envelopes are opened simultaneously and, while it is claimed that cost is only one of many factors, where the selection is undertaken by those without good understanding of the engineering issues, money becomes the only objective means of assessment and hence dominates the award.

The appointment of the first expertise to the project is a vital, possibly the most vital, moment for a project. Upon this appointment will hang the means of appointing others and the conditions for their appointment. A lesser engineer, accepting responsibilities beyond his capacity, will ensure that risks will be placed on others, regardless of the means for their control, and that antagonistic relationships will emerge from the start. This feature will remain at the heart of all that will tend to go wrong — and probably will go wrong.

The initial engineering appointment will require the Client to be competent to provide reliable answers to many questions in order to establish the criteria whereby the optimal project is to be judged. Many such issues will be interlinked and, unless a sound overall strategy is being developed, it will be difficult to evolve a sound project strategy.

I recall below two experiences for major road tunnel projects which introduced unnecessary problems as a result of decisions being taken prematurely without the benefit of sound engineering advice:

a) The Act under which the Corporation of Glasgow had authority to build the Clyde (Road) Tunnel prescribed an inadequate diameter for the tunnel, doubtless based on obsolete examples. The initial approaches were also far too cramped in the endeavour to link with an existing road system. The latter defect could be eliminated as the project proceeded but the size of the tunnel remains a permanent problem which design could only partially mitigate.

b) Tunnels to reduce the road traffic problems for the city of Bath started from a curious situation whereby their planning was the prerogative of the City Planner, construction the responsibility of the City Engineer. City Engineer and City Planner did not communicate with each other. Moreover the planning consultants were advising on tunnel costs without understanding the issues. As a consequence the engineering designer's problems started from selection of a tunnel for through traffic confined too close to the centre of the city since its cost per unit length had been over-estimated by a factor of around two. This tunnel was intended to be complemented by a short in-city tunnel whose costs had been under-estimated by about the same factor. There was an obvious solution: to return to basics and devise a scheme based on sound cost estimates to keep through traffic well clear of the city centre. This was politically impractical, the scheme foundered and Bath remains a city of chronic chaotic traffic problems.

4 Design and investigation

The responsibilities for tunnel design and its supporting investigations, including investigations of the ground, should never be separated. The absence of a common interlinked strategy in these respects will introduce avoidable problems. The designer uses the investigation, as it unfolds, to define the project, with increasing confidence and precision, as to its nature, line and, most important, scheme of construction. Site investigation (s.i.) is designed accordingly, in an iterative manner, addressing issues of uncertainty with greatest bearing on the design. As part of this operation, the designer should at all times be indicating what is expected to be found, of relevance to the project, so that discovery of the unexpected may lead to an immediate reaction, with possible change in the priorities of the site investigation. Such a situation emphasises the need for the designer to have overall control of the s.i. campaign, identifying requirements from it. The geotechnical engineer translates these requirements into the techniques for their attainment. In this way, the designer acts as the 'client' for the s.i., reflecting, in this capacity, the interest overall of the project Client.

I have written elsewhere [5]:

> *'As the principal hazard in tunnelling springs from the behaviour of the ground, the process of site investigation clearly has a leading role. Far too often a site investigation is undertaken as a ritual without thought for the special circumstances of the locality or for the particular information required for tunnelling. Suggested guidelines .... may be summarized as follows.*
>
> - *Assess relevant regional geology.*
> - *Identify factors of special importance to tunnelling.*
> - *Design site investigation (possibly in more than one phase if layout has to be established) adequately to establish ground characteristics and to explore doubtful geological aspects.*
> - *Reassess continuously during site investigation whether or not results suggest any unexpected features.*
> - *Use geological expertise, colour photography and core preservation techniques so that the value of the site investigation is adequately realized.'*

The most vital features to be explored will determine the nature of the investigation. This fact may have subtler forms. Suppose, for example, that the depth to which deep weathering of igneous rocks occurs may be a feature of significance. The weathered zone may be expected to have limited width, decreasing with depth. In consequence, the probability of encounter by boreholes will depend on their spacing and orientation; furthermore, the probability of encounter will reduce with depth. Such investigation will not determine the maximum depth of weathering; it will simply establish the maximum depth at which weathering has been proved. Adequate data may encourage attempt at predicting a maximum depth at sufficiently high odds, but the 'tail' of extreme probability prediction is notoriously fickle, in the absence of an understanding of the underlying physical laws.

While there may be occasions in which doubt remains concerning the preferred method of tunnelling at the completion of pre-Tender s.i., there should be sufficient information on which to base an informed estimate of project cost. Of course uncertainty will remain and the uncertainty needs to be taken into account, whether as a contingency or as an allowance to be included in the Tender. It is not clever to lead to a favourable interpretation of the data when alternative interpretations have not been eliminated. Settlement of differences by lawyers costs far more than agreement within the Contract by engineers.

The extent to which the design should be developed at this stage must depend on circumstances. An innovative solution for a tunnel in dependable geological setting may demand a thorough analysis combined with specification of the manner in which the work is to be undertaken. This was the case for the Heathrow Cargo Tunnel, a 10.3 m. internal diameter road tunnel, lined with expanded rings of precast lining elements, in the stiff London clay beneath London Heathrow Airport. The conditions and method for construction were here laid down rigorously to maximise ground control and minimise ground loss, kept within a factor of 0.3%.. Where, on the other hand, a general scheme of tunnelling is established with expectation of possible variation to suit the ground or the preference of a contractor, the design will be related to criteria for choice, using an observational approach both to choose the precise needs at each point and, also probably, to modify the scheme in the light of construction experience.

Again, I have remarked [5]:

> *'If there were some way of benefiting from accumulated global experience, tunnelling mishaps would be rare because few have had no precursor. Unfortunately there is no notation for communicating judgement so, although the factors may be well enough known, their incidence in a given situation has largely to be weighed by personal experience. What it should be possible to do is to accumulate case histories specific to certain situations and geological sequences for better briefing of those concerned. For example, such information could well form an important part of the engineering geologist's report for a comparable project in comparable ground.'*

Risk increases when the design changes after the s.i. has been undertaken for a different method. For example, a tunnel designed for drill-and-blast, for which the varied characteristics for a series of folded rocks could be readily tolerated, had its line shifted after completion of s.i.. This would not have had serious effects but for the fact that a road header was subsequently preferred by the Contractor. The uncertainties of interpretation of the s.i. were overlooked so the tunnel encountered a fold of rock whose strength was beyond the capacity of the machine.

It is fortunate that usually the ground/lining composite structure is not constrained from relaxing into an equilibrium state, however remote this may be from the design assumptions. However, the good fortune that may thus have attended designs based on simple loading assumptions can be overstretched by the following circumstances:

(a) when the equilibrium state of the loading is incompatible with the stiffness or the degree of compressibility of the lining in relation to the ground

(b) the unexpected presence of swelling or squeezing ground.

(c) cavities or inhomogeneities which lead to unacceptably uneven loading.

(d) differential ground movements

(e) inadequate consideration of the construction process and its intermediate stages.

Present-day designers too readily turn to numerical methods of presentation before determining the specific questions to be answered. A differentiation should be made between 'numerical analysis', in which a specific predetermined problem is solved numerically, and 'numerical modelling' in which data are fed into a program for compilation into a presentation compatible with the data. The first has a specific purposeful objective; the second entails a limited cognitive contribution by the operator, who becomes a passive observer of the results.

There have been instances where, to obtain convergence for an unstable problem, special features have been introduced into the model without apparent awareness that these may change the problem to be solved, effectively altering the physical properties of the materials involved. In the absence of a prior simple analysis of the issues and mechanisms, such a variation of the design criteria remains undetected. A new program should always be 'validated' by applying it to a model for which an analytical solution is available ('verification' should also be undertaken, implying that the problem to be solved has been correctly constituted). Those who engage in numerical modelling without reflecting on the physical nature of the problem may lose the inherent scepticism, which should always remain the mark of the engineer. Solutions are from time to time counter-intuitive, and the engineer should then ask why. The engineer studying tunnel stability is interested to observe stress patterns in the ground and in the lining. These then point to critical areas for more detailed analysis, particularly where the program has no automatic remeshing capability in zones of high strain. The structural analyst, on the other hand, will choose a program which incorporates a failure or yield function for the ground or structural component, based on the stress tensor, such as von Mises or Tresca. All this is really far too sophisticated for the spatially variable, naturally uncertain, construction-affected, real world of the tunnel. The results, presented in multiple colours and endless print-outs to six or more significant figures, provide an air of assurance and verisimilitude to a practically untested conclusion. There is a place for such complexity in exploring specific phenomena of particular situations, for instance the strikingly non-linear relationships for stress and strain for over-consolidated clays, but not for the robust character expected of a tunnel lining, adaptable to the natural and man-induced variability of the materials.

So much effort is applied in these ways to the completed tunnel that the intermediate state, which may well be more vulnerable, tends to be overlooked. If the finished tunnel has received the full (and often excessive) treatment of 3-D non-linear analysis for rock and concrete lining, the logic of a similar type of analysis for the unquestionably 3-D intermediate stages of construction of a tunnel junction becomes conveniently ignored, for a condition of yet greater complexity. In many instances we need only to call up the lower-bound theory of plasticity for masonry arches, expressed thus by Heyman [4]:

> *'If a thrust-line can be found, for the complete arch, which is in equilibrium with the external loading (including self weight) and which lies everywhere within the masonry of the arch ring, then the arch is safe.'*

For an intermediate stage of construction this is an acceptable criterion but it needs to be applied with much care to a permanent concrete tunnel structure. It is a common fallacy to suppose that an arch (or dome) inscribed in a thick concrete slab or plug, its thickness determined for direct and bending stress, may then be accepted by ignoring, 'wishing-out-of-place', the remainder of the mass concrete. Masonry is flexible, at least at the joints; mass concrete is not. For the arch to develop to accept the load, it needs to yield and deflect, the 'rise' of the arch will slightly flatten (for an external load), the abutments, real or imaginary, slightly spread. The thick slab cannot deform in this way so, unless the abutments are very stiff, some other mechanism must intervene. For a slab, this will usually be by cracking towards the centre. Comfort will then be illusory in supposing that this will provide a stable structure, leaving aside the cracking, acting as a three-pin arch. The danger now is that stress concentrations at hinge points, at the edge of the slab, will be more extreme than for a continuously deforming curved arch. Spalling and cracking become probable, with the release of tensile stress in an explosive manner, a possibility under increasing load or decreasing end support.

In principle, the design process should be kept as simple as possible, recognising the basic principles for stability, with an eye to safety and practicability in construction. It is important to note that such considerations may entail specific features of construction, which should never have become separated from the concept of the design. It is usually in tunnelling not too difficult to ascertain which limit of variable physical features should be accepted to establish a general 'worst case'. Such an analysis will also point to the benefits of a flexible lining except in very weak ground where the stiffness is essential to avoid collapse.

The beneficial advances in tunnel design occur through a series of 'learning loops' illustrated by Figure 1. The important issue is to ensure that continuity in feed-back is assured, or the next design does not profit from the experience of its predecessors.

Muir Wood [6] proposed two factors for the representation of a lining by its compressibility ratio, R_c , and stiffness ratio (i.e. to 'elliptical' distortion), R_s, in relation to the surrounding ground.

These simple relationships permit the load sharing between ground and lining to be estimated from the appropriate equations. For non-linear moduli, at least the

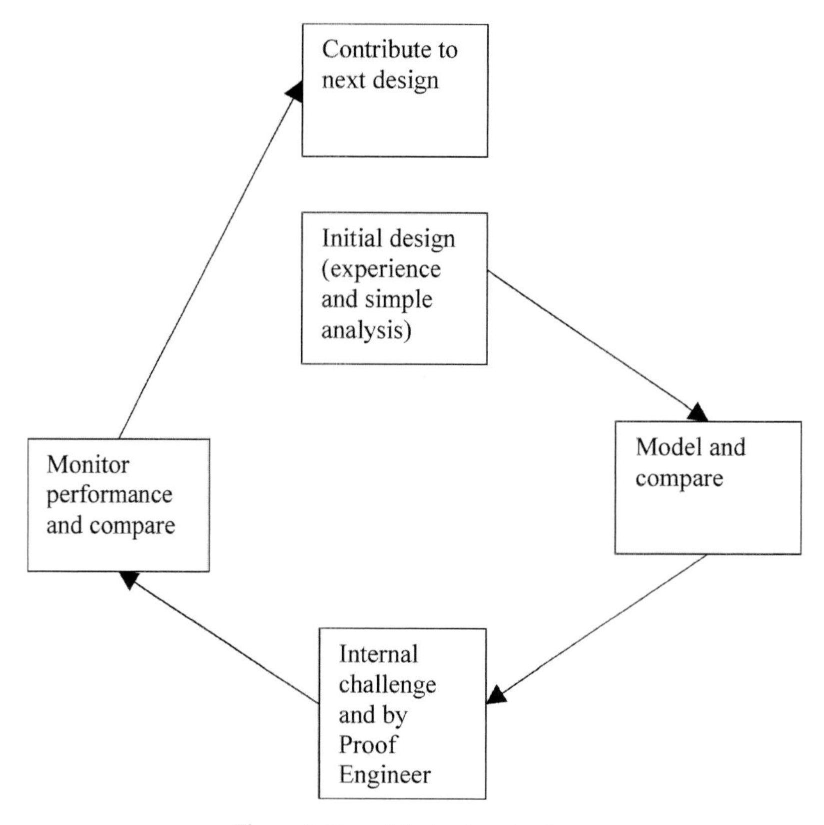

Figure 1: Tunnel design learning loop

limits may be estimated by sensitivity analysis. For stiff ground and for rock it will generally be found that the lining has slight effect on the 'elliptical' deformation of the ground around a circular tunnel so the more flexible the lining, the lower the bending stress to be withstood. The precast lining with rolling joints between segments exemplifies this objective.

As a practical issue of design this feature implies that analysis of behaviour of ground deformation near to the tunnel face may be simplified by assuming no effect from the lining to elliptical deformation, the lining simply resisting in part the inward convergence of the ground. It is then possible to apply a very simple first check on the lining quality and thickness. The ratio between tunnel radius and lining thickness is represented by $r/t = m$; the ratio between lining and ground modulus, $E_l/E = n$, major initial principal stress in ground p_v and ratio between principal stresses, K.If Poisson's ratio for ground and lining is assumed as 0.2, in the absence of shear stress between ground and lining, for the 'wished-in-place' tunnel, the maximum circumferential bending and mean direct compressive stresses are found [2], [6], approximately to be respectively:

$$f_{bc} = \pm 2(1 - K)p_v n/m \quad (1)$$

$$f_a = [(1 + K)p_v nm]/(2n + 1.7m) \quad (2)$$

By equating d($f_{bc} + f_a$)/dm to zero, the value of $m$is found to produce lowest maximum stress in the lining. Alternatively the least value of $m$is found for a given value of $f_{bc} + f_a$, checking that $f_a - f_{bc}$ >O if cracking is to be avoided. Additionally, as suggested by Muir Wood [6], the assumption may be made that 50% convergence and deformation have occurred, prior to provision of support, as a reasonable approximation for a precast lining behind a shield or for a full ring of sprayed concrete, each being provided within a distance to the face corresponding approximately to r. Such a simple approach needs discernment, with allowance for time-dependent increase in loading for a permanent lining, and could not be applied, for example, to squeezing ground.

A point of confusion concerns the adequacy of two-dimensional analysis for tunnel support and lining. In simple terms, the needs may be set out thus:

(i)	initial support or lining provided close to the face	3-D(?)
(ii)	lining provided at a distance from the face	2-D
(iii)	hydrostatic pressure applied to a lining	2-D
(iv)	long-term effects	2-D
(v)	critical intermediate stages (e.g. junctions)	3-D

Where the third dimension needs to be taken into account, it is likely that an axisymmetrical or similar simplification will much reduce the complexity without destroying the rigour. We need constantly to remind ourselves that the uncertainty of our data does not justify complexity for its own sake — or to impress others. Why therefore do we find concern expressed for analysis of a permanent lining in 3-D or, for the intermediate state, the adoption of HME (Hypothetical Modulus of Elasticity) which, its users admit, has no theoretical basis? This expedient is introduced to combine the effects of convergence between the tunnel face and the (shotcrete) support with the early-age stress/strain characteristics of the shotcrete. The problem with such an approach is that, being empirical , it may be estimated from a set of data relating to a specific set of conditions. These, for a particular size of tunnel in a particular geological structure, may include the rate of advance, face support and the distance from the face of the completion of the ring of the support. If any factor varies, there is no theoretical basis for adjusting the HME. Simply, the approach is non-predictive. The '50% rule' described above, supported by observational improvement, is frequently adequate. A minor mystery of HME concerns whether one uses the same value of HME for uniform convergence as for ellipticity, i.e. for R_c and for R_s, and, if so, why.

Where a tunnel is subjected to external pressure of ground-water, this loading will develop relatively slowly during the course of construction or, relatively rapidly, after temporary drains are sealed. In either instance, this is clearly a 2-D problem for analysis. The effect of buoyancy needs to be taken into account with the need for high standards of contact grouting of the crown if design assumptions are to be valid. For a non-circular tunnel the hydrostatic loading will tend to cause distortion; simple analysis permits the effects to be estimated within reasonable limits. If an inherently waterproof lining is required, the criterion for cracking of the lining will evidently be more severe than for the same lining protected by grouting or by an external membrane.

A widely used means of representing ground/lining interaction is through the use of springs. Springs do not well reproduce the interface forces and the value to select for their stiffness, even for the simplest elastic analysis, is not clear. It is commonly stated that the stiffness is inversely proportionate to the radius of the interface, derived from consideration of the relationship $u/\sigma_r = r(1+\nu)/E$. for the association between radial strain and stress for a cylindrical hole in an elastic solid. It is incorrect to apply such a relationship where the interface is non-circular, where the radial stress varies around the interface or where shear occurs across this interface.

In my experience, the introduction of the structural engineer to tunnel design, without the constraining hand of the experienced tunneller, has had a malign influence. The experienced tunnel designer knows that the time/space dependent problems of tunnel design and of the design of tunnel construction are sufficiently complex to warrant a transparent design process whereby the influence of particular assumptions is ever evident and readily checked. Simple analytical or, on occasion, numerical, techniques serve this purpose. If possible economies appear to justify more complex numerical modelling, this may subsequently be indulged. What I find inexcusable is the growing habit, particularly encouraged by structural engineers without practical experience of tunnelling, to engage immediately in complex numerical modelling (with output incapable of useful interpretation) without previously ascertaining, through sketches and analysis, the mechanisms of failure to be investigated.

5 The proof engineer

As in warfare, so in tunnelling, it is imperative that a single mind, the 'conductor' be identifiable as having overall responsibility for the project strategy, as it passes through the several interactive stages of planning, investigation, design and construction. Any confusion of this principle tends to cause distraction. I place the Proof Engineer in tunnel design in this category, where the appointment creates the illusion of providing an external check of the overall functions of the designer.

The concept of Prüfungsingenieur, originating in Germany, was imported to Britain as 'Design Checker' (later more commonly described as Proof Engineer) for major bridge structures following problems with box-girder bridges. I fully accept the desirability for external checking of a complex structure of this nature, for which the checker has access to all design data: material properties; loading; wind; temperature; seismicity. I do find considerable problems when the Proof Engineer enters tunnel design. In the late 1960's I was asked to advise Government in Britain on the intention to extend the function of Design Checker to tunnelling. I emphasised in my report that the design process must then include site investigation and its interpretation, including the discarded hypotheses, the application to the chosen technique of tunnel construction and the expected quality of control. These are all factors to contribute to the assessment of a satisfactory design overall. Unless, therefore, the Design Checker was present and engaged throughout all such activities and decision-making, acting as a 'shadow designer', his testimony could do no more than confirm that the process of analysis was acceptable for the stated assumptions. I concluded therefore that the degree of assurance to be provided by a Proof Engineer for tunnelling was illusory. It might even discourage innovation. Experience since that time only confirms my viewpoint. Proof Engineering moreover, with the analysis of the design removed from other vital aspects, encourages the work to be undertaken by structural engineers without experience in tunnelling. Two factors appear to have encouraged the wider adoption of Proof Engineering in tunnelling:

- The increasing complexity of the design analysis, for which perhaps it is both cause and effect.
- The increasing practice for the project designer to be employed by the Contractor, arousing suspicions that safety may be jeopardised by cutting corners.

I believe that the Proof Engineer would be better employed:

(i) In undertaking simple analyses of failure modes and mechanisms in order to pose particular questions on these issues to the designer.

(ii) Taking a broad view of possible problems, including the comparative review recommended for the engineering geologist (section 4 above), to enquire whether each has been adequately addressed or whether, on the contrary, sound reasoning leads to its elimination.

The German word 'Gestalt' (form) has been broadened internationally to equate to something near to 'holistic'. I am advocating that the designer and the Proof Engineer should be expected to exercise such an approach in preference to the narrow role of checking the structural analysis, which is much more likely to be correct that its assumptions and interpretation. The Employer who engages a Proof Engineer might then expect to gain value for money. He should not expect such a valuable service from one who has been engaged at least cost for a function of uncertain and inadequate content.

6 Procurement 2 — appointment of the contractor

The seeds of project failure, sown by a lawyer-advised Client and by the segregated functions of the designer, begin to germinate at the time of Tender. It is manifestly absurd to heap on a Tenderer unknown risks beyond his control, of a magnitude sufficient, with reasonable contingencies, to turn profit to loss. What can then be the Tenderer's game plan in such circumstances? To make a reasonable allowance and then to rely on litigation to recover the balance is a risky and expensive strategy. Assume the chance of success is $1/n$ and the legal costs a fraction p of the award for success, where p is possibly 0.6-0.8. Hence the expectation of reward from litigation may be represented as $(1 - p)/n$, not a very favourable option. Alternatively, does the Tenderer make a pessimistic forecast of project risk knowing that this will probably eliminate his Tender. The attempt by the Client to turn uncertainty into certainty is very high and the very attempt to do so will prevent attempts subsequently to mitigate cost. The reason for this is that a Fixed Cost Contract will be voided if its conditions are subsequently changed, regardless of whether or not these changes made the project easier to perform for the altered circumstances.

The worst possible start to the tendering process is for the Tenderer to find that the information on site investigation is made available reluctantly in such a manner as to satisfy the legal obligation for access, but no more. The data and interpretation from site investigation provide a key resource for the project, possibly of yet greater interest for the Contractor than the Designer. Why then the apparent reluctance to make it freely available? This stems from a mistaken, and long disproved, view that an adversarial posture on behalf of the Client will achieve best results.

While most tunnel engineers will support free access to s.i. data, the wider availability of interpretative reports is less readily accepted. Interpretative reports directed towards the application of data to the project form an inherent part of the valuable resource. These reports will have been prepared by those with special skills and with far greater time to absorb the material than the Tenderer. The provision of these reports may well be accompanied by a note that they are 'without warranty' (although why such a qualification should accompany s.i. data as it often does is incomprehensible). The principle goes hand-in-hand with the concept of 'reference conditions', discussed in Section 8 below, whereby the basis may be established for assumptions vital to the preparation of a reasonably priced scheme for undertaking the project.

A valuable part of the interpretative report should include an assessment of the geotechnical risks of the project, drawing upon experience from comparable projects (comparable but never the same) and describing measures already taken to identify and manage the risks. Where risks may be left for resolution during construction, the practicability, in terms of time and of adequate control, of applying observational techniques should be discussed. This type of approach helps the Tenderer to choose whether to prefer a system tolerant to the risk or one which depends on adequate prior mitigation of the risk.

7 Value engineering

'Value Engineering' is the title given to a procedure whereby savings from a change proposed by the Contractor are shared between Contractor and Client. For relatively minor variation, the practice is harmless and may encourage initiatives for desirable innovation. The origin of the term is associated with more radical variation and here there are obvious dangers. In the United States there was traditionally a break between project design up to tender and the subsequent acceptance of overall responsibility by Contractor. The absence of continuity much discouraged innovation. In the event of any problem, lawyers would pounce upon the notion of a 'novel, untried method' as indicating irresponsibility of the designer. Value engineering was seen as a solution to this disruption since the Contractor would then have no one else to blame for design problems. The Tender would thus be awarded against a well-tried conservative design, leaving innovation, with its reward of a handsome profit, to the Contractor. In this way the Client lost half the benefit of an innovative design, often an appreciable fraction of the project cost. Far more important, the initial planning and investigation would not be related to the changed scheme. The Client would find the proposal sufficiently attractive to accept it without regard to the possible difficulties, the associated indirect increases in cost and the insufficiencies of s.i.. Moreover, as the practice became more prevalent, so was the initial design seen as entailing little responsibility for the designer since it would not survive for the project, encouraging superficial attention to the initial design and its supporting studies.

The prevention of design continuity through the tendering stage by International Development Banks has resulted in Value Engineering becoming accepted by FIDIC (Fédération Internationale des Ingénieurs Conseils) [3] for the Conditions of Contract for Construction as a measure for partial compensation. As already stated, this for tunnelling provides no solution to the real problem of discontinuity between pre-Tender and post-Tender design. Where proposals for Value Engineering are made, these merit close examination to ensure that they are technically sound and that supplementary s.i. and design that may be required are taken fully into account in respect of time and money. It would be far more to the interest of the Client, and his Banker, for the innovative feature to be established from the outset.

Among other examples, I recall a tunnel for irrigation in Egypt for which two designs were prepared, one based on an observational approach and hence requiring full continuity in design, the second at an increase in cost of about 50% using more traditional methods. The conditions for the first were unacceptable to the authorities, the second was too expensive so an inferior surface solution was preferred.

8 Reference conditions

The term 'Reference Conditions' for tunnelling was first coined for a report by CIRIA in 1978 [1] as a component part of risk management and risk sharing between Client and Contractor. Reference Conditions relate to issues of significant uncertainty at the time of Tender, providing a basis for tendering, possibly also a basis for measuring variations. While the simplest application concerns specific aspects of geological interpretation of available information (e.g. whether or not a tunnel would be entirely contained within a particular type of ground suitable for tunnelling), the report [1] gives examples of wider usage in associating the class of ground with physical properties and with practical adequacy of particular forms of support. This latter example may be applied, for example, to 'zoning' of the ground, relating means for assessing quality to terms of payment. The principle has wide application, as a factor towards low project cost. For a start, it helps to avoid costs of subsequent litigation on the issue of what was, and what was not, to be expected of the ground in relation to 'tunnelability'. Reference conditions address such issues more broadly than the later US proposal for a 'Geotechnical Baseline Report'.

As an example of the value of Reference Conditions, the design of the Cargo Tunnel (1966-68) at Heathrow Airport, referred to in Section 4 above, depended on a top cover of about 1.2 metres of London clay. Rigorous researches had been undertaken but, with the technologies of the time, it was not possible to be certain that this condition was assured in the most critical situations, including beneath the airport runways. In consequence, this risk was removed from the Contractor, with provisions for the contingencies needed to overcome any deficiency. These were not required and the tunnel was constructed at a notably low cost.

9 The observational method of design

The Observational Method (OM) is a powerful design tool for the economic and efficient control of uncertainty. The principle is abused by those who claim to be relying on OM in the absence of a specific and timely means for its application. OM is a design method and has, in each instance, itself to be designed. The rigid rules set down by Peck [9] for when, and when not, to resort to the OM are unnecessarily restrictive. The essential features for tunnelling may be described as:

1. Devise a conceptual model of ground and tunnel behaviour.
2. Select conditions for design and construct to this design.
3. Predict critical features from (2) for observation.
4. Observe and compare observations against (3). Assess whether differences call for foreseen supplementary measures or whether variation of (1) is needed.
5. React appropriately to (4), with any supplementary measures applied toa predetermined scheme and timing..
6. Repeat (1) to (5) for control and for varying circumstances.

The selection of the standards for (2) is a vital feature of the method. In some circumstances, it may be seen as economic for any supplementary measures (e.g. additional support) to be exceptional. Elsewhere, it may alternatively be considered that safety and economy are better achieved by reliance on observation to indicate where supplementary work may be needed for a proportion of the tunnel.

OM may also be applied to 'designing down' rather than 'designing up'. This implies, for tunnel support for example, that initially the degree of uncertainty calls for an over-provision of support with the expectation that observation (on the lines of 1-6 above) will permit reduction as the uncertainty is reduced. The OM should never be used on a basis of 'What, if?' without full preparation for the consequences.

Observations of movements of survey points around a tunnel are sometimes related to 'triggers' of the amount of measured displacement. What is not so commonly appreciated is that the 'triggers' should consider combinations of displacements. For example, if there are points *a, b, c d.....* around a tunnel ring, and if tangential and radial displacements are designated as t and $r$respectively, the important features for 'triggers' are $r(a) + r(c) - 2r(b)$ etc, representing approximate change in curvature, hence bending stress, and $t(b) - t(a)$ etc approximately representing compression or, more likely, partial failure of the ring.

10 Conclusions

Following the title of my contribution, it would have been too easy to provide a catalogue of tunnel disasters, with the cause attributable to poor workmanship or, more rarely, to inadequate design. What I have attempted is, from personal experience, to dig deeper. Many factors , some which may be deemed beyond the normal control of the tunnel engineer, may contribute to a specific mishap. We are concerned with what has been termed 'A climate of risk' [10]. I will illustrate two examples:

(a) the methane explosion at Abbeystead in 1984 on the Lune-Wyre water transfer tunnel.

(b) The collapse of NATM tunnels for Heathrow Express railway station tunnels in 1994.

For the former incident, failure to heed the significance of the confidential nature of a nearby borehole contributed to disregard of the risk of methane entry. For the latter, an unfamiliar contract system imposing design responsibilities on the Contractor was applied without ensuring that the implications were understood and appropriate resources provided.

I also indicate a few of the simple means for exploring the adequacy of a solution. An insidious trend at the present time is for the investigation and design process to proceed mechanically, s.i. providing numbers for rock classification systems, without regard for the specific features and their variation. This is followed by a leap into numerical modelling of such complexity that output is unrelated to possible failure mechanisms, the real justification for analysis.

While we learn from the obvious immediate causes of each tunnel accident, it is my belief that the underlying causes are inadequately investigated and that these are the real culprits for the popular view that tunnelling is a risky business.

Bibliography

[1] Construction Industry Research and Information Association: Tunnelling — improved contract practices. Report 79, CIRIA, London 1978

[2] Curtis, D.J: correspondence on [6]. *Géotechnique*, Vol 26 pp 231-237, 1976

[3] FIDIC: Conditions of Contract for Construction, First Edition, 1999.

[4] Heyman, J: The Masonry Arch, Ellis Horwood, Chichester, 1982

[5] Muir Wood, A.M: Tunnel hazards: UK experience, Hazards in tunnelling and on falsework, Institution of Civil Engineers, London, pp 149-164, 1975

[6]] Muir Wood, A.M: The circular tunnel in elastic ground, *Géotechnique*, Vol 25 pp 115-127, 1975

[7] Muir Wood, A.M: The first road tunnel, World Road Association (PIARC), Paris, 1995

[8] Muir Wood, A.M: Tunnelling: management by *design*, Spon, London, 2000

[9] Peck, R.B: Advantages and limitations of the observational method in applied soil mechanics, Géotechnique, Vol 19, pp 171-187, 1969

[10] Pugsley, A.G: The safety of structures, Arnold, London, 1966

Application of Design-Build Contracts to Tunnel Construction

Robert A. Robinson

Shannon & Wilson, Inc., Seattle, Washington, USA

e-mail: RAR@shanwil.com

Abstract:

Many public agencies, pressured to reduce design and construction costs, unhappy with cost overruns and schedule extensions associated with many standard design-bid-build contracts, are searching for alternative cost-effective contracting methods. One of the methods being considered and tried is design-build contracting, used successfully on several large civil projects. Consequently, many owners have elected to try design-build contracting on underground projects in the hopes that there will be less risk of claims and legal entanglements with the single source responsibility that is part of the design-build process. Other owners are assessing design-build as a means for accelerating the tunnel project delivery process while potentially reduc-ing owner risks and ultimate project costs. The Federal Transportation Administration (Tunnels & Tunneling 1999) elected to implement design-build contracting approaches on five major demonstration projects. Sev-eral European owners have also experimented with design-build formats on major tunnel projects. This paper will present case history data for several design-build tunnel projects, noting changes in cost and schedule, during the course of the project. This paper will also discuss likely pros and cons, and key considerations for the owner, the contractor and the designer in a design-build contracting approach for tunnel construction.

1. Design build basics

The civil industry has utilized design-build contracting to successfully construct a wide variety of buildings, plants, and other facilities. Design-build contracting accounted for about 15% or an estimated $70 billion of the U.S. construction industry in 1999 and 2000. This is an increase of $16 billion from the $54 billion reported in 1990 in Engineering News Record.

The Engineer's Joint Contract Documents Committee (EJCDC 1995) provides these definitions of the design-build method of contracting:

- A single company providing both the design professionals and construction services from within its own forces.
- A joint venture of an engineering firm(s) and a contractor(s).
- Either an engineering firm or a contractor providing their services, and subcontracting the complimentary necessary services through an appropriate subagreement.

Design-build contracting procedures are best suited to design and construction projects that are repetitive in nature and where the opportunity for unforeseen conditions (geotechnical or subsurface utilities) or impacts on adjacent property owners (third party issues) are minimal or nonexistent. At the other end of the construction spectrum, design-build has been used successfully for a variety of complex or high-tech projects requiring highly specialized construction techniques and specialty design and construction firms, and where third party issues are negligible or non-existant. These include "clean rooms" for high-tech assembly areas, petrochemical facilities, hydroelectric plants, and steel mills (Loulakis 1992), all with well-defined final performance requirements and conditions.

Typically, the design-build process is thought to provide these attributes:

Owner developed preliminary design - Ideally, the owner's engineer will develop a preliminary design that incorporates the essential project requirements, owner's preferences for design details and configurations, and provides sufficient assessment of ground conditions and third-party impacts for the design-build team to develop a realistic bid and schedule and to provide sufficient information for the development of a final design.

Unified design and construction team - The teaming of the final designer and contractor provides a single point of contact and responsibility to the owner. The design-build format gives the contractor more control over the work and provides increased opportunities for the contractor to be innovative with respect to strategies for both reducing cost and accelerating schedule (Brierly et al. 1998). It also allows the designer and contractor to interact closely during construction and react quickly to the excavated and exposed ground conditions or perceived cost and schedule savings with alternate construction means and methods.

Accelerated schedule - Reduces the length of the overall design and construction process by combining the competition and selection phases for the final designer and the contractor into one unified selection process for the design-build team. Overlapping design and construction efforts allows for simultaneous design and construction, for a "design as you go" approach, thus reducing overall schedule. Procurement of long-lead items such as tunnel boring machines can proceed in parallel with the final design. The design-build format may also potentially reduce the time and effort required by the owner to administer the work by reducing the number of contracts.

Award using best value procurement process - The owner may chose to award the contract to the most qualified team based on objective selection criteria focusing on the talent and experience of key team members, work programs the team will develop and implement, preliminary design and construction concepts, and the estimated cost of their work.

However, the design-build submittal and selection process is considerably more costly for the proposing teams than traditional design-bid-build, due in part to the large amount of information and the level of design detail generally required for submittal by the project owners to assess the various design-build teams. The design-build teams often find it necessary to carry the owner's preliminary design to a 50 to 60% level in order to develop design and construction approaches, assess construction impacts, and develop a reasonable construction bid estimate.

To encourage a larger number of proposing teams, some owners have elected to include a stipend or honorarium payment for the losing bidders in a design-build contest. This stipend is intended to partially reimburse the losing bidders for the large amount of effort that is necessary to develop the extensive submittal documents that are required for most design-build procurements. Stipends have generally ranged from 0.05% to 0.3% of the estimated total design and construction cost of the project. Most design-build teams would argue that these stipends, although helpful, do not come close to covering all of their costs (let alone providing a profit) for developing their proposed design and construction approaches, as required for most design-build proposals. Many owners, inexperienced with stipends and the cost of a design-build proposal, consider stipends to be unnecessary for promoting a healthy and competitive bidding environment.

2. Traditional design-bid-build contracting

Most major public sector tunnel projects in the U.S. have been contracted using traditional design-bid-build contracting. In this approach, the owner contracts with the designers to develop feasibility studies, environmental impact statements, preliminary design, and final design. A comprehensive site investigation program is implemented to assess ground conditions, followed by the development of a design, cost estimate, and set of plans and specifications. In the iterative process of developing the final design with repeated reviews and feed-back from the owner, the design concepts are fleshed-out, third-party impacts and issues are addressed and resolved, permit requirements are identified, right of entry agreements are signed, construction impacts are predicted and mediated with design refinements, and public meetings and interaction are accomplished.

Following this progressive design process, the owner then competitively bids the fully explored and designed project. The "low-bid" contractor generally sublets portions of the designed project to a variety of specialized subcontractors. The subcontractors can review the bid documents to fully ascertain the nature and extent of their involvement, as well as conflicts, schedule impediments, and other issues that might affect their costs.

Under the traditional design-bid-build approach, the contractor's primary goal is to economically implement the plans and specifications and complete the project on schedule and under budget in order to receive a reasonable profit for the construction team's efforts. The contractor is not obligated to interpret or second-guess the intent of the designer, nor the condition of the ground. Ambiguities or errors in design or interpretation of the explorations will likely result in additional compensation for additional work and "changed conditions".

The design engineer's goal is to produce a functional and satisfactorily completed project within the owner's budget, with minimal changes or claims, while meeting the intent and goals of the design. This may at times be at apparent odds with the contractor's goal of successfully completing a project while earning a reasonable profit. The contractor's meeting his goals should in fact be a significant concern of the designer and owner, since only contractors who can successfully and satisfactorily complete a project while earning a reasonable profit can afford to go after future tunnel work.

Traditional design-bid-build contracting provides these attributes:

- Strong owner control of all aspects of the project though construction;
- Clearly divides and defines the roles of the designer versus the contractor;
- Resolves most of the third-party (utilities, buildings, highways) issues prior to construction;
- Defines the most of the geologic and environmental issues and impacts prior to final design;
- Provides abundant opportunities for public involvement;
- Promotes a high quality of design and construction for better long-term performance of a completed project;
- Resolves many of the possible design questions and the owner's design concerns with an iterative design process prior to construction;
- Is familiar territory for the owner, designer, and contractor.

Therefore, an owner must weigh and compare the risks of working with a relatively new and untried contractual approach versus refining his use of the traditional design-bid-build approach. The owner's level of risk in design-bid-build contracts can be managed and reduced by employing risk sharing strategies developed over the last 30 years to suit the tunneling industry (USNCTT 1974) and that include the use of dispute review boards, escrowed bid documents, geotechnical baseline reports, as well as thorough alignment investigations (USNCTT 1984).

The traditional design-bid-build contracting method has been repeatedly promoted by the American Consulting Engineers Council as preferable to design-build contracting, as follows: "The American Consulting Engineers Council strongly believes that the use of the traditional design and separate bid/built project delivery system is in the best interest of the owner as well as protecting the health, safety and welfare of the general public. This traditional system provides the checks and balances necessary to give the owner and the public the greatest degree of assurance that the project is the most appropriate solution for the circumstances through the use of a qualifications-based selection process and direct owner representation by the design professional" (ACEC 1992).

3. Design-build – pros & cons

Due to a general lack of experience with design-build contracts for tunnel construction, there are few owners or contractors familiar with the pros and cons of this contracting method. A few of the more salient issues, concerns and perceptions are discussed in the following paragraphs.

Reduced level of owner control. Owners often resist the loss of control inherent in the concept of design-build and react throughout the project as if it is standard design-bid-build. However, the owner's ability for continuous review and full control over all of the nuances of the final design in the design-build process is greatly reduced.

Reduced checks and balances. One of the most significant drawbacks to the owner is that in joining the designer and contractor together, many of the checks and balances present in the traditional contracting method are eliminated. These checks and balances help to protect the owner from an inferior project.

Uncertain level of design. The level of design appropriate to design-build is uncertain. Some design-build RFPs are based on a conceptual design level of maybe 15%, whereas other RFPs have been carried to a 60% design level, with accompanying plans and specifications. Likewise, the level of geotechnical investigations and third-party property and utility assessments varies considerably from next to nothing, to an 80% to 90% evaluation. There are pros and cons to both of these approaches. Certainly, the minimalist approach provides the greatest potential for innovation by the contractor, although it also provides the greatest risk for changed conditions, and the attendant cost and schedule impacts. Whereas, the maximalist design approach significantly reduces the potential for changed conditions and delays due to design changes or permitting requirements, but reduces the opportunity for innovation on the part of the contractor, while maximizing the owner's control over the nature of the end product.

Geotechnical considerations. The applicable geotechnical conditions can only be resolved through a comprehensive exploration program implemented incrementally during the iterative phases of conceptual, preliminary, intermediate, and final design. However, under a design-build format, ground conditions may not be thoroughly explored and consequently the negotiated construction costs will be based on an incomplete understanding of these conditions. If geotechnical conditions and issues are resolved with additional explorations during the design phase of a design-build contract, then they may involve some additional costs for changed conditions. However, if significant differences in geotechnical conditions are not revealed until the construction phase, then major claims for changed conditions resulting in increased project cost and construction schedule may occur.

Third party. Many of the third-party and geotechnical issues that impact the outcome of an urban tunnel project are resolved during the iterative phases of intermediate and final design. However, under the design-build format, the multitude of third party right-of-entry, permitting, and utility issues may not even be recognized, let alone resolved during the preliminary design phase. If they are discovered and resolved during the design phase of a design-build contract, then they may involve some additional costs for changed conditions. However, if significant differences in third-party conditions are not revealed until the construction phase, then major claims for changed conditions resulting in increased project cost and construction schedule may occur.

3.1 Owner's perspective

A perception common to many owners seems to be that the design-build process will shift much of the cost and schedule risk to the contractor, and significantly reduce project schedule and possibly the overall project cost. Some owners have come to associate traditional design-bid-build with extended project design and construction times, disputes over changed conditions, and prolonged and costly litigation. In traditional design-bid-build, the owner has been held responsible for unanticipated changes in ground conditions, unknown third-party utilities, and often for construction delays due to third-party permitting requirements. For design-build, the owners appear to retain much of this same risk for third-party issues and changes in geotechnical conditions.

Abundant case histories from design-build applications to above ground facilities indicate that the owner also does not realize any significant cost savings, but some possible savings in schedule, provided that the owner can restrain their desire to actively control the project. Design-build was never intended for nor has it proven capable of routinely reducing construction costs. In fact, as shown in later sections, the preponderance of the evidence from tunnel projects to date is that design-build generally results in increased costs and schedule time beyond that contracted by the owner.

Most design-build, at least as it applies to tunnel work, appears to involve a negotiation of the final cost of the project after the final design-build team has been selected. At this point, there is very little incentive for the design-build team to minimize their prices, since they are the selected team. However, there may be a tendency to increase prices as the owner adjusts the design during the negotiation period, based on ideas provided by the various competing teams.

Owners and their design engineers are used to frequent and detailed reviews of several iterations of the design and construction documents. It is through this process that many of the problems and issues of a project are revealed and resolved during the various stages of design. However, unless the design-build documents detail out the levels, frequency, and general scope of submittals and reviews, it will be difficult to enforce such frequent submittals and reviews. Furthermore, if the reviews result in substantial changes to the design-build team's approach, then the owner will likely be expected to pay additional costs. This could easily lead to significant disagreements and conflicts between the owner and contractor team.

In design-build, the owner now finds himself on the opposite side of the table from the final designer, who now works for the contractor. This often makes it extremely difficult for the owner to find out exactly what the final design will look like, except in a piecemeal fashion as various portions of the design are being developed to support the construction schedule of the contractor. The owner may have little opportunity to directly interact with the designer without the intervention, and control of the contractor. Naturally, the contractor is interested in reducing, as much as possible, the complexity and cost of a project, whereas the owner may have design goals, possibly not adequately detailed and explained in the contract documents, that he wishes to impart to the designer as the project progresses (as would be the case in normal design-bid-build).

Lastly, the Owner may believe his site inspection staffing needs can be significantly reduced, since the design-build team now has a greater responsibility for quality assurance (QA) and quality control (QC). However, yielding this activity to the design-build team further removes the owner from the project. In tunneling, the best opportunities for observation and recording of ground conditions and quality of the work are afforded during the various phases of construction, rather than observation of the completed tunnel.

3.2 Contractor's perspective

Many contractors believe that design-build offers an opportunity to maximize profits while retaining the current balance of risk with the owner. Many contractors also appear to relish the idea of having significantly more control over the progress, details, construction means and methods, and character of the final product. There may also be a greater tendency for contractors to try new and innovative construction approaches that could save money and time, and for a more productive partnership between the contractor and designer to develop these innovative construction approaches.

Design-build teams may attempt to shed much of the risk for unanticipated or changed ground and third-party conditions back to the owner by including, as part of their proposal, a long list of assumptions, guidelines, issues, concerns, exclusions and understandings. This list is generally intended to serve as "baselines" for establishing changed condition claims or for limiting the contractor's liability exposure. Several pages of baseline assumptions from each competing team make comparison of the various technical proposals and costs extremely difficult, and have resulted in some design-build projects never going to contract.

Contractors that set up and lead a design-build team must interact directly with the designer. Due to the presence of numerous uncertainties, and the wide range of possible design and construction solutions, the contractor must often learn to cope with frequent design uncertainty and revisions. This is particularly true if there is considerable latitude in the design requirements and/or a scarcity of information on geotechnical conditions and adjacent third-party utilities and structures. Contractors are often not aware of the amount of lead time, labor hours, and labor costs required to complete a final design and prepare the specifications and plans.

The contractor must also provide a more comprehensive proposal submittal than is normally required for traditional design-bid-build contracting. A disproportionally large amount of this proposal information must be provided by the designer, who may have been provided with very limited geotechnical and third-party information. If a stipend is provided, then the contractor may elect to pay some portion of the designer's base salaries with a partial multiplier.

The proposal submittals often contain unique construction means, methods, and materials. Consequently, the design-build bidders have often expressed concern over the possible release of their proprietary approaches to possible competitors.

Public owners and design-build proposers may find it difficult to prevent release of proprietary proposal information through various forms of the Freedom of Information Act.

For some design-build proposals, utilities and overlying or adjacent property owners have had little or no contact with the project owner. Consequently, initial contacts with third-parties made by the design-build team to test out alternative construction options has resulted in considerable consternation for all parties. Unfortunately, where a design is only carried to a 15 to 30% level by the owner, third-party issues are usually not fully addressed by the owner at bid time.

Design-build often shifts a greater proportion of the quality control and quality assurance issues to the design-build team. However, the contractors are often ill-prepared and ill-disposed toward the extensive inspection, testing and reporting necessary for a thorough QA/QC program. A design-build team's goals of constructing a project quickly, efficiently, and cheaply do not generally mesh well with the normal goals of a comprehensive QA/QC program, or the normal agency requirements for thorough QA/QC documentation.

3.3 Designer's perspective

The designer has traditionally worked for and developed the design through its various phases to support the requirements of the owner. This has traditionally been a collaborative and iterative approach, with the design developing and maturing as additional data is obtained and the goals and needs of the project are refined, often under public scrutiny.

Under the design-build approach, the designer, and his design related subcontractors now work for the contractor. This change in allegiance requires that the designer and his subcontractors now promote and protect the interests of the design-build team. These interests include:

1. The development of the least expensive construction approach (not necessarily the most cost-effective from a long-term operation and maintenance approach);
2. Minimization of the exploration and design phases of the work to accommodate rapid turn around of construction documents;
3. Reduction of the amount of construction monitoring and review;
4. Minimization of the amount of design and construction documentation. Furthermore, the owner has now lost direct access to the final designer, and may have very little control, except through the preliminary specifications, on the final product.

Design engineers and contractors need to develop formulas for equitable sharing of the risk from a problem project and profits from a successful project. Often there is little or no owner payment for design work and, consequently, the team has little cash to reimburse the engineers for their design efforts. There is also a strong push to segment the design so that portions of the project can get under construction to enhance the teams cash flow. Contractors may require that the designer reduce their billing rates in the early phases of the project until cash flow improves by meeting construction milestones.

The designers and contractors in the design-build team recognize that they may face significantly increased exposure to liabilities that were not present in more traditional design and construction methods. Designers may be required by the owner or the contractor to guarantee that their design will meet defined performance criteria, resulting in increased liability potential over and above normal negligence definitions (Loulakis 1992). Some design-build projects have even been expanded to include a period of maintenance and operation. The I-15 project in Salt Lake City, Utah, included an extended maintenance requirement. The Tolt River water treatment project near Seattle, Washington, requires the team to maintain and operate the plant for several years before turning it over to Seattle Public Utilities.

4. Design-build – Case histories

Few design-build tunnel project case histories have been published in the literature. However, these few relevant published case histories do shed light on the successes and failures, pros and cons of the design-build contracting approach for tunnel projects. From these projects we can glean several lessons to incorporate into future design-build contracts. Selected recent tunnel projects that have been contracted utilizing design-build methods include:

4.1 Channel tunnel, England to France (Lemley 1991)

The Channel Tunnel project was conceived over a century ago, but it took nearly a full century to realize that dream as tunneling techniques caught up with the needs and requirements of the project. The project was implemented with the award of a concession to design, build, equip, and commission the cross Channel project in 1986. The tunnel was completed in 1993, with full operation commencing in the summer of 1995. The design-build team, Transmanche Link JV, consisted of a consortium of five British and five French firms to construct a total of 57 km, portal to portal, of 4.8 m diameter service tunnel and 7.7 m diameter twin railway tunnels in a scheduled 33 months. Over the course of the design-build construction project, the estimated price of tunnel construction climbed from $3.8 billion in 1981, to $8 billion in 1988. This upward creep was related to inflation, improved geotechnical data, and changes in owner requirements. All parties recognized that there were major risks, not the least of which were geotechnical. Consequently, the risk was shared between the owner and contractor using a target cost reimbursable contract for the tunnel portions of the project.

An extensive geotechnical program (Varley et al. 1996) was undertaken in three phases prior to letting of the contract and in two phases during construction. Prior to the contract, there were 97 over-water and 34 land borings. During construction, an additional 19 over-water borings and 34 land borings were drilled. Over 4,000 km of geophysical surveys were performed on land and sea.

Despite the large number of very costly over-water borings and extensive geophysical surveys, there were geologic surprises that resulted in delays, schedule adjustments, claims, and additional compensation. In the design-build contractor's proposal, a comprehensive discussion of anticipated ground conditions and behavior was presented. This listing basically served as a baseline for comparison with actual ground conditions and ground behavior for the resolution of claims for differing ground conditions as allowed under the contract. Claims revolved around: significantly larger amounts of rock overbreak which contributed to major difficulties in erecting the segmental liner system, local areas of high groundwater inflow, and unidentified zones of folding and fracturing that adversely impacted tunneling (Mansfield 1996).

4.2 Whittier tunnel rehabilitation, Alaska (Moses et al. 2000)

The 4.2-km-long Anton Anderson Tunnel was originally constructed during World War II to transport goods and people to a military base established at Whittier, Alaska. However, it has been a bottleneck for car traffic in and out of Whittier, since cars could only enter or leave the town via flatbed railcars through the tunnel. Consequently, a preliminary design was developed HDR Inc for the Alaska Department of Transportation for conversion to the first U.S. tunnel to provide dual highway and railroad use, making it the longest rail and highway tunnel in North America, and the longest highway tunnel in the U.S. Four design-build teams proposed, but two dropped out prior to the final submittal and selection phase. Although the owner's estimate was $35 million, after negotiations the project was awarded on the basis of "best value" for $57.3 million to a design-build-operate contract team consisting of Kiewit Construction Co., with design provided by Hatch Mott McDonald and five local Anchorage firms. Final cost of the project is estimated at about $61 million, a little more than a 5% increase over the negotiated price. The contract was completed on schedule despite weather delays and tunnel access scheduling complications with the operating Alaska Railroad.

No geotechnical explorations were necessary since the existing tunnel was more than 99% unlined. However, the almost total rock exposure throughout the tunnel provided little opportunity for a claim for changed geologic conditions.

Discussions with various team members suggest that this was a successful project overall. However, there were some problems with implementation of the design-build contract approach. The very linear nature of the project made staging and organization a critical issue for the success of the job. Difficulties in coordinating with the Alaska Railroad's freight and passenger service schedule resulted in irregular short and long work windows. Quality control was often a point of contention between owner and contractor. Several iterations of various design components were required and the owner, by contract, had a very long review period (up to 3 weeks), which fortunately was rarely used completely, but could have caused significant delays in construction. Environmental permitting with the U.S. Forest Service did result in some delays.

4.3 Copenhagen subway system, Denmark (Reina 2000)

The current project, begun in 1996 and scheduled for completion in 2002, consists of 14 km of rail line bid at $350 million. However, the project has $175 million in claims and will be two years over schedule. The project includes 8.3 km of twin 4.9 m tunnels, 4.2 km of track elevated on embankments and structures, 0.9 km of at-grade track, and numerous stations including six stations reaching nearly 20 m deep. The winning team, CONMET is an international consortium including: Carilion plc (U.K.), SAE International (France), Astaldi SpA (Italy), Bachy Soletanch Ltd. (England), Ilbau GmbH (Austria), NCC Denmark A/S, and ILF Consulting Engineers (Austria).

Reportedly, the owner was reluctant to relinquish control, adding to construction delays and costs, lengthy submittal reviews, and an estimated 300% increase in drawings over the contractor's original estimate. The large increase in number of drawings may have also related to the Contractor's optimistic prediction of number of required drawings, if they were not itemized or otherwise described in the bid documents. Stringent zero settlement criteria and excessively abrasive silts, sands and gravels contributed to heavy wear and high maintenance requirements for tunnel equipment; and stringent controls on locomotive exhaust have contributed to increased costs and schedule delays. Also, the international flavor of the team may have created difficulties due to cultural differences in their approach to the work and the distance between the various team members.

4.4 Tren Urbano Transit System, Puerto Rico (Gay et al. 1999)

The new Tren Urbano transit system consists of 17.2 km of track alignment, 16 stations, and 1.5 km of underground alignment. The Rio Piedras Section 7 includes open cut, cut-and-cover, earth pressure balance machined (EPBM) running tunnels, and NATM and stacked drift excavation methods for cross-overs and stations. The design-build team selected for the underground section consists of Kiewit Construction Co., Kenny Construction Co., H.B. Zachary Co., CMA Architects and Engineers, Sverdrup, Jacobs Associates, and Woodward Clyde. The KKZ/CMA Team won the project for $225,600,000, which, although not the lowest price, was determined to be the best value of the three bidders.

A geotechnical program consisting of 22 borings and two pumping tests were undertaken by the owner prior to award of the contract. Fifteen additional borings were drilled by the design-build team to better define the alluvial clays, sands, and silts, and the groundwater conditions on the project.

Many lessons were learned on the Rio Piedras design-build project by the various participants. These included: 1) improved awareness and documentation is needed of each member's role and responsibilities including the required levels of quality control and team management, 2) the designer must modify his approach in order to coordinate and segment design with construction for earlier construction progress payments, 3) improved awareness of the owner and designer that design is on the critical path and that quick efficient document control is essential to maintain schedule, 4) a recognition by the owner and design-build team that design costs for design-build are very comparable to traditional design-bid-build, 5) the designer must adjust his specification writing and quality control (QC) requirements to expedite construction, and 6) the normal payment on the basis of constructed elements of the project can lead to serious cash flow problems for the team during the early phases of the project.

Although this project is complete, the actual cost of the project and its success in meeting the schedule and in providing the owner with the quality of work that was desired won't be known for a year or more. However, discussions with various project personnel have indicated that there were problems with localized excessive and damaging settlements, old undocumented utilities, greater levels of design effort to meet the owner's needs, and a much greater level of QC labor than was anticipated. All of these issues will likely lead to increased cost. As of September 2001, contract costs for the entire 17.2 km alignment had increased by nearly 70% to $1.9 billion and the schedule has slipped by 2 years, due to increases in scope, higher than estimated construction costs, and differing site condition claims. A representative of the U.S. Department of Transportation has suggested that "A traditional design-build contract would not have had so many claims" (ENR 2001).

4.5 High Speed Rail Link Germany

A 177 km long high speed rail link is now nearing completion from Nurnberg to Munich in Germany. Cost for the entire line is estimated at $2.04 billion (DM 4,320 million) and completion is scheduled for late 2000 or early 2001. This alignment includes nine tunnels, totaling about 25.6 km and ranging from 650 m to 7,700 m long. NATM construction methods will be used to excavate the 12.9 m I.D. double-track tunnels. The owner, Deutsche Bahn AG, experimented with the use of design-build contracts for this project and divided the alignment into several design-build sections. Negotiations with the winner of each section took up to 6 months, due to price differences, design approach differences, and baseline issues presented in the proposals.

In Germany, as in the U.S., the owner is generally acknowledged to retain ownership of the ground. All geotechnical explorations were undertaken by the owner and the data was collected along the alignment and presented to the various bidders in data reports. Borings were typically spaced about 300 m apart. In describing ground conditions, care was taken to avoid discussing construction means and methods in relation to the ground, since the construction approach was to be determined by the design-build teams.

The German owners experienced many of the same difficulties with design-build that have been experienced in the U.S. Difficult ground conditions have been claimed to be partially unanticipated and have contributed to claims; however, delays in gaining and granting permits have caused the greatest delays to the project. The owner's engineers voiced concern over the quality of the design submittals, quality of the construction, inadequate time for thorough submittal reviews, lack of owner access to and interaction with the designer, and higher numbers of claims. The contractors indicated concern over excessive time for review by the owners, excessive owner involvement in details of the project, and insufficient preliminary design details. For the traditional design-bid-build, claims were generally within 3% of the bid price, but for design-build the claims appear to be about 25% of the negotiated bid price. Some of the owner's representatives were not enthusiastic about using design-build on future tunnel projects in Germany.

4.6 "Link" Light Rail, Seattle, Washington (Gildner, 1999)

The entire $3.6 billion project consists of 40 km of light rail alignment including 10 km of twin-bore tunnel. The north 8.4 km of the alignment will consist of up to 6 m diameter tunnels, up to 70 m deep along with three deep (up to 65 m) mined stations, one 40 m deep cut and cover station, a shallow crossing beneath Interstate 5, a deep crossing beneath Portage Bay, and passage beneath numerous streets, utilities, and buildings up to 30 stories high. The north corridor tunnel portion of the project was designated to be design-build. Three teams qualified, but one withdrew during the formal proposal and cost estimating phase. The Modern Transit Constructors (MTC) team was selected as providing the "best value" (a combination of technical qualifications and cost). The MTC team consisted of Modern Continental Construction, S.A. Healy Construction/Impregilo of Italy, Dumez of France, Parsons Transportation, D2 Consult, and Goldberg Zoino Associates.

When Sound Transit selected the design-build contract approach they also established a philosophy of risk allocation based on the fundamentals dealing with "ownership of the ground" as distinguished from "ownership of the means and methods". This approach is reflected in a substantial geotechnical exploration program. A phased exploration program, consisting of 137 borings, spaced an average of 100 m apart along the tunnel and with 6 to 9 borings at each station location was undertaken during the feasibility and preliminary design phase of the project. An additional four borings were drilled at the suggestion of the two design-build competitors. After selection of MTC, an additional 15 borings were initiated based on discussions with the design-build team and minor revisions to the alignment. All exploration data was presented in a Geotechnical Data Report (GDR) and interpreted in a Geotechnical Characterization Report (GCR). A Tender Geotechnical Baseline Report (TGBR) was prepared, which included definitions of baseline conditions such as boulder quantities, nature of soil unit conditions, but not ground behavior, since this would largely be determined by the contractor's selection of means and methods for tunnel and station excavation and support.

Unfortunately, the selected design-build team could not develop a cost estimate and construction approach that met the goals of Sound Transit. The tentative agreement between the selected team and Sound Transit was terminated.

4.7 Case history conclusions

Of these six projects, the Whittier Tunnel Rehabilitation project appears to have been the most successful from a cost and schedule perspective. This success likely relates to the roughly 99% exposed geology in the mostly unlined existing tunnel, and relatively few third-party impacts. The most critical third-party issues involved working around the scheduled freight/auto carrier train service during the enlargement of the tunnel, and installation of a roadway surface and safety and traffic control facilities to permit both rail and automobile traffic in the tunnel.

5. Where's the risk?

Over the last 25 years the tunneling industry (owners, contractors and designers) have evolved a unique risk sharing program to help to reduce the number and severity of claims and disputes (USNCTT 1974). This risk sharing approach included the development of:

1. The presentation of all factual field and laboratory data in a Geotechnical Data Report (GDR)
2. A geotechnical baseline report (GBR) to establish a level "playing field" for bidders and for reference during construction;
3. Escrowed bid documents to provide information on the contractor's assumptions for unit prices, quantities, and means and methods of construction;
4. A 3-person Disputes Review Board (DRB) consisting of reputable peers from the tunnel construction industry.

This risk sharing approach has proven to be successful, when correctly applied by experienced and knowledgeable owners, designers, geotechnical engineers, and contractors for traditional design-bid-build tunnel construction. However, no testing and calibration of this system of risk sharing has been accomplished for design-build contracting. Consequently, considerable experimentation is still required to determine how best to use these established risk sharing principles.

A critical and central issue for any tunnel contract is the question of who "owns" the site conditions. It is of critical importance that ownership of the risks and benefits of a particular site be clearly distributed between the owner and design-builder in the early stages of the project (Jaffe and Goode 1992). Some owners have attempted to shift all responsibility and ownership of the project environment to the contractor. This harkens back to the pre-risk sharing days of the heavy construction industry when owner's attempted to abrogate any responsibility for ground conditions, or changes therefrom, that might occur in the course of a project. This was often accomplished by including disclaimers on all exploration boring logs. For these projects, minimal interpretation was provided and consequently no geologic profiles, sections or other interpretations were commonly provided. This contracting approach led to numerous major claims for changed ground conditions and development of the risk sharing approaches presented in the previous paragraphs.

Loulakis (1992) notes that a design-build contract should include a differing site condition (DSC) clause that provides for the owner retaining the financial risk for unanticipated site conditions. The incorporation of a DSC clause allows the bidders to submit a minimal bid that does not incorporate contingencies for unexpected conditions. In implementing a DSC clause, it behooves the owner to assess the site conditions to a level necessary to reduce the risk for encountering unanticipated conditions during the design or construction phases of the work. One of the largest areas of potential risk is associated with variations in the ground conditions. It has become a basic tenant of tunnel construction contracting that the project owner retains overall ownership of the site conditions, including geologic conditions and variability, soil and groundwater contamination, and imprecisely located or describe third-party utilities.

A modified GBR was used on the Sound Transit North Corridor project, that established baseline geotechnical conditions utilizing data from borings spaced about 100 m apart. Baselines were established for the geologic and groundwater conditions, soil properties, and boulder and fracture frequency. The selected contractor was required to amend this Tender GBR with the agreement of the owner.

A Disputes Review Board (DRB) is most effective when a well defined set of contract documents, and GBR is in place. Without these two prerequisites, it will be difficult for the DRB to establish the bidding parameters that provide a baseline from which to determine if a changed condition has occurred.

The bid documents of the selected design-build team may mimic the role of escrow bid documents, provided that enough detail is presented in the documents. The owner and contractor will need to thoroughly assess and review these documents, keeping in mind that they may be used by a DRB to assess the validity of a claim. Contractors have expressed a concern over the possible accessibility of escrow documents to competing teams. After review by the owner, these documents should be left in the possession of a third-party, such as a bank, and the contractor should retain ownership of the documents. The documents of losing bidders should remain their property and be returned to them after the selection and negotiation with the winning design-build team. However, when a stipend is paid to the loosing bidders, then some owners may maintain that they now "own" the losing bidders documents, even though the owner has not paid full labor rates for their preparation.

6. Third-party and geotechnical information for design-build

Very little information has been published on the expected or anticipated levels of site evaluation and exploration that an owner should provide to prospective bidders for a design-build project. For traditional design-bid-build contracts, experience has shown that for both the owner and the contractor, the risk is significantly reduced by presenting the bidders with all of the geotechnical information collected (USNCTT 1984). However, very little has been published regarding the amount of geotechnical, utility, foundation, and historical construction information that should be provided to design-build bidders. A wide range of levels of geotechnical and site utility and foundation investigations have been utilized for design-build projects, ranging from little or no investigation for projects such as the Whittier Tunnel Modification, to a thorough, nearly 100% complete design exploration program for the Channel Tunnel, Tren Urbano, and Seattle Light Rail projects.

6.1 Failure without exploration data

Some owners have attempted to let design-build contracts with little or no information on existing utilities, foundations, permits, and/or subsurface investigations, arguing that the design-builder assumes responsibility for all conditions relative to the project site. A recent design-build procurement for the rehabilitation of an approximately 100 year old tunnel presented very little data on subsurface utilities and relied entirely on subsurface data from adjacent properties collected more than 30 years ago. Since the owner's engineer prepared less than a 10% level of design, none of the third-party issues, such as utility realignments and impacts on an adjacent freeway, had been resolved prior to the bidding phase. The lack of utility data prompted several of the bidders to contact local utility owners, who were as surprised as the bidders that numerous relocations were required and that utility companies had not already been contacted by the project owner. Due to the very preliminary nature of the owner supplied design, a lack of third-party agreements, and lack of current geotechnical data, the various proposed approaches and bids were very diverse and included extensive lists of bidder's concerns and baseline assumptions. The very diverse bids and approaches made it difficult to compare the various proposals. Consequently, the owner elected to cancel this project, pending more design work and agreements with third-party stakeholders.

6.2 Suggested level of geotechnical explorations

For design-build, the level of exploration should be similar to the traditional design-bid-build except that the program would be phased to permit input and suggestions for additional explorations from the design-build teams. Experience has shown that any information not disclosed to the bidders will eventually be revealed if any related claims litigation is pursued. Therefore, as with traditional design-bid-build, the exploration data should be presented in a GDR and summarized and interpreted in a GBR. A Geotechnical Interpretive Report may be prepared for the owner's designer to provide them with the necessary geotechnical interpretations to develop their preliminary design, engineer's estimate, and preliminary plans and specification guidelines. The content of these various documents have been well described and documented by a number of authors (USNCTT 1984 and UTRC 1997). However, the use of these formats for design-build contracts has not generally been addressed in the literature.

Most claims or cost increases for tunnel projects tend to revolve around unforeseen changes in the ground conditions (soil, rock, groundwater, methane gas, etc.). The U.S. National Committee on Tunneling Technology (USNCTT 1988), reviewed 87 tunnel case histories, and found that overall claims averaged about 30% of the engineer's estimate and contractor's bid price. About 60% of the tunnel projects experienced claims involving unanticipated or changed ground conditions. Over 95% of these claims were for relatively substantial dollars, and that about 30% of the claims were settled for the full amount of the claim. Where a substantial exploration program amounting to roughly a meter of boring per meter of alignment is undertaken, the claims are generally less than 20% of the bid price. Where roughly 1.5 m of borings per meter of alignment were performed, then claims were generally less than 10% of the bid price. However, where the length of borings is only about half the length of the alignment, claims ranged up to about 30 to 40% of bid price. Therefore, it benefits an owner to undertake a thorough investigation in order to reduce the potential for such claims.

There are some significant differences and considerations in presenting geotechnical exploration data for design-build versus design-bid-build. Since a design-build team is fully responsible for final design and for selecting their construction means and methods, it is not appropriate nor in the owner's best interest to present interpretations of construction behavior that might define or limit construction means and methods. Consequently, baseline statements presented in an owner developed baseline report should clearly note the construction condition, if relevant, that the baseline relates to, such as "flowing ground in an unsupported heading", and should also note that this assumption in no way implies approval or is meant to constrain the design-builder. Ultimately, the design-build teams should be asked to provide their own interpretations of the exploration and laboratory data, since the contractor has a much greater responsibility for selection of the means and methods of construction. The owner should have an opportunity to comment on and amend the contractor's baseline interpretations.

Also during the proposal phase, the various bidders should be asked for their review comments and input on the content and extent of the explorations. Additional explorations may be suggested by the bidders, dependent on their design and construction concerns. Furthermore, the selected design-build team may, through their further assessment of the project conditions, propose additional borings and tests. When possible, and depending upon schedule, it is desirable to perform these additional explorations prior to the design-build teams final design scope and construction bid. This will help to reduce claims for changed conditions later in the project.

7. Conclusions

Design-build has only been used for the construction of a small number of large tunnel projects in the United States and Europe in the last 10 to 15 years. Consequently, this contracting technique is still being refined and tested in the underground environment. The primary advantage of design-build is a potential savings in schedule, which the owner pays for with reduced control over the final product and potentially higher negotiated final bid prices. Thus far, many of the major design-build tunnel projects have been significantly over schedule and have experienced cost increases of 25% to 100%.

7.1 Murphy's Laws live underground

Tunnel construction frequently reveals the unexpected and few, if any, tunnels have been constructed without some surprises. Many of the same risks exist for both the design-bid-build or design-build approaches. These risks include: 1) the potential for changed geologic and groundwater conditions, 2) potential occurrences of contaminated groundwater or soil along a long linear alignment, and 3) third-party impacts such as unknown utilities, settlement induced damage to structures, and delays for property procurement and permit acquisition. Since it is a general tenet of U.S. and many European construction practice that the project owner retains ownership of the ground (including third-party issues), then these risks are not easily transferred or eliminated by shifting to the design-build contracting process. A fourth potential risk in design-build projects is the risk that the owner's 10 to 30% design may contain significant design flaws that will be exposed by the more thorough design process of the design-build team and result in significant additional costs and schedule delays.

7.2 Risk reduction contracting

To reduce these risks, whether a tunnel project is constructed using design-build or the more standard design-bid-build approach, the owner is encouraged to undertake a thorough geotechnical investigation, combined with a complete assessment of utilities and structures along the alignments and a determination of the permit requirements that will impact design and construction approaches, schedule, and cost. The geotechnical data should be presented in a factual GDR. The interpreted conditions upon which the bidders are to rely along with other geotechnical baseline issues should be presented in a GBR, which should avoid discussion of means and methods for design and construction. The contract documents should contain a DSC clause in recognition of the possibility of unrecognized ground conditions that may warrant a change in construction approach and the associated project cost and schedule. Ideally, the bid documents of the successful bidder should be retained and stored with a third-party, such as a bank, and referred to when resolving disputes. Disputes should be presented to a DRB established jointly by the Owner and Design-Build Team. Utilizing these up-front precautions, should help to appreciably lower the risk in using any of the contracting approaches.

References

[1] American Consulting Engineers Council (ACEC). 1992. Board of Directors. October.

[2] Brierley, G. and Smith, G. 1998. Going Under? How about Urban Design/Build! *World Tunneling*. November. p. 42-46.

[3] Engineer's Joint Contract Documents Committee (EJCDC). 1995. Guide to Use of EJCDC Design/Build Documents. Issued by American Consulting Engineers Council, National Society of Professional Engineers, and American Society of Civil Engineers.

[4] Engineering News Record (ENR). 2001. Rising Costs, Slipping Schedule Have Tren *Urbano in Hot Water*. September 17. p. 15.

[5] Gay, M, G. Rippentrop, W.H. Hansmire, and V.S. Romero. 1999. Tunneling on the Tren Urbano Project, San Juan, Puerto Rico. 1999 Rapid Excavation and Tunneling Proceedings.

[6] Gildner, J.P and Borst, A.J. 1999. Sound Move-Seattle's Light Rail Continues Underground. 1999 Rapid Excavation and Tunneling Proceedings.

[7] Jaffe, M.E. and C. Goode. 1992. Chapter 9 – Allocation of Risks Between Designer and Builder. Design-Build Contracting Handbook. Edited by R. F. Cushman and K.S. Taub. Wiley Law Publications.

[8] Lemley, J.K. 1991. Channel Tunnel – Overview and Contractural Arrangement. 1991 Rapid Excavation and Tunneling Conference. p. 739-749.

[9] Loulakis, M.C. 1992. Chapter 1 - Single Point Responsibility in Design Build Contracting. Design-Build Contracting Handbook. Edited by R.F. Cushman and K.S. Taub. Wiley Law Publications.

[10] Mansfield, A.J. 1996. Chapter 31- Disputes, Arbitration and the Geotechnical Engineer. Engineering Geology of the Channel Tunnel, C.S. Harris, M.B. Hart, P.M. Varley and C.D. Warren, Editors, Thomas Telford House. p. 467-471.

[11] Moses, T., P. Witt, and F. Frandina, 2000. Two-In-One Tunnel. *Civil Engineering*. April. p. 48-53.

[12] PSMJ Resources, Inc., 1995 PSMJ's Book of Design Build Contracts.

[13] Reina, P. 2000. Tunnel Vision. Design-Build. McGraw Hill Construction Information Group, December.

[14] Tunnels & Tunneling, North America. 1999. *Design-Build to Control Costs?*. Vol. 2. November. p. 5.

[15] Underground Technology Research Council (UTRC), Technical Committee for Geotechnical Reports 1997. Geotechnical Baseline Reports for Underground Construction, ASCE.

[16] U.S. National Committee on Tunneling Technology (USNCTT). 1974. Better Contracting Practices for Underground Projects. National Academy of Science.

[17] U.S. National Committee on Tunneling Technology (USNCTT). 1984. Geotechnical Site Investigations for Underground Projects. National Academy of Science.

[18] Varley, P.M., C.D. Warren, W.J. Rankin, and C.S. Harris. 1996. Chapter 8 – Site Investigations. Engineering Geology of the Channel Tunnel, C.S. Harris, M.B. Hart, P.M. Varley and C.D. Warren, Editors, Thomas Telford House. p. 88-117

Cost and schedule management for major tunnel projects with reference to the Vereina tunnel and the Gotthard base tunnel

Felix Amberg[1], Bruno Röthlisberger[1]

[1] Amberg Engineering AG, Rheinstrasse 4, P.O. Box 64, CH-7320 Sargans, Switzerland, Tel. +41 81 725 31 13, Fax +41 81 725 31 02, e-mail: famberg@amberg.ch, URL: www.amberg.ch

1. Introduction

Cost and time management are decisively important factors for all major projects, and this statement is equally true of underground construction ventures which are currently being planned or executed in Austria or Switzerland. Especially in the underground construction sector, projects have often been completed with substantial additional costs in the past, making it clear that efficient cost management is a necessity.

Projects currently under way in Austria include the Brenner Base Tunnel, the Wienerwald (Vienna Woods) Tunnel and the tunnels in the Perschling Valley, while in Switzerland the Gotthard and Lötschberg lines are currently under construction by AlpTransit. In Spain, there are the forthcoming major projects for the high-speed railway network and in France there is the Mont-Cenis Tunnel, etc. However, major underground projects are not limited to mountainous regions - expanding cities and agglomerations also require underground infrastructures in order to cope with increasing traffic flows. Projects such as the Madrid Metro, the Citytunnel project in Sweden and the Betuwe Line in the Netherlands may serve as examples here.

Especially in present times when public-sector resources are in short supply, engineers have a duty to make economical use of the available funds. There is no doubt that efficient cost and time controlling which is appropriate to the project has a crucial role to play in this respect.

Using some current examples from Switzerland, we shall demonstrate which means are being used for this purpose at present.

However, one point must be clearly emphasised at this stage: no matter how much the controlling is optimised, it will never compensate for incorrectly applied technology, an inadequate project or slipshod execution. Nevertheless, properly functioning cost and time management - as a management tool for project handling - can ensure that those responsible for the project or the clients/owners are informed promptly if costs exceed or indeed fall below the budgeted amounts.

Cost management or cost controlling also involves diligent administrative work in order to track the development of costs on a major project with the minimum possible time lag, i.e. in parallel with the progress of the actual construction work.

If it appears that costs are going to exceed the budgeted levels, one action to be taken is certainly to have the project reviewed by the project partner or the body responsible for supervising the execution. In a situation of this sort, the client should have the options of redimensioning the project, of blocking certain parts of the work, or - in the event of a disaster - of being able to apply the "emergency brakes" in good time. Conversely, if costs are evolving positively, sections of the building work which have deferred or parts of the project which have been redimensioned must be "unblocked" without delay.

As an example of the basic task of cost controlling, let us quote the Swiss **Neat Controlling Instructions**:

Cost controlling aims to ensure that those carrying out the construction keep to the cost benchmarks (target values). This should guarantee universal cost transparency in all cost stages, and early identification of cost variances.

Cost controlling should be implemented as appropriate to the specific times and stages, and also on a universal basis. Moreover, the cost controlling data are used as the basis for project-related financial management.

The basis for perfectly executed cost control is a realistic set of proposed costs with a clearly defined price basis. From this key date onwards, the "grey inflation" must be defined with the help of an inflation rule which is appropriate to the sector or sectors involved.

When construction starts, tasks are executed on the basis of countersigned work contracts. Each individual work contract stipulates the key date (base date) for calculating inflation. In Switzerland, the OIV inflation method (based on so-called KBOB (Conference of Federal Construction Organisations) indices) is used for all work contracts for tunnels, including the Gotthard and Lötschberg Base Tunnels. More will be said on this subject later.

Additions to work contracts must be incorporated into the performance schedule of the contract on the same price basis as the original work contract to which they apply.

Finally, two examples from a bygone era of cost control for tunnel projects should indicate the crucial importance of cost controlling. This importance becomes especially evident - with exemplary clarity - whenever fundamental mistakes are made.

- **The Furka Tunnel**

  In 1971, the Federal Parliament approved a credit of CHF 75 million for the completion of the Furka Tunnel (length: 15.4 km).

  Construction work was started on both the portals in 1972.

  Thereafter, a total of 3 follow-up credits amounting to CHF 150 million were approved by the Federal Parliament. On each occasion, serious consideration was also given to abandoning the project.

  Before any thought was given to effective and efficient cost controlling in those days, the geology had to be analysed again, the project had to be revised and drastic adjustments had to be made to the contractor's installations. Based on realistic geology which was not guided by wishful thinking, the project was revised to bring it into line with this geology and the available financing. This again resulted in adjustments to the contractor's installations.

  An enormous amount of manual labour was then expended to build up a cost control system on this basis, and the system was then successfully deployed as a management tool for the remainder of the construction period.

- **Galleria Centovallina**

  The underground entrance of the Centovalli Railway into the town of Locarno was approved by the Federal Parliament in 1985 with a credit of CHF 57 million.

  There was an accumulation of negative influences on this venture: the construction project was slimmed down so as to keep it below CHF 57 million, geological conditions proved to be worse than expected, the construction boom in the Sopraceneri region went into overdrive, and supervision of the construction work was incompetent. In 1989, these factors literally led to a standstill on the construction site. In the same way as happened with the Furka Tunnel but with EDP incorporated to a greater extent, the project had to be revised and an appropriate cost control system had to be developed for it.

These two projects - the Furka and the Galleria Centovallina - had many features in common which were ultimately responsible for the substantial overshooting of budgeted costs:

- both projects were the subject of major political arguments during their planning phases. In response to political pressure, the cost estimates for both the projects were forced right down in order to achieve political acceptance, i.e.:
- the geology was interpreted favourably or even optimistically;
- the project was played down in every respect;
- those involved in the project were sometimes vastly overtaxed by the tasks assigned to them;
- no cost controlling as such was undertaken, and it had to be implemented at great expense during the actual construction period.

However, it has also been possible to complete very many major projects within the scope of the proposed costs, or even below them.

Using the Vereina Tunnel as an example, ladies and gentlemen, we shall now present a cost controlling system which has enabled a demanding construction project to be completed below the proposed costs. Before we do so, however, let us single out one other special feature of cost controlling for construction projects that involve long durations, as is usually the case when major tunnels are being built. This relates to inflation. During the period in which long-term construction projects are completed, there will inevitably be an increase in costs due to influences beyond the construction project as such. The planners and contractors involved in the construction project must be compensated fairly and appropriately for these additional costs, which they are unable to influence. Various approaches are available for this purpose, and these will be explained at a later stage.

2. Cost controlling

2.1 The Vereina project

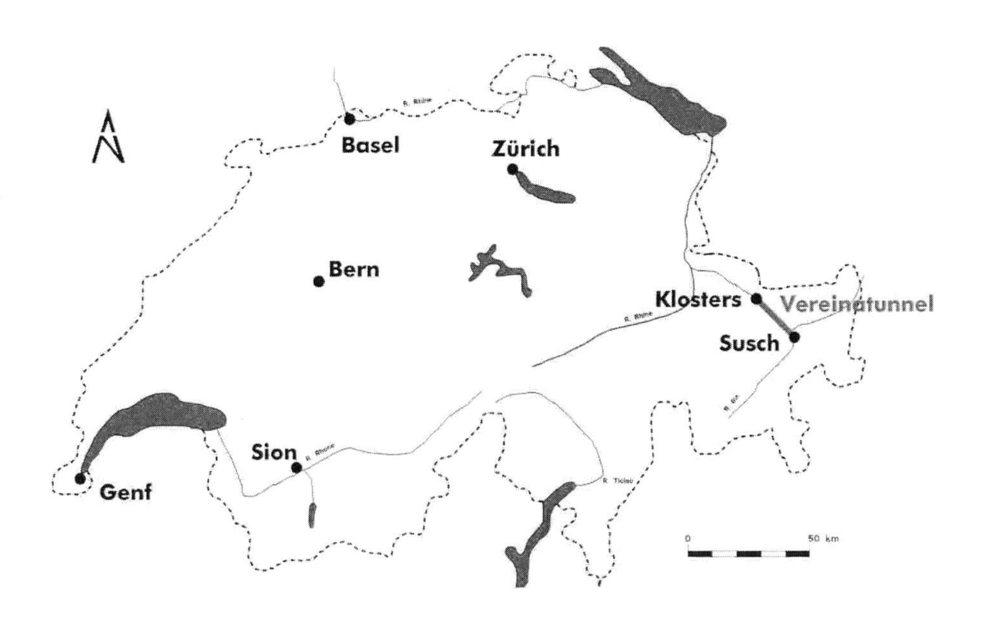

Figure 1: Situation

The Vereina Project comprised the adaptation of the existing railway network of the Rhetic Railways in eastern Switzerland. The project was located 1,200 m above sea level, and it included bridges, stations and car transportation facilities on trains for vehicles up to 4 m high.

At the heart of the Vereina Line is the Vereina Tunnel (length: 19,050 m). The overburden is as much as 1,500 m. The tunnel climbs at a gradient of 1.4 % from Klosters in the north to a point 1,463 m above sea level, and then it drops down to the south portal at Susch with a gradient of 0.5 %.

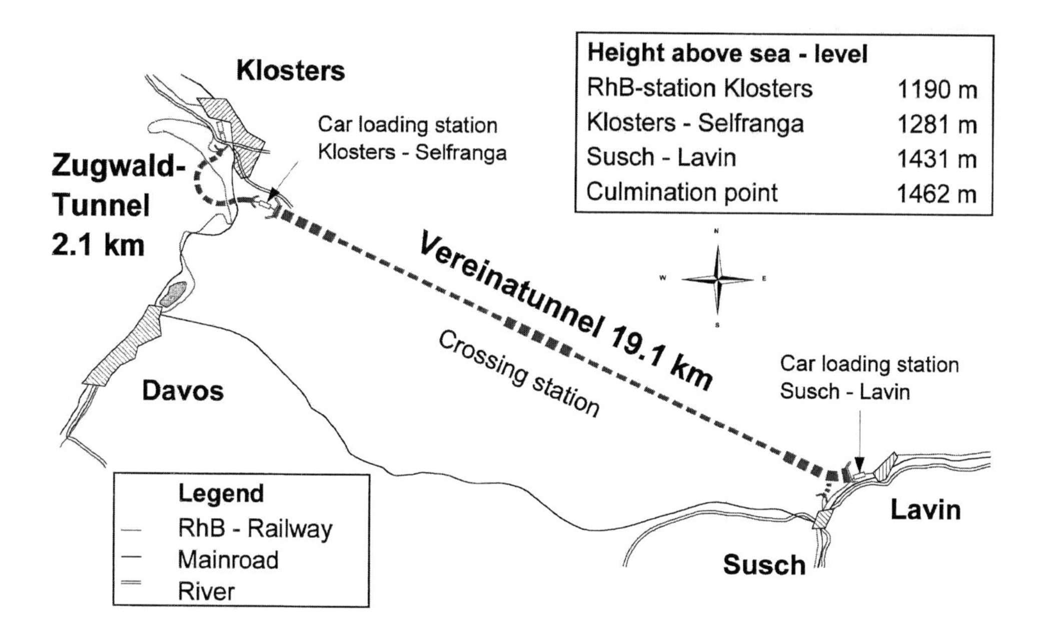

Figure 2: Project overview

The Vereina Tunnel is basically a single-track tunnel, but twin-track and triple-track sections in the middle and at both ends (length: approx. 2 km) allow trains to pass one other. The driving work was carried out by drilling and blasting in the twin-track and triple-track sections (with cross-sections from 70 to 85 m^2 and 135 m^2 respectively) and the same method was used for the single-track section on the south side (with a cross-section of 39 to 42 m^2); a TBM was used for the single-track section on the north side (with a cross-section of 46 m^2).

The Vereina Project involved overall construction costs of US$ 360 million (price base: 1987). Work on the north side of the Vereina Tunnel began in April 1991, and a start was made on the south side in October of the same year. The TBM heading work in the Vereina Tunnel started in May 1995 and was finished with the breakthrough in March 1997. An open TBM was used, with rock bolts, sprayed concrete, steel arches and meshes as rock support. After the breakthrough, the central crossing station was excavated, the tunnel lining was finished and the tracks were laid. In 1999, the railway installations were completed, test runs were carried out and the tunnel went into operation on November 19, 1999 - more than half a year ahead of schedule. What is more, the costs were considerably less than estimated.

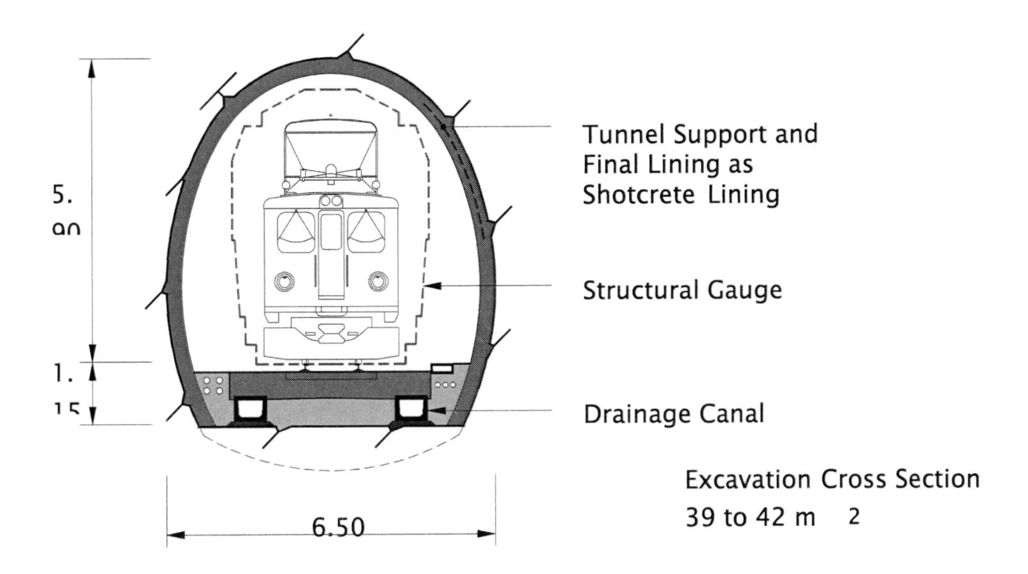

Figure 3: Single-track section, Vereina south lot

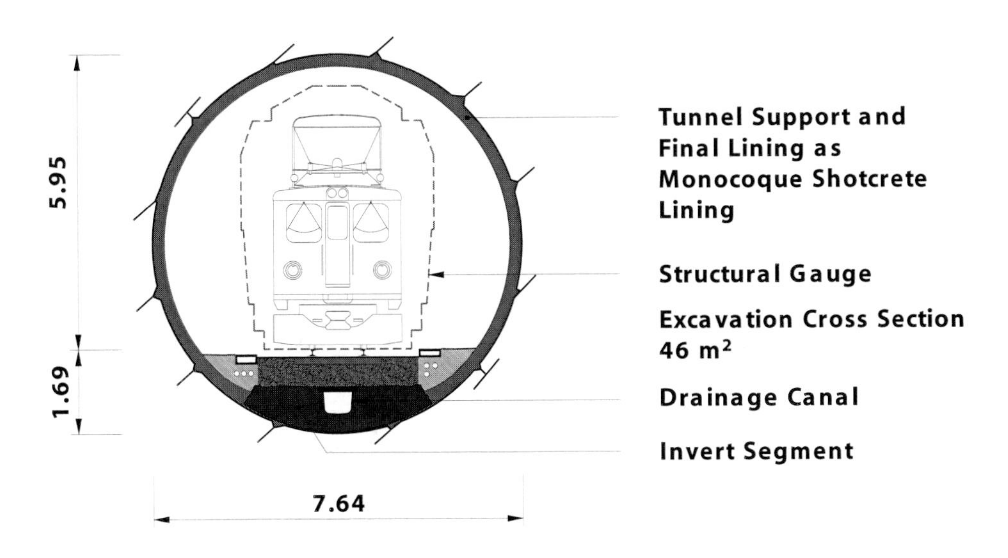

Figure 4: Single-track section, Vereina north lot

2.2 Basic principles

For the Vereina Tunnel, the basis for cost controlling was also a cost estimate, partially based on quoted prices with a clearly defined price basis. From this key date onwards, the "grey inflation" had to be specified using an inflation regulation that was appropriate to the sector. When construction started, work was carried out on the basis of countersigned work contracts, each of which stipulated the key date for calculating inflation. KBOB indices were used for all the tunnelling work contracts for the Vereina Tunnel.

Cost estimate

The cost estimate is drawn up by the planning engineer. The way the estimate is determined and its accuracy are also adapted to the respective planning status.

In pre-project status, the accuracy is +- 25% and the costs are usually determined with the help of experience-based values. During the construction project phase (which forms the basis for the invitation to tender), the accuracy of cost determination is better than +- 10%. In order to achieve this level of accuracy, detailed calculations are required for the amounts of the major cost items in particular (excavation, rock support, invert, lining, insulation, installations, etc), taking account of current material prices, wages and machinery costs. This in turn means that the project must be handled in a manner that is suitable for this purpose.

Inflation regulations appropriate to the sector

An 'inflation regulation appropriate to the sector' means that the settling of the payments for the inflation should take account of underground construction sites in general as a specific category of its own within the construction industry. In terms of inflation, it matters a great deal whether (for example) the tunnel is driven by means of drilling and blasting with a large number of people employed, or whether a highly mechanised TBM drive is used calling for considerably less staff but requiring a large number of expensive machines instead. The development of inflation for wages may be subject to completely different natural laws and tendencies than those applying to machinery, spare parts and fuel. An inflation regulation that is appropriate to the sector attempts to take account of this fact.

Grey inflation

"Grey inflation", as it is known, is used to adjust cost estimates that were compiled on a specified date so as to bring them into line with new key dates, without having to recalculate all the details of the estimate. Grey inflation is also used if the construction work is not started immediately after the work contract is signed. This can happen, for example, if legal procedures have not yet been completed and the work cannot be started. If a phase of this sort carries on for a lengthy period, the contract is subject to inflation at a rate which is not precisely known because no work has actually been carried out.

KBOB indices

KBOB is the abbreviation for the 'Conference of Federal Construction Organisations' (or 'Konferenz der Bauorgane des Bundes' in German): this government body specifies the inflation indices, taking account of the general development of inflation. Indices for wages, materials and machinery, etc are individually studied and specified. The effect of this procedure is that clear base values which are binding on all parties are applicable for inflation calculations.

Groups of similar grey inflation on the Vereina Tunnel project

The total costs of the Vereina Line were divided into 4 inflation groups in order to calculate the grey inflation.

Cost group 1	Planning / construction management / ancillary project services with fee rates as per KBOB, subdivided into fee categories A to G
Cost groups 3, 4 and 51	Underground and overground construction work, also superstructure, using the Zurich Index for Residential Construction Costs (this is a collective index which is calculated periodically on the basis of the KBOB indices)
Cost group 5	Technical rail installations, using ASM wage indices
Cost group 7	Rolling stock, using ASM wage indices and wholesale price indices for metal and metal products

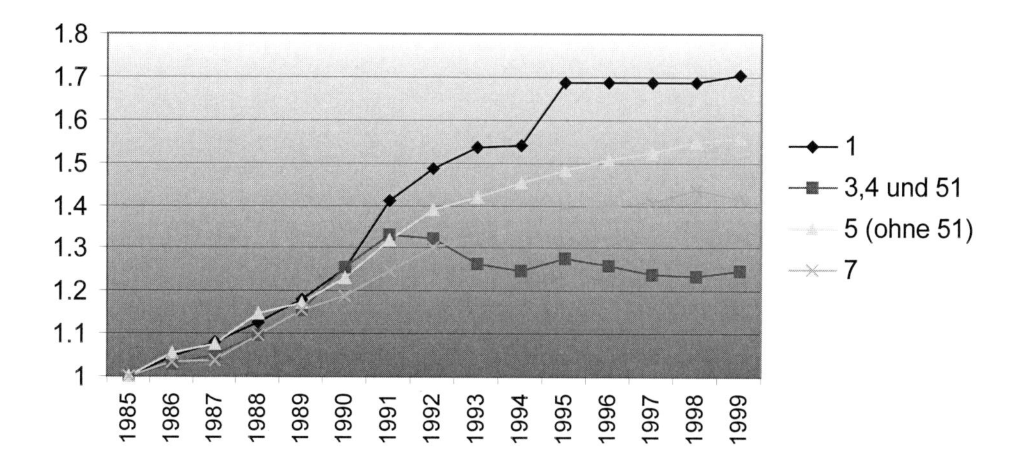

Figure 5: Grey Inflation as per the BAV (Federal Transport Authority) disposition

As the chart shows, inflation developed at rates of between 26% and 69% from 1985 until 1996.

Indexed inflation while construction is in progress

Contractors submitting projects appended the basic elements as per SIA Recommendation 121, "Charges for Price Changes using the Object Index Method" to each bid for the Vereina Tunnel. As part of the bidding procedure, the contractors had to examine form 1021/1 and enter the envisaged sector code on form 1021/3. In each case, the 0-quotas were adjusted during the negotiations with contractors.

Recommendation 121 of the Swiss Association of Engineers and Architects (SIA) describes a method for calculating inflation on construction projects. This method represents a simplification of the so-called "proof of quantities method". The basis of this method is that inflation is compensated according to quantities used in a specified period and the inflation occurring therein.

The 'Object Index Method' simplifies the effort required to determine inflation by formalising the inflation calculation and relating it directly to the part payments.

In construction lot T4, of course, the calculation was undertaken using the distribution for a mechanical driving operation which is typical for the sector.

VEREINALINIE

OBJEKT - INDEX STAND FEBRUAR 1993 FUER DAS BAULOS T4

SIA 1021/3 Ermittlung des Objekt-Indexes

Objekt / Arbeit :	Zugwald- + Vereinatunnel Los 4	Bauherr :	Rhätische Bahn
Unternehmer :	Stuag/ Züblin/Frutiger/Jäger/Vetsch/Bordoli	Angebot vom :	29.05.90
Beilage zur Rechnung Nr. vom :		Periode vom :	bis :

Kostenarten		Anteile		Index- oder Preisstand			Ueber-wälzung	Objekt-Index-Anteil
		in 1000 Fr.	in% von P	Basis Datum (Stichtag)	Periode vom : 01.02.93 bis :	Aende-rung		
P	Angebotspreis / Vertragspreis, ohne Abzug für Skonto		100.00					
F	Abzug für Risiko, Verdienst / Verlust, WUST		8.12					
S / R	Selbstkosten		91.88	01.01.90	28.02.93	%	%	%
O	Kostenarten mit objektabhängiger Verteilung (objektabhängige Kosten)							4*7*8
1 Index- Bez.	2 Kostenarten	3	4	5	6	7	8	9
2.44	EL ENERGIE TBM	2'319.8	1.16	237.3	264.9			
05.121.1	BAUHOLZ	177.4	0.09	173.0	175.5			
7.4	KUNSTSTOFFWAREN	5'624.8	2.82	140.3	144.7			
08.S.12	SPRENGSTOFFE	280.7	0.14	146.1	161.6			
08.S.13	ZUENDMITTEL	174.7	0.09	447.0	524.9			
08.BTC 2	BAUTECH. CHEMIKALIEN	8'676.2	4.36	226.4	263.6			
09.111	BETONKIES	834.1	0.42	353.4	433.5			
09.112	SAND	1'176.6	0.59	335.0	408.8			
09.115	ZEMENT	5'076.7	2.55	194.1	224.0			
09.132	ZEMENT- und BETONWAREN	13'992.2	7.02	260.2	314.6			
10.121.33	BETON-ARMIERUNGS- STAHL	1'159.4	0.58	99.8	56.4			
10.121.34	ARMIERUNGSNETZE	1'310.4	0.66	136.9	131.1			
10.121.4	PROFILEISEN	2'714.5	0.15	144.2	104.8			
10.121.6	STAHLBLECHE	298.7	0.15	167.7	131.1			
10.152	MITTL + KL SCHMIEDSTUECKE	6'022.7	3.02	112.6	113.6			
ASTAG	STRASSENTRANSPORTE	768.1	0.39	100.0	113.2			
SBB / RHB	BAHNTRANSPORTE	1'993.6	0.64	100.0	111.7			
09. X	Baumaterialien	1'272.5	0.64	202.9	222.5			
O	Total O	53'873.1	25.47					
Kostenarten mit spartentypischer Verteilung		Spartenschlüssel,S = 100						
Sp / Cat	Total Sp / Cat = S / R - O	100	66.41					
	Personalkosten	66	43.83	100.0			95	
	Kapitalkosten (Zinsen)	4	2.66				0 / Art. 5	
	Amortisation	10	6.49				0 / Art. 5	
	Ersatz- und Verschleissteile, Betriebsmaterial	5	3.32	112.6			0 / 50 / 100	
	Dieseltreibstoff 02.22	3	1.94	264.7			100	
	Mineralschmieröl 02.5	2	1.33	217.9			100	
	Elektrische Energie 02.44	2	1.33				0	
	Uebrige Kosten	8	5.31				0	
Total der Anteile ohne Teuerungsüberwälzung, inkl. 5% Anteil von Personalkosten			29.42	Objekt - Index %				

Massgebender Betrag der in der Rechnungsperiode geleisteten Arbeiten (ohne Abzug für Skonto)	Fr.	Preisänderung ±	Fr.
		WUST %	Fr.
		Total Rechnung	Fr.

Instanzen	Unternehmer	Oertliche Bauleitung	Oberbauleitung
Unterschriften			
Datum			

2.3 Recording the development of costs

Within each work contract, the cost comparisons were calculated back using the price basis for the work contract in question. For the project as a whole, the year 1985 was used as the cost basis.

As a fundamental principle, any cost controlling system on a project in progress is only as good as the quality of the daily quantity assessment. For this reason, it was an essential requirement on all the construction lots of the Vereina Tunnel for the quantities of both all the excavation and rock support works and the final lining work to be assessed by mutual agreement with countersignatures.

Any pending issues regarding quantity assessment or accounting were quickly cleared up at construction meetings. On occasion, quantities missing in the work contracts were incorporated into the contracts by means of supplementary bids, which also meant that these quantities were included in the cost controlling system. This was the only possible reach the target to enable settlement of the accounts "net of all claims" for the construction lots 30 days after the contractors withdrew from the construction site.

The following graph shows the set up of the cost controlling system in graphic form.

Row 1) Costs estimation including additional project components, price basis:1985

Row 2) Costs for electric current (set at disposition of the client)

Row 3) Internal Costs (part of the specific contract)

Row 4) Costs of Work contracts lot T1 to T5 including cost-plus work, unforeseen and miscellaneous items

Row 5) Awarded cost amount for construction lot T4, price basis:1990

Row 6) Cost for sub projects parts (here lot T4a a and T4b)

Row 7) Costs in a specific sub project by sub categories

Row 8) Costs of sub categories by main working

Row 9) Effective Costs of driving, rock support and final lining work

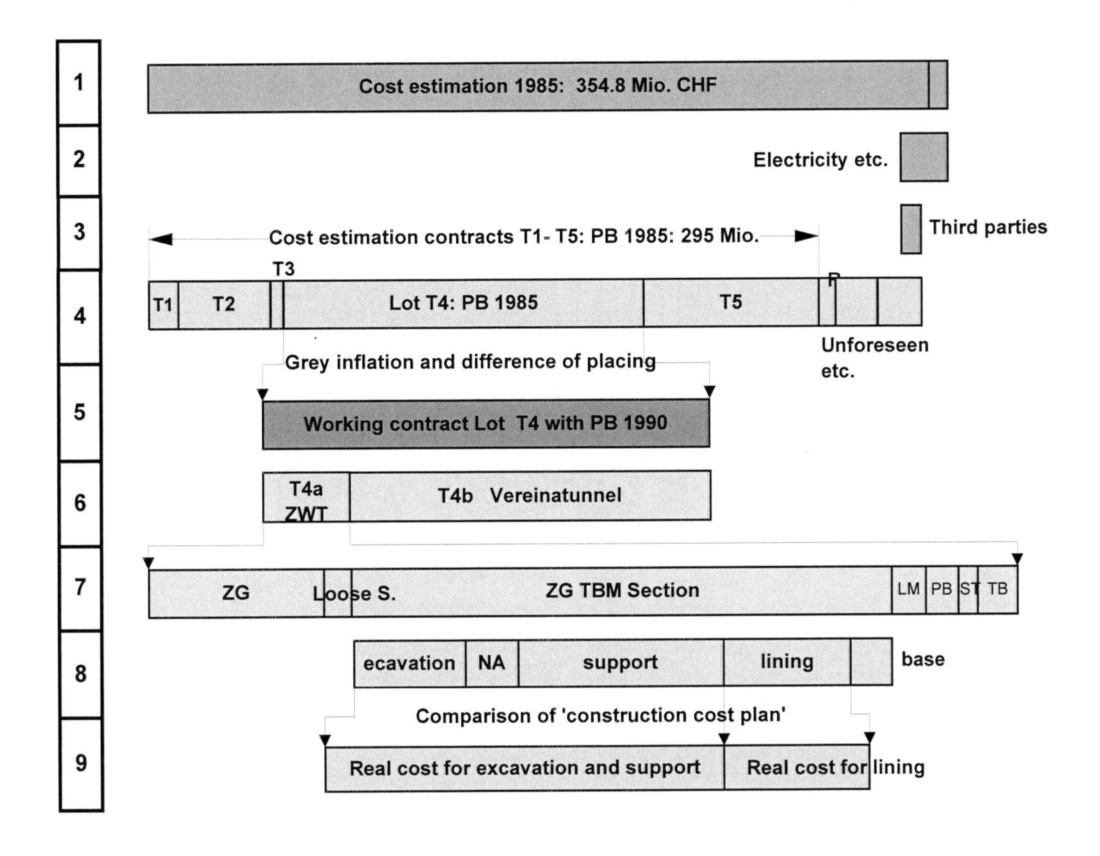

2.3.1 Cost controlling for driving work

The costs of driving and rock support work are clearly the decisive factor in the development of costs on tunnel construction sites. They also represent the largest cost risk. Therefore special attention must be paid to an adequate cost estimation in the planning phase and to a strict cost controlling during the carrying out of the works.

In construction lot T4b, the mechanical driving work for the Vereina Tunnel, excavation classes I-VII and attrition classes VI-V5 were assigned to the geological formations in the longitudinal geological profile. A summary of this distribution is shown below. The corresponding boring classes A-E were also assigned to the excavation classes.

21 %	2,214 m	I	**Summary of excavation classes for construction lot T4b**
56 %	5,740 m	II	
14 %	1,492 m	III	
5 %	478 m	IV	
2 %	218 m	V	
1 %	123 m	VI	
1 %	120 m	VII	
10 %	1,092 m	VI	**Summary of attrition classes for construction lot T4b**
20 %	2,029 m	V2	
35 %	3,631 m	V3	
20 %	2,105 m	V4	
15 %	1,528 m	V5	

	A	B	C	D	E	
I	130 m	970 m	500 m	314 m	300 m	**Summary of excavation and boring classes for construction lot T4b**
II	470 m	650 m	2,350 m	1,150 m	1,120 m	
II	240 m	225 m	402 m	525 m	100 m	
IV	70 m	80 m	220 m	70 m	38 m	
V	60 m	60 m	80 m	18 m	---	
VI	30 m	50 m	43 m	---	---	
VII	38 m	42 m	40 m	---	---	
	1,038 m 10 %	2,077 m 20 %	3,635 m 35 %	2,077 m 20 %	1,558 m 15 %	**Total of boring classes**

The entire tunnel was split into sections of 200 m for cost controlling purposes.

The driving costs, with the preliminary dimensioning bases and the work contract prices were calculated for each of these sections.

The following subdivision was made:

- excavation costs
- attrition costs
- rock support costs, such as rock bolts, reinforcement meshes, inbuilt steel, base tubbing
- special measures such as preliminary drilling, injection
- pre-sealing

A form adapted for the purposes of the construction lot was used to record the daily quantities each day, after which these data were initially transferred to the construction record plan. The amount of rock support work (based on sections of 10 m) was used to define the excavation classes, and the penetration measurements were used to define the boring classes (see page 292).

Weekly cost controlling

Up-to-date processed figures from the driving operations are a decisively important management tool for cost controlling; in the case of the Vereina Tunnel, these were the North and South drives. No purpose at all is served by collating the driving costs weeks or months afterwards. This makes it impossible for the site supervision management to exert any influence whatsoever.

In the case of the Vereina Tunnel, the driving reports for the preceding week were available every week, i.e. by 12.00 noon every Monday, enabling the site supervision management/senior construction Management to influence the development of costs immediately.

The example of construction lot T5 shows the driving performance of 55.0 m achieved in week 43 in 1996 (see page 293). The average excavation class for the reporting week is entered in the second block. The third block compares the costs in tunnel section 6,910 to 6,970, i.e. over 60 continuous meters. The effective costs consist of the daily quantities, and the work contract costs are an average value for the single-track section.

The bottom block also summarises the costs for this section.

Vereinatunnel, TBM–Strecke Tm 5'300 bis 5'500

Ausbruchspreise (Var. 1)

Ausbruchspreis in Fr./m'	Ausbruchsklasse						
	I	II	III	IV	V	VI	VII
Bohrklasse A	1'785.–	1'925.–	2'200.–	2'690.–	3'970.–	7'500.–	16'830.–
Bohrklasse B	2'040.–	2'240.–	2'550.–	3'260.–	5'000.–	9'020.–	16'830.–
Bohrklasse C	2'085.–	2'295.–	2'630.–	3'425.–	5'000.–	9'020.–	16'830.–
Bohrklasse D	2'600.–	2'760.–	2'930.–	4'460.–	6'265.–	16'600.–	–
Bohrklasse E	2'815.–	2'995.–	3'200.–	4'880.–	8'790.–	21'650.–	–

Verschleisskosten (Var. 2)

Verschleissklasse	V1	V2	V3	V4	V5
Verschleisskosten in Fr./m'	197.–	394.–	661.–	985.–	1'478.–

Mittlere Laufmeterkosten gem. Werkvertrag

TBM–Profil 1.VN	Tm 5'300–5'500
TBM–Vortrieb nach Ausbruchsklassen	2'548.–
Werkzeugverschleiss	749.–
Ergänzungsleistungen u. aussergew. Vortrieb	185.–
Ausbruch	**3'482.–**
Felssicherung inkl. Tübbing	**2'561.–**
Bohrungen, Injektionen etc.	374.–
Ortsbetongewölbe armiert	112.–
Sondermassnahmen	**486.–**
Entwässerung/Abdichtung	**207.–**
	6'736.–

Abschnittkosten Tm 5'300 bis Tm 5'500

TBM–Profil 1.VN	Effektiv	Werkvertrag	Differenz
Ausbruch	627'347.–	696'400.–	– 69'053.–
Felssicherung	344'278.–	512'200.–	142'470.–
Tübbing/Sohle	310'392.–		
Sondermassnahmen	––	97'200.–	– 97'200.–
Entwässerung/Abdichtung	––	41'400.–	– 41'400.–
Abschnittkosten Total	1'282'017.–	1'347'200.–	– 65'183.–

Driving costs.

VEREINATUNNEL SÜD **BAULOS T5**

Vortriebsbericht Woche 43 / 1996

06.11.96

Vortrieb 1-Spur-Strecke im Vollprofil

Wochenleistung			
Vortriebsstand Montag	28.10.96	07 Uhr:	Tm 7042.0
Vortriebsstand Montag	21.10.96	07 Uhr:	Tm 6987.0
Wochenleistung			Lfm 55.0

Vortriebsleistung pro Arbeitstag (AT)	
Anzahl ganze Arbeitstage der Betriebswoche	5 AT
Mittlere erreichte Vortriebsleistung	11.0 m / AT
Mittlere Ausbruchsklasse: A II	
Vertragliche Vortriebsleistung für obige Ausbruchsklassen	10.45 m / AT

Kostenkontrolle Ausbruch / Felsicherung

Betrachtete Tunnelstrecke: Tm 6910 bis Tm 6970 d.h 60 Lfm

Vollprofil Profil 1. VS	Effektive Kosten	Werkvertrag (1)	Differenz
Ausbruch	124'680.-	167'040.-	-42'360.-
normale Felssicherung	61'029.-	94'020.-	-32'991.-
Zwischentotal	185'709.-	261'060.-	-75'351.-
Sondermassnahmen	4'320.-	27'300.-	-22'980.-
Entwässerung / Abdichtung		6'420.-	-6'420.-
Total für 60 Lfm	190'029.-	294'780.-	-104'751.-
Mittlere Laufmeterkosten (2)	3'167.-	4'913.-	-1'746.-

(1) Die angegebenen Werkvertragskosten basieren auf dem Mittelwert der ganzen 1-Spur-Strecke

Der Vortrieb der betrachteten Tunnelstrecke erfolgte in den Ausbruchsklassen A III. Die effektiven mittleren Laufmeterkosten (2) stehen im Vergleich zu den Werkvertragskosten von Fr. 3'769.- für diese Ausbruchsklasse.

Bisheriges Bauteil-Total	19'706'776.-	23'838'970.-	-4'132'194.-
Allfällige Korrektur			
Total für 60 Lfm	190'029.-	294'780.-	-104'751.-
Neues Bauteil-Total	19'896'805.-	24'133'750.-	-4'236'945.-
Mittlere Laufmeterkosten	4'001.-	4'853.-	-852.-

Driving reports for the preceding week 43.

2.3.2 Cost controlling for the final lining work

The Vereina Tunnel was built as a monococque structure with shotcrete, meaning that:

- the rock support represents a load-bearing component of the tunnel lining, and it must be taken into account when determining the thickness of the final lining
- the thickness of the lining is variable throughout the entire length of the tunnel, and it forms one monolithic structure together with the rock support.

In order to determine the dimensions for the final lining work, use was made of the geological records, the rock support measures from L1 and L2 which were documented in construction record plans, and also the evaluations of the carried out rock mechanic measurements.

200 m section plans were used to show the resources and costs specified for the final lining work. The daily recording of quantities and the weekly mass quantity checks in the relevant sections made it possible to carry out a very accurate review of the quantities and hence the costs as well.

2.4 Quarterly cost controlling report

The quarterly cost controlling report was compiled for the attention of the client, the BAV (Federal Transport Authority) and the Federal Financial Controlling Service, with comments as appropriate. Also, cost comparisons for the individual construction lots were made between effectively charged costs as of the key date and costs as per work contract. Any difference between the work contract total and the engineers project cost estimation was taken into account separately as a constant.

All the cost information in the quarterly report was calculated back to the 1985 price basis using the grey inflation index.

The costs of excavation and rock support in the two driving operations (North and South) developed as follows (see following two graphs):

In construction lot T4b the drive proceeded as far as Tm (tunnel meter) 4,600 with substantially higher excavation classes. Consequently, this resulted in additional costs of several millions.

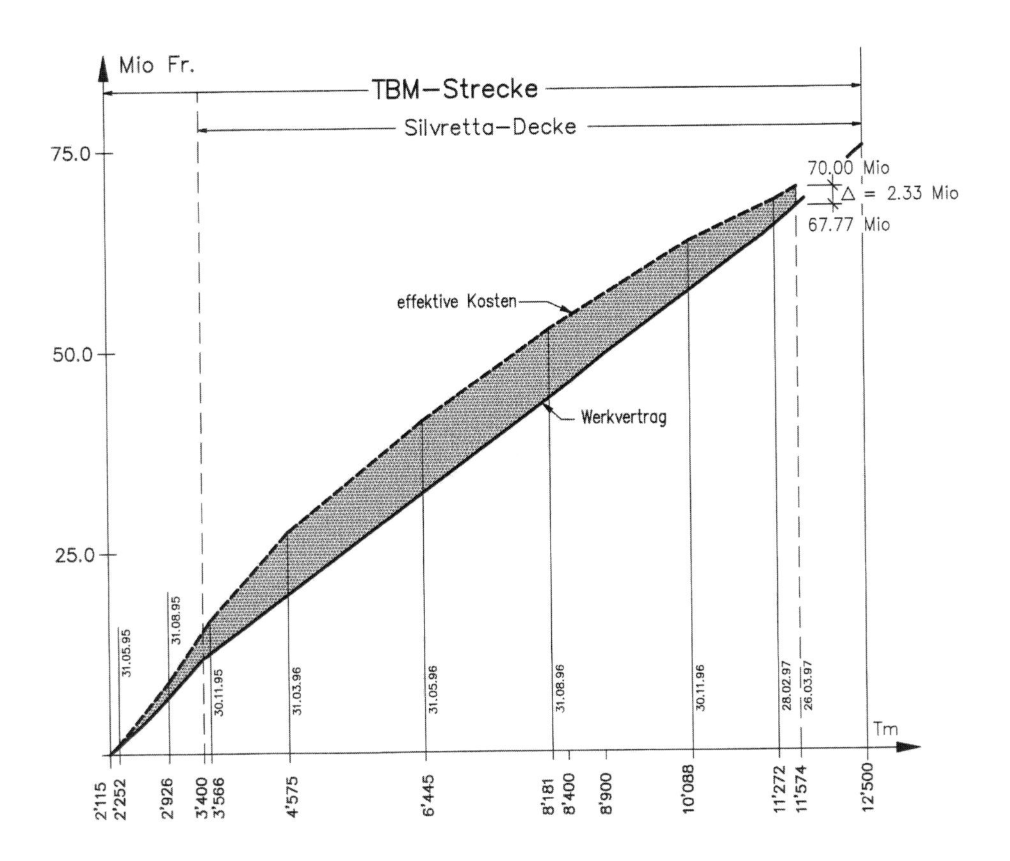

Once the Silvretta crystalline was reached, the driving costs developed in parallel with the forecast. Thanks to better rock conditions between Tm 8,000 and 9,400, it was also possible to reduce the additional costs reported on 31.8.1996 to some extent.

In construction lot T5, in the single-track Vereina South section, optimal geology meant that construction was possible under more favourable conditions each month than those in the work contract forecast.

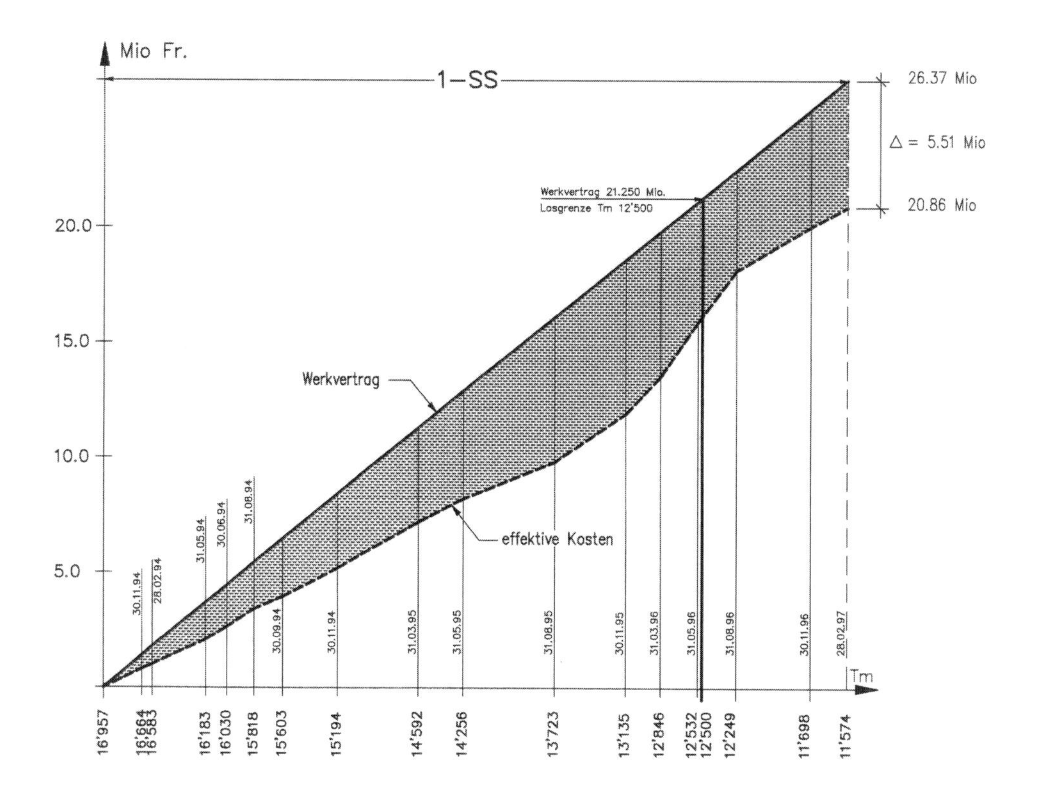

From Tm 13,723 onwards, the higher excavation classes had an immediate effect on performance, but most important of all, they led to higher costs.

73% of the total costs of CHF 340 million, price basis:1985, had been used up as of 31.8.1996 (see graph below). According to the work contracts, a further 27 %, or CHF 79 million, were still available for use. Since no really major surprises were to be expected in the remaining 2 kilometres of driving work, it was already possible at this point to be confident that the tunnel construction work could at least be completed within the scope of the proposed schedule of costs, or the total credit of CHF 340 million.

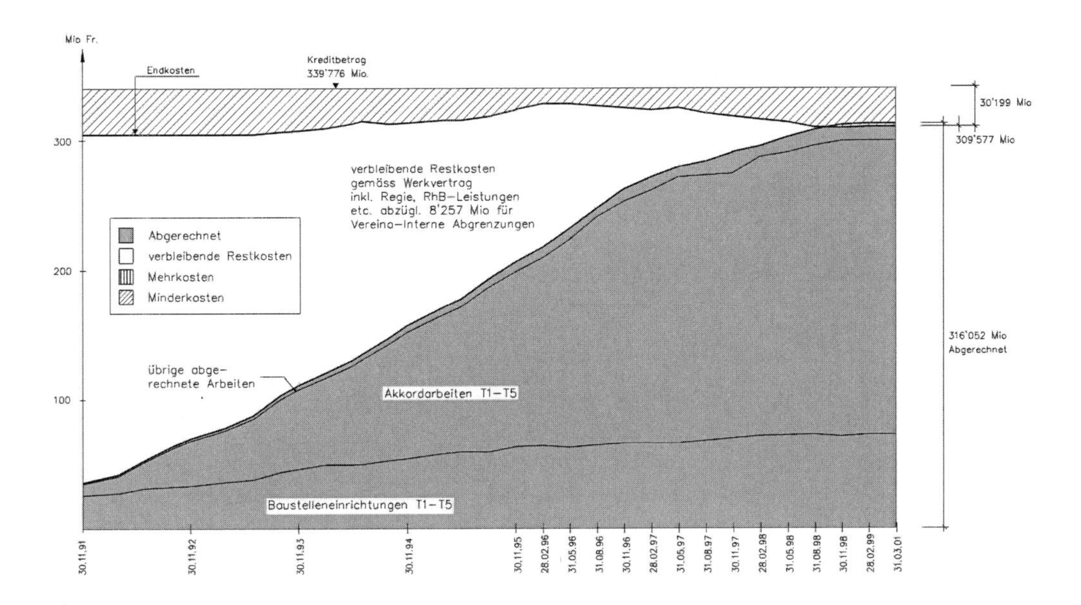

2.5 Summary

- The basic preconditions for functional cost controlling on the Vereina Tunnel were:
  - a clear price basis as the starting point
  - a lucidly defined inflation regulation for grey inflation and construction-related inflation
  - daily, reciprocally agreed recording of quantities on site
  - weekly cost controlling as a management tool for the construction management
  - quarterly balance-sheets with forecasts.
- However, cost controlling of the sort undertaken for the Vereina Tunnel - involving a great deal of manual work - is also very costly. As the table below shows, about 0.75 persons had to be deployed for cost controlling:

	per month	per quarter	per year
Construction manager for:			
- weekly costs	20 h	80 h	320 h
- quarterly costs	10 h	40 h	120 h
Draughtsman for cost controlling plan	20 h	80 h	320 h
Project engineer for:			
- weekly reports	10 h	40 h	120 h
- quarterly reports		50 h	400 h
		Total per year	**1,280 h**

3. Schedule management

3.1 An introduction to schedule management

Increasing requirements for the completion of construction projects on schedule are compelling all those involved to devote more attention to operational planning and monitoring of these projects. In particular, the site supervision management team has the central responsibility for this requirement.

Various scheduling tools for planning, project design, execution and commissioning have proven their merits in practice. These tools are presented below.

3.1.1 Objectives

The scheduling and monitoring system should:

- ensure compliance with the interim and final deadlines (commissioning) as well as other major key dates, in accordance with the project objectives;

- indicate the effects of any individual scheduling variances on the project as a whole and its implementation (in technical and financial terms);
- enable controlling actions (technical and financial) to be initiated in good time;
- indicate the scheduling bases and marginal conditions for the project design and the execution of the individual building works and construction lots;
- supply deadline information of various densities, matched to the needs of the individuals and bodies involved in the project (concerning status, variances, forecasts).

3.1.2 Scheduling methods

The varying effectiveness of scheduling methods calls for clarity about the way the method is to be used and the objectives to be achieved with it when a selection is being made.

The project, or one phase of it, must be clearly delineated. Depending on the status of the project's progression, this may involve a description (e.g. objectives) or plans which are already available (e.g. the construction project). All the existing marginal conditions must also be recorded in addition.

The individuals and bodies involved in the envisaged planning area must be known. The best overview is provided by an organisation chart which authoritatively defines the various information flows. The listed participants indicate who can supply information about which areas, and who requires which information.

An assessment of what is to be recorded by the scheduling plan, and for whom (an overview for the client, coordination information for the project manager, execution programmes for contractors, etc.), can be used as the basis for making a choice which is appropriate to the plan's purpose. The following sections present the most commonly used methods of scheduling.

3.1.2.1 Bar chart

The bar chart is the simplest method of planning. Even though new techniques have become available, this is still the most frequently used planning aid in the construction sector at present.

The handling of operations and procedures over time is presented in a time grid on the abscissa. This grid is plotted according to the time units (days/weeks/months) selected for the operations, and units of time and/or calendar dates can both be used for this purpose.

The header column lists the operations (brief descriptions). They can be listed in chronological order or according to responsibility.

The durations of the operations are entered as bars in the specified grid, using the grid scale, beginning from the planned start date.

A bar chart can be used expediently if the following benefits and drawbacks are taken into account:

Benefits	**Drawbacks**
well established	no contexts
easy to understand	no indications of urgency
Clear	limited scope (as regards size and parameters)

Bar charts are suitable for

- smaller, limited projects (sub-projects)
- determining load diagrams
- summarised presentations (management information)

3.1.2.2 Line chart / path-time chart

A line chart is suitable for planning and monitoring continuing operations of the sort which occur in underground driving work.

In a presentation based on coordinates,

- the construction task is plotted to scale on the abscissa; and
- a time grid is drawn on the ordinate, with the selected time divisions (days, weeks, months) taking account of the marginal conditions (public holidays, vacations).

The individual operations are entered as lines; the performance rate can be read from the gradient of the line.

In most cases, several operations will follow one another, and their performance rates may differ in this case. If there are subsequent operations that are accomplished more quickly, it should be ensured that they do not come too close to their predecessor (critical approach). Operations of this sort should be started later, or they should be interrupted.

For each point in time, the line chart can be used to read what progress has been made with the individual operations, or when a certain point has been reached with a particular operation. This makes the line chart highly suitable for representing tunnel driving operations.

The line chart has the following benefits and drawbacks:

Benefits	**Drawbacks**
easy to understand	dependencies are not explicit
link between place/quantity and time	no identification of urgent operations
clear as regards different performance rates (critical approach)	

Line charts are suitable for

Execution planning and monitoring of linear construction sites (construction of roads, tunnels, canals, galleries, pipelines and railways).

3.1.2.3 Network planning method

The network planning method is the newest of the planning aids which are available nowadays. It was originally developed in the USA for major industrial and marine projects, where it was known as the "Critical Path Method" (CPM) and the "Program Evaluation and Review Technique" (PERT), but a form of the network planning method that is suitable for the construction sector has since developed over the years. Inclusion of the relevant influencing variables leads to a gradual build-up of the network plan.

In the first step, the operations to be planned are arranged in the correct logical relationships to one another. The graphic 'map' leads to a network.

The second step consists of recording the times for each operation, and the calculation based on this information which in turn generates all the timings, the time reserves and the longest-time or critical path.

There are various ways of presenting network plans. In the example shown below, the operations are shown as nodes (boxes). The arrows represent the positional relationships between the operations.

The relationships between the individual operations may be characterised as follows: end-start, start-end, start-start, end-end. If necessary, times are also defined for this relationship - for example, the waiting period between concreting and the time when it is possible to walk on the concrete.

Theoretical example:

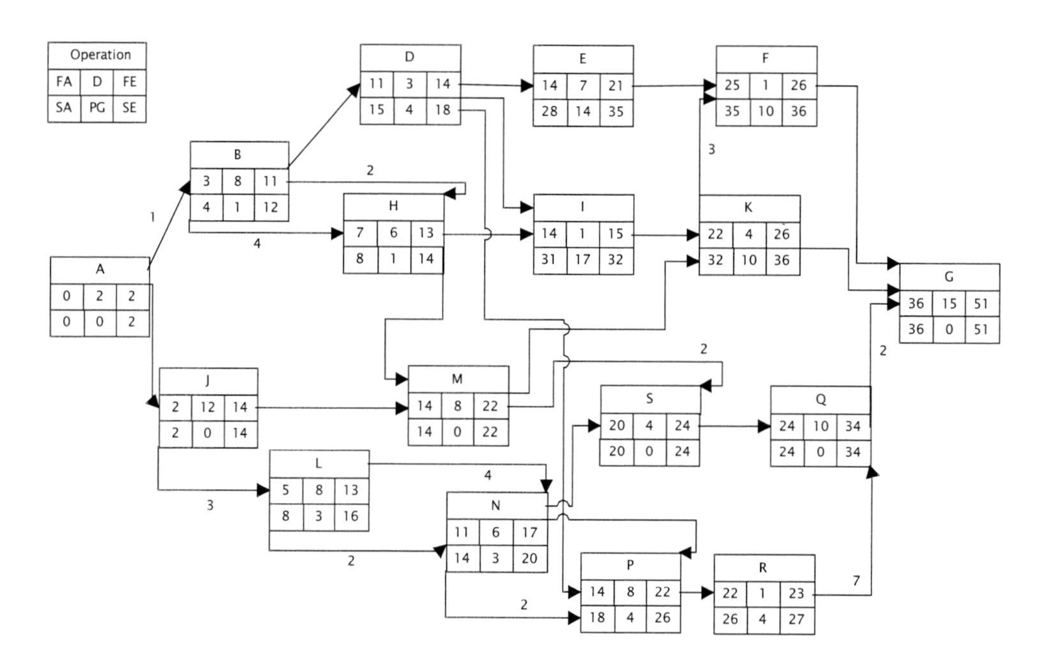

Operation: designation of the activity
FA: earliest start
D: duration of the operation
FE: earliest end
SA: latest start
SE: latest end

In a further step, the working days can be converted into specific dates.

The network plan has not been able to establish its position in the construction sector. As a general rule, it is costly to maintain and is only used in special cases.

3.2 Scheduling for the Gotthard base tunnel

3.2.1 The Gotthard base tunnel

The Gotthard Base tunnel is part of Switzerland's ambitious AlpTransit project.

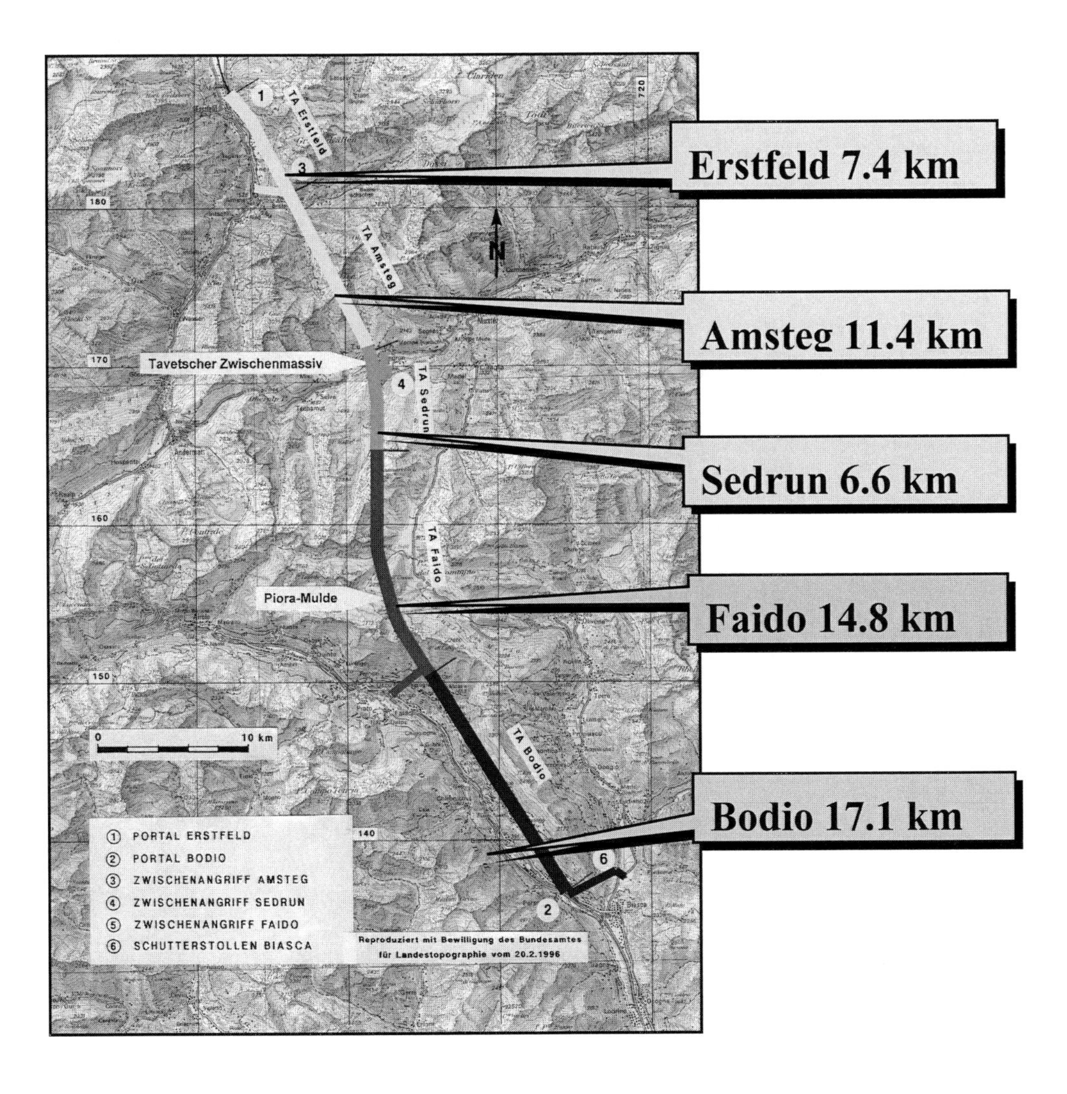

Figure 6: Situation

On the Gotthard axis, not only the Base Tunnel but also the entire axis has had to be planned and designed so that high-speed passenger trains can travel at speeds of between 200 km/h and 250 km/h, with cargo trains travelling at 140 km/h. This has resulted in a maximum permissible gradient of 1%, and the tunnel portals have to be built at an elevation of 600 – 800 m below those for the existing Gotthard Railway Tunnel.

For the Base Tunnel, geological conditions had to be taken into account so that difficult geological formations could be avoided or traversed with minimal expansion. Other points to be considered included the overburden, and compliance with safety distances from reservoirs.

The location that was originally planned for the northern portal was at Amsteg, but it was later relocated approximately 8 km to the north for reasons related to traffic and regional planning. The tunnel now has its northern portal near Erstfeld and its southern portal near Bodio, and it covers a length of slightly less than 57 km.

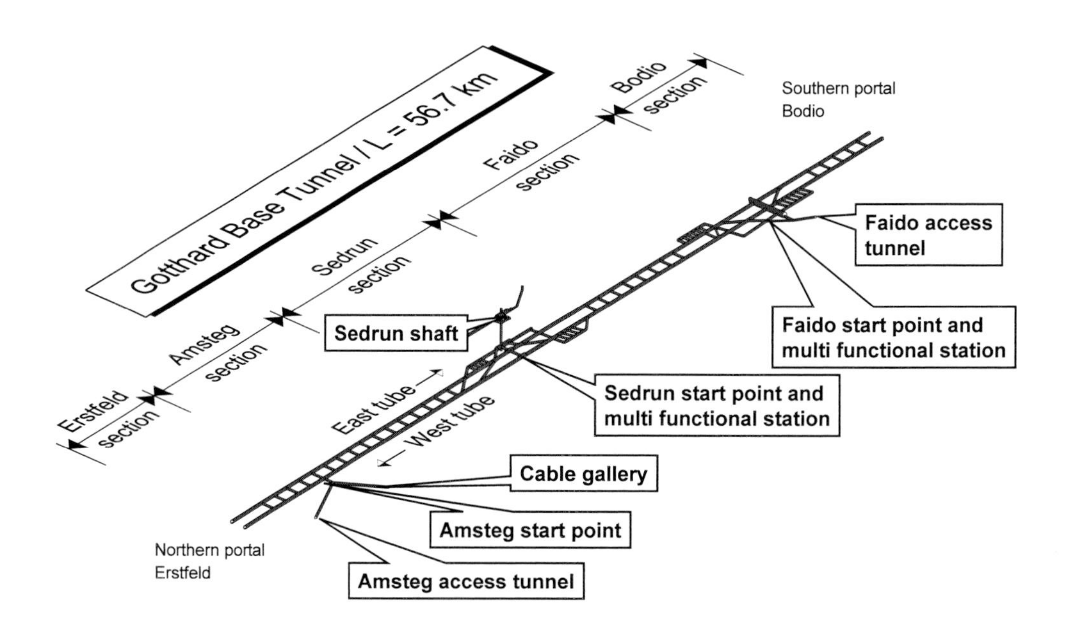

Figure 7: Tunnel layout

Two parallel single-track tunnels were selected as the system for the Gotthard Base Tunnel. This choice was based on intensive studies of constructional aspects, traffic capacity, facility maintenance, safety, construction and operational costs, together with a thorough analysis of risks related to the costs and duration of the construction.

Given that the tunnel is 57 km long, special consideration has had to be given to efficient construction and adequate ventilation when traffic is passing through it. Intermediate points of attack were determined in order to achieve an acceptable overall construction time. As well as functioning as additional points for the tunnel heading work, they will also be used as intakes and outlets for ventilation once traffic starts to use the tunnel.

The intermediate points of attack were topographically positioned so as to divide the tunnel into sections of roughly equal length. This meant that geologically difficult sections could be tackled during the earlier stages of the construction work on the tunnel.

The following intermediate points of attack were selected for the Gotthard Base Tunnel:

- Erstfeld: northern portal
- Amsteg: a horizontal access tunnel, approx.1.2 km long
- Sedrun: a blind shaft, 800 m deep and 8.0 m in diameter, accessed through a horizontal tunnel approximately 1 km long . An additional shaft extending from the head of the shaft to the surface will provide ventilation when the tunnel is operational.
- Faido: a tunnel with a length of 2.7 km, inclined at a gradient of approximately 12% with a height difference of 330 m.
- Bodio: southern portal

53 km of the tunnel's total length of 57 km consists of three major gneiss zones (the Aar Massif, the Gotthard Massif and the Penninic Gneiss Zone). In the main, these zones can be regarded as favourable for tunnelling purposes, all the more so as the zones with a significant overburden have an almost vertical dip, so they will be tunnelled through perpendicularly.

Technically difficult sections of the tunnel are located at the Tavetscher Intermediate Massif and in two younger sedimentary zones (the Urseren Garvera Zone and the Piora Basin).

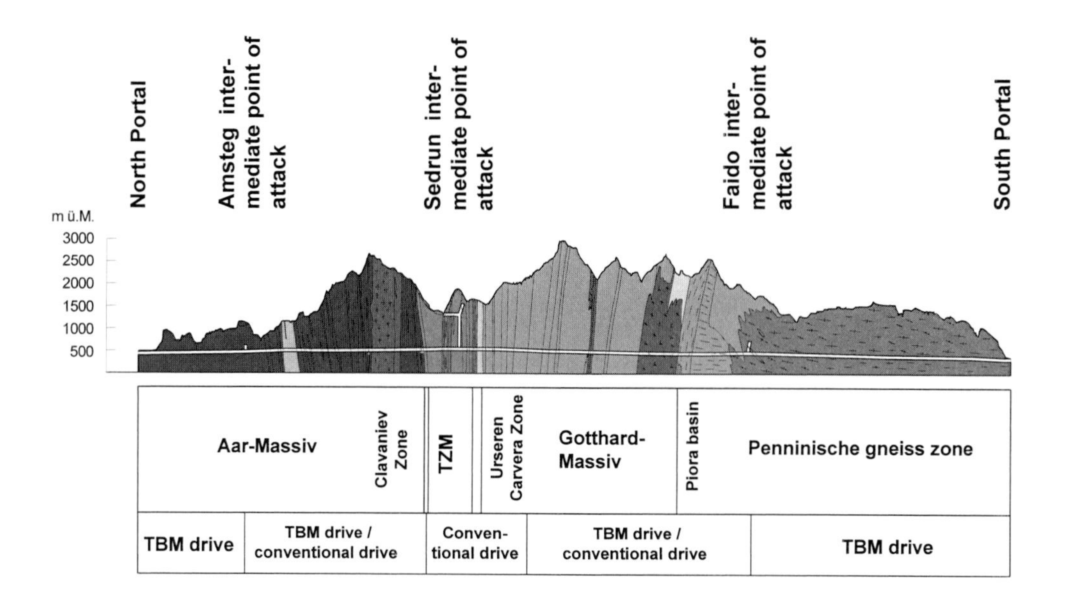

Figure 8: Longitudinal geological section

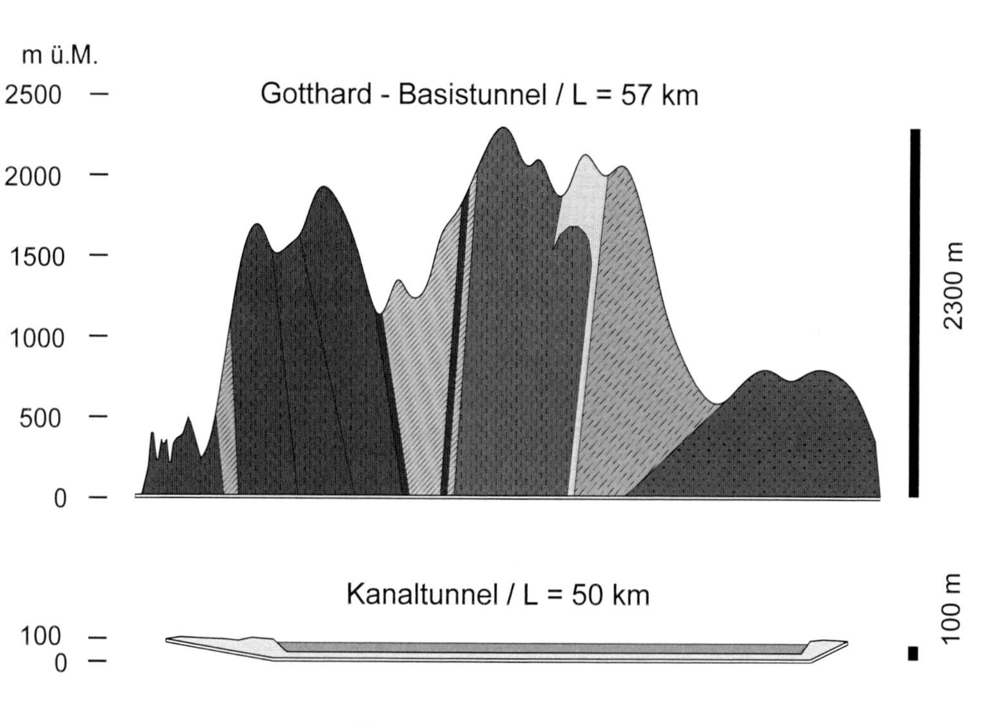

Figure 9: Overburden

The overburden will be extremely high over substantial sections of the tunnel. 35 km of the tunnel will have an overburden of more than 1000 m, 20 km will have more than 1500 m, and 5 km will exceed 2000 m. The maximum overburden is as much as 2300 m.

The standard cross-section is based on a double-shell lining system with an inner cast concrete lining. The rock support consists of shotcrete reinforced with mesh or fibres, rock bolts and steel arches where necessary.

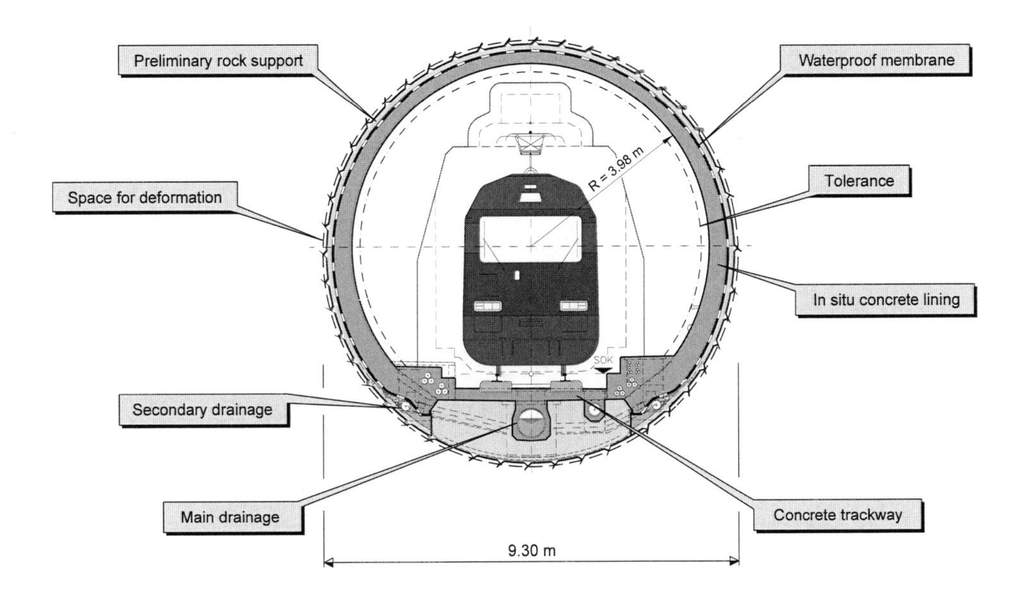

Figure 10: Cross-section of standard TBM drive

The insulation between the rock support and the inner lining consists of a membrane seal attached to longitudinal and perpendicular vault drainage strips that lead to the main drainage tube at regular intervals. These drainage strips are made of nap membranes. Special constructional features will minimize the sintering of the drains.

The chart shows the construction programme for the period between the start of construction of the tunnel tubes and the installation of the technical equipment in the tunnel. It is easy to see that the time-critical section of the tunnel is located between Faido and Sedrun, where the breakthrough is expected at the end of 2007 or early in 2008.

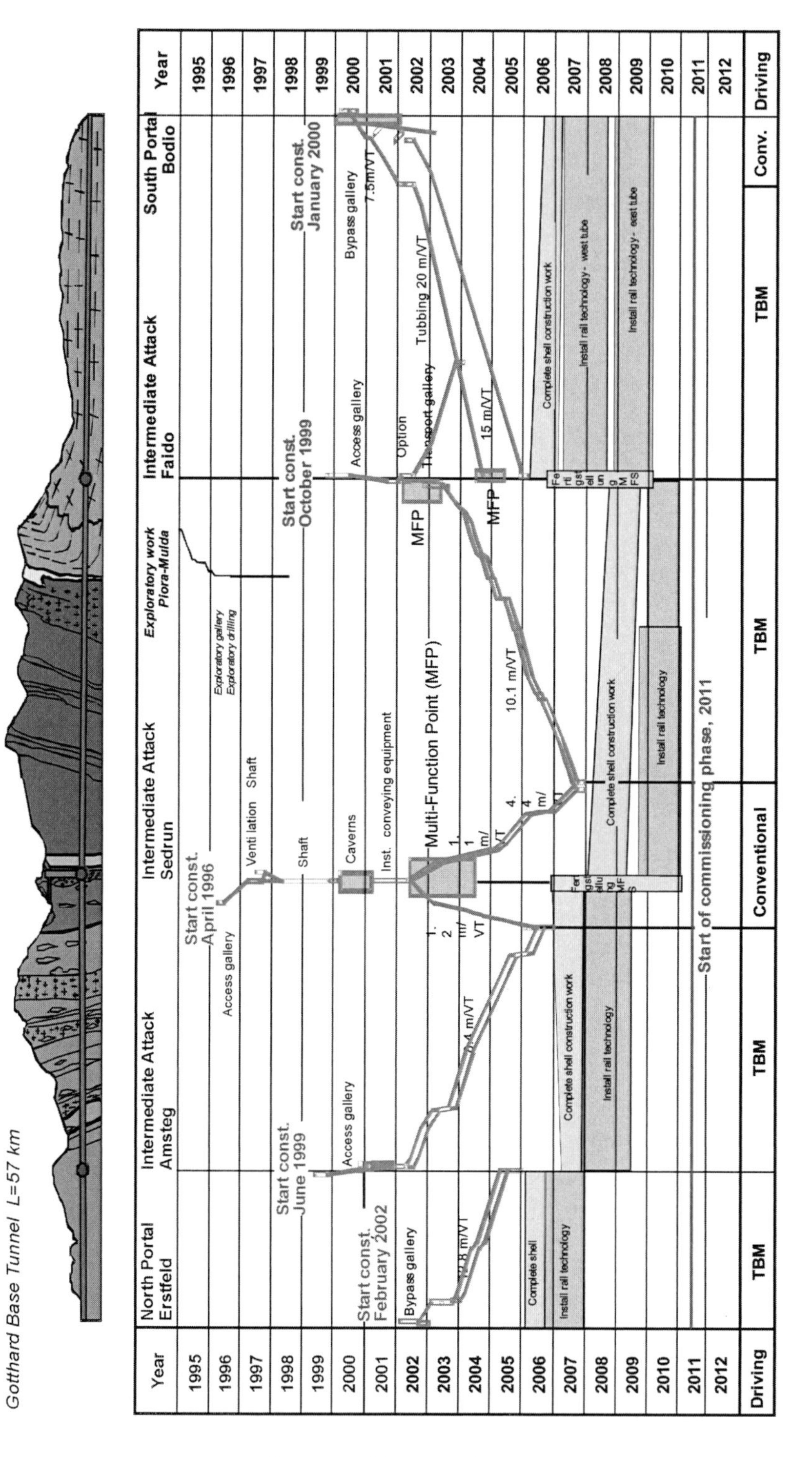

Figure 11: TBM Schedule

3.2.2 Deadline controlling at BAV (Federal Transport Authority) level

Using the example of the Gotthard Base Tunnel, we shall now show how the senior State authority - the Federal Transport Authority (BAV) in the Department (Ministry) of the Environment, Transport, Energy and Communication - and the client, Alptransit Gotthard AG (ATG) have structured the schedule controlling system.

In 2000, an "Instruction on the Basic Principles of Project Control, Project Supervision and Reporting in connection with the Overall AlpTransit Project" (Version 3.00 dated 30.11.1998) was brought into force by the Federal Department of the Environment, Transport, Energy and Communication (UVEK).

The purpose of this NEAT Controlling Instruction (NCW), as it is known, is as follows:

"By means of this instruction, the UVEK intends to support the overall AlpTransit project and to ensure that the competent Federal agencies can efficiently perform the following tasks in particular:

- *project controlling, together with the provision and management of the financing for the overall AlpTransit project;*
- *project supervision, coordination between the constructors (clients) and reporting on the overall AlpTransit project.*

Project information will be required, standards will be specified and random samples will be taken as the main means of achieving these objectives."

Section 11 of the NCW contains the following instructions on schedule controlling:

Article 1 Objective

Schedule controlling aims to ensure that the constructors (client) keep to the scheduling benchmarks (target values). In this way, universal transparency should be guaranteed regarding deadlines in all phases of the project, and early recognition of variances from deadlines should be ensured.

Article 2 Procedure

Schedule Controlling is underpinned by the scheduling benchmarks (target values). Each half year, the scheduling situation (actual values), the scheduling forecasts (planned values) and the scheduling benchmarks are compared with each other and with the values for the preceding period, also taking account of scheduling changes as per the concept in section 19.

Article 3 Scheduling plans

The scheduling plans basically show phases at levels 1 to 4 of the project structure plan. Sub-phases are used at level 5 (project groups).

Article 4 Scheduling benchmarks (target values)

The target values for the deadlines are specified in the current scheduling benchmarks.

Article 5 Scheduling situation (actual values) and scheduling forecast (planned values)

The constructors (client) will inform the BAV about the following items every half year:

- the scheduling situation (actual values) including the current levels of completion, and times ahead of or behind schedule;
- the scheduling forecast (planned values) including the remaining duration, and
- comparisons and explanations as per Article 2 of this section.

Article 6 New scheduling plans

The constructors (client) shall inform the BAV immediately of:

- new scheduling plans, including variants, and comparisons with earlier scheduling plans;
- areas of critical importance for scheduling, and any important scheduling changes in the future;
- major changes to the scheduling reserves.

Article 7 Schedule monitoring

Within their areas of responsibility, the constructors (clients) shall implement universal schedule monitoring for all the elements of the project structure plan, in all phases of the project. The structures of the scheduling plans in project structure plan levels 2 (works) to 5 (project groups) shall also be retained for this purpose.

3.2.3 Schedule controlling at ATG level

As the constructor, ATG has implemented the NEAT Controlling instruction in section 10 of its Project Manual: "Implementation of the BAV's NEAT Controlling Instruction for Scheduling".

The "AlpTransit Gotthard Project Manual" was completely revised in 2000 and the new version has been published with the title "ATG Guideline".

The overview which follows provides a summary of how schedule controlling has been implemented at ATG until now. It shows which scheduling plans had to be compiled, for what purposes, addressed to whom and in which form.

	Level	Purpose, recipients	Form	Compiled by
0	Overall project, AT Gotthard	Rough information for senior levels of authorities	DIN A4 landscape	ATG
1	Works on Gotthard axis and connection to eastern Switzerland	Rough information for authorities (status report) Working tool for ATG	Bar chart	ATG
2	Section (North, GBT, South, East)	Controlling at ATG Working tool Section Managers Basis for stages 0 to1	Bar chart	Section Managers
3	Coordination scheduling plans, stage: PAR project or construction project	Working tool for Project Managers Basis for stage 2	Bar chart	Project Managers
4	Work and construction programme, phase plan	Working tool for special assignments Basis for stage 3	Bar chart Path-time chart, etc.	Project Managers or construction firm

At present, the NCW is also being revised at BAV level, as is the implementation of the NCW at ATG level with regard to the execution phase. The new instructions are in the final streamlining phase at BAV/ATG/BLS level. These instructions will also contain new regulations for schedule controlling.

3.3 Schedule management by the local construction management

3.3.1 Bases, contractual agreement

3.3.1.1 SIA standard 103, regulations for services and fees for civil engineers

SIA LHO 103 describes the rights and obligations of the contractual parties for assignments commissioned from civil engineers and specialists in related professions. This standard forms the basis for all engineers' contracts in Switzerland.

Article 4 of the standard describes the engineer's services for each phase of the construction process. It defines the following basic deadline-related services for the execution phase:

Senior site supervision management:

- monitoring deadlines, costs and credits up to final inspection and testing;
- arranging for action in case of technical, financial or scheduling variances in liaison with the local site supervision management.

Local site supervision management:

- monitoring execution of the project in accordance with the description;
- drawing up detailed programmes for technical execution and time sequence, together with the contractor and suppliers;
- periodic reporting on the execution and progress of the construction work, and the development of costs;
- preparing action in case of technical, financial or scheduling variances.

3.3.1.2 IG-GBTS Engineering Contract dated 17.10.1997

The Engineering Contract between IG-GBTS (JV of Engineering firms responsible for the design and site supervision of the 3 southern lots of the Gotthard Base

Tunnel) and ATG includes the basic services set forth in SIA LHO 103. The following are mentioned as further services in addition to those mentioned above:

- monitoring deadlines and costs according to requirements specified by ATG

3.3.1.3 Work Contract / Special Provisions

The Work Contract and the Special Provisions regulate all the contractual elements with the contractor. As far as deadlines are concerned, these are:

- the contractual construction programme with binding deadlines (milestones)
- the target construction times for each excavation and drilling class
- adjustments to set periods, longer availability

3.3.2 Construction programme in the bid

In connection with the invitation to tender, the contractors were issued with scheduling programmes to serve as the basis for calculating their bids.

These scheduling programmes were drawn up on the basis of the division of the tunnel into excavation and boring classes, which in turn were based on the "longitudinal geology/construction technology profile".

The "longitudinal geology/construction technology profile" divides the full length of the tunnel into sections with the same excavation class, boring class, type of rock support and lining type, on the basis of forecast rock types, assumed water inflow, geotechnical hazard patterns, etc.

The illustration shows the complete programme as valid when bids for the main Faido and Bodio lots were submitted (TBM drive variant):

Bases for construction times, lot 452

The construction times for the excavation support types were calculated by the project engineers. They were based firstly on the geological classification mentioned and secondly on the following principles (Special Provisions, page 190ff):

Working hours

- The construction programmes for the Sedrun, Faido and Bodio sub-sections are based on 24-hour-operation for 340 working days per year, for the underground work.
- The completion of the raw work of the tunnel and the installation of technical rail equipment is calculated at 320 driving days or installation days respectively.

Tunnel driving

- To take account of the learning phase, the TBM driving work on the first 1,500 meters of tunnel is calculated with 3 months' more driving time than would be necessary to head through the same section in normal operation.

Fault zones

- Obstructions to the driving work due to the forecasted fault zones are differentiated according to the following levels of difficulty affecting the driving rates:
  - local injection work overhead, in the roof area
    3.5 m/VT (driving day)
  - systematic injection work overhead, full extent
    2.5 m/VT (driving day)
  - pre-injection of bypasses driven by blasting
    2mths/12m

Assuming 4/3 shift operation and based on the information regarding probable geological conditions, this yields average driving performance rates of between 10.4 and 17 m per driving day in the individual geological formations of the Faido sub-section.

If the traversing of the fault zones and the learning phase at the start of the driving work are also included, then an average performance of approx. 10.2 m/VT (driving day) is obtained for the Faido sub-section.

3.3.3 Work contract construction programme / contractual construction times

The construction programme which was drawn up in connection with the construction project formed the basis for defining the credits, but it does not govern schedule management on site.

A so-called contractual construction programme, or target programme, was drawn up in conjunction with the contractor in the course of the contract negotiations. This programme is based on the contractor's target construction times which are defined in the work contract, and it forms the joint basis for schedule monitoring between the construction management and the contracting company.

According to this programme, the contractor must state the target construction times as part of his bid. These times are based on the installation selected by the contractor and the envisaged working sequences and procedures. The target construction times also govern the calculation of extensions to set periods and the calculation of longer availability periods.

As a general rule, the contractor must submit the following time calculations with his bid, as a basis for calculating the target programme:

- construction times for all excavation and boring classes
- included times for interruptions to driving work
- effect on timing in case of delays due to difficulties with water
- included times for heading through fault zones
- included times for drilling and injection work
- included times for measurements regarding rock mechanics

An example of target construction times is given below:

MFS, Single-Track Tunnel, East Tube SPV — EST SPV

Target const. time (as per SIA 198 Annexe 6)
Font A = to be completed as work contract document (contractor must provide information in coloured boxes)
Font B = to be completed as accounting document

Target const. time, accounting const. time, periods		Work contract:		
Const. section: Single-track tunnel, east (temp. 1. tube), blasting drive EST to ensure deadline		Monthly production time	in WD	26.9
Work phase: rising drive		Interruptions to prod. time		
Work hours: shifts / WD	3	New Year	in WD	21
Hours / shift	8	Works holidays Easter	in WD	7
		Works holidays Summer	in WD	14
Inspection:		No. of New Years from start of SPV EST until Bodio-Faido breakthrough	x	2
Hours / Inspection shift	6	No. of Easters from start of SPV EST until Bodio-Faido breakthrough	x	2
		No. of summers from start of SPV EST until Bodio-Faido breakthrough	x	2

Work category	Unit	Unit per WD	Work contract Qty	Work contract Target const. time in WD	*Accounting Qty*	*Accounting const. time WD*
Start const. TA Faidc	Date	-	-	04.03.02	-	
Start SPV EST (Tm 139.172)	Date	-	-	13.01.03	-	
Period from starting const. Faido to start SPV EST	Days (incl. pub. holidays)		-	315.00	-	

Driving:						
EC I	m	0.0	0			
EC II	m	6.0	1527	254.5		
EC III	m	6.0	742	123.7		
EC IV	m	4.0	74	18.5		
EC V	m	2.4	103	42.9		
EC VI	m	1.8	16	8.9		
EC VII	m	0.8	10	12.5		
Total:	m		2472	461.0		
Driving interrupt'ns / pre-exploration:						
- drilling campaigns, systematic pre-drilling *)	x	4.0	14	3.5		
- seismic pre-exploration campaigns *)	x	3.0	10	3.3		
Difficulties due to water:						
- rising: 10 ... 20 l/s (reduction factor 0.2) ****)	Gr.hrs.	-	120	1.3		
20 ... 40 l/s (reduction factor 0.4) ****)	Gr.hrs.	-	60	1.3		
40 ... 60 l/s (reduction factor 0.6) ****)	Gr.hrs.	-	40	1.3		
60 ... 80 l/s (reduction factor 0.8) ****)	Gr.hrs.	-	10	0.4		
80 ... 100 l/s (reduction factor 0.9) ****)	Gr.hrs.	-	10	0.5		
Cutting through fault zones (cost-plus)	WD	-	6	60		
Other difficulties						
- blasting cross cut	x	1.0	6	6.0		
- site open days (12 h each)	x	2	1	0.5		
Drilling and injection work						
- injection and drainage drilling ****)	Gr.hrs.	-	84	4.7		
- injections ****)	Gr.hrs.	-	60	3.3		
Rock mech. measurements (Driving interruption per meas- x-section due to drilling and moving extensometers, probe extensometers**)						
- EST for crevasses, loosening or brittle fractures						
- meas. x-section no. 1 (as per VB IIIC88)	x	2	1	0.5		
- meas. x-section no. 2 (as per VB IIIC88)	x	2	1	0.5		
- meas. x-section no. 3 (as per VB IIIC88)	x	2	1	0.5		
- EST in fault zones						
- meas. x-section no. 4 (as per VB IIIC88)	x	2	1	0.5		
- meas. x-section no. 5 (as per VB IIIC88)	x	2	1	0.5		
Tunnel measurem't (Driving interruption for tunnel measurement as per III B5)						
- minor check	x	3	2	0.7		
- major check	x	1	1	1		
Other interruptions ***)	Gr.hrs.	-	35	1.5		
Net driving time	WD			552.9		

Total work phases (from start of driving EST SPV) incl.period from start const. to start driving EST		867.9	-	
Period from end EST SPV to start TBM driving EST (incl. vacations)		172.1		
Total driving work phases incl. period from end EST SPV to breakthrough Bodio Faido incl. period from start const. to start driving EST SPV		1040.0	-	
Interruptions to prod. time:			-	
- New Year		42.0	-	
- Works holidays, Easter		14.0		
- Works holidays, summer		28.0	-	
Total const. time	**in WD**	**1124.0**	-	
Final deadline = start const. date + total const. time)	**in months**	**37.3**	-	
Difference - accounting const. time / target const. time	in WD	-	-	
	in months	-	-	

All payments are made on the basis of contractor's info in the target const. time tables
*) The unit nos. and hours set out in the perf. schedule for seismic exploration and pre-drilling were included in the calculation as follows: 30% in west tube, 70% in east tube.
**) The no. of rock mech. measurement x-sections is assumed as about half in the east tube and half in the west tube
***) Payment for interruptions is made as per perf. sched. section 263, item 511.500
****) For conversion into WD, the divisor is reduced by the daily inspection time [Gr.hrs/((hours/WD)-inspection)]

3.3.4 Factors influencing the deadlines / construction programme

The statement in the NCW that schedule controlling should ensure that the constructor (client) can keep to the scheduling benchmarks is rather unfortunately phrased, given that controlling can only ever pinpoint a situation that is currently prevailing.

Adherence to the schedule is primarily ensured by the correct decisions on site, which - of course - may be reached not least on the basis of variances that are revealed in the process of controlling.

The driving performance rates govern the scheduling situation of an underground construction site. Once the breakthrough has been achieved and no further expansion is required, the work on the final lining generally proceeds according to the planned programme. This is mainly because the greatest factor of uncertainty - the geology - can then be excluded.

The major factors influencing the driving rates for a mechanical driving operation with an open TBM and shotcrete rock support are especially dependent on:

1. the time-related effects of the introduction of the excavation support on the excavating operation (excavation class/excavation support class),
2. the net driving performance rate of the TBM (penetration) and
3. the shutdown times.

Rock support work is defined on the spot, according to the geological findings in each case. The extent to which large-scale support work leads to delays in the heading is linked to the corresponding drive rate. If the rock support work can "keep pace" with the boring, no delays will occur. But if the support work is so extensive that it impedes the driving, this will influence the entire programme.

The driving rate is fundamentally dependent on the geology. Especially in the case of a driving operation with an open TBM, it can only be influenced to a limited extent once the machine has been installed and is operational.

Shutdown times mainly occur for maintenance work on the machinery. With a 4/3 shift operation, the third shift is normally used as a maintenance shift.

3.3.5 Schedule management by the local construction management

3.3.5.1 Recording deadlines and performance

The chief site manager is responsible for monitoring the construction programme.

The status of work and the daily performance rates are recorded each day by the individual local construction managers. These data form the basis for schedule monitoring as well as cost recording.

The work status is recorded with the help of daily quantities sheets, daily reports and daily journals. In the case of Faido, a daily quantity sheet and a daily journal are compiled for each lot, with differing structures depending on the particular lot.

Contents of the daily journal:

- staffing level
- weather / temperature
- performance rates (status of driving operations, daily performance rates)
- special events
- special instructions
- visitors

Contents of the daily quantity sheet:

- type of rock support
- type of profile
- daily driving rate
- daily quantities

3.3.5.2 Deadline monitoring and comparison with the work contract

For clarity's sake, the daily performance rates are transferred into tables and a path-time chart.

An example from the access gallery (Lot 451) is explained below.

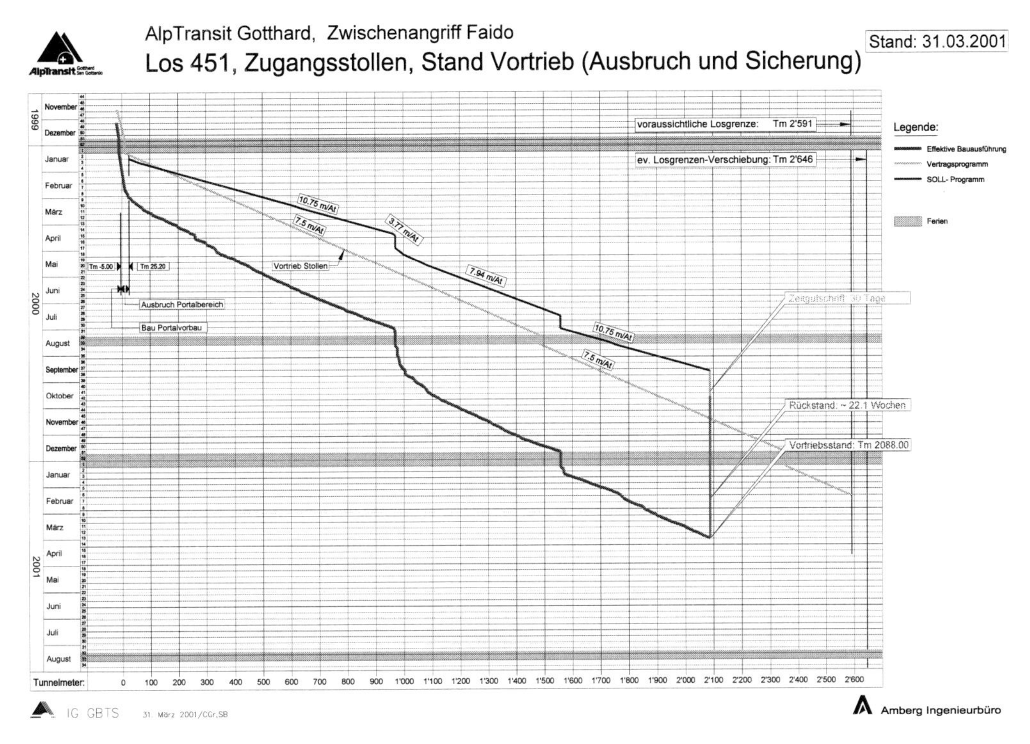

Lot 451, Access Gallery, Status of Driving (Excavation and Support Work) - 31 March 2001

The red line indicates the effective status (effective performance rates and effective types of excavation support). The SIA refers to the actual programme and the actual construction time.

The light blue line shows the work contract programme (forecast excavation and drilling classes as per the construction project, and target construction times as per the contractor's work contract). The SIA refers to the target programme and the target construction time here.

The dark blue line traces the target programme (types of excavation support effectively headed and target construction times, contractor's work contract).

The line for the construction programme in the bid is no longer shown. It runs about 3 months after the contract programme. The construction work started ahead of schedule by these three months. It is usual practice to shift the contract programme to the effective starting date.

3.3.5.3 Variance analyses

The variances can be interpreted as follows:

Total variance	=	**variance on excavation support types + performance rate variance**

The variance between the contract programme (light blue) and the target programme (dark blue) is the variance on excavation support types. This shows the variance due to the difference between the planned and the effective distribution of excavation support types.

The variance between the effective execution of construction work (red) and the target programme (dark blue) is a performance rate variance. This shows the variance which arises due to non-compliance with the contractual target construction times and the time credits. Failure to keep to the contractual performance rates may be due to various causes and is not necessarily the fault of the contractor.

The variance between the effective execution of the construction work (red) and the contractual programme (light blue) shows the total variance.

The performance variance is broken down into a time credit, and a period behind or ahead of the scheduled construction time.

Performance variance	=	**time credit + period behind construction time**

The time credit and the period behind the scheduled construction time are now explained with the help of a specific example:

Comparison: TARGET-ACTUAL construction time in the driving operation, construction lot 451 as of 26.04. 2001

Driving status: Tm 2194.00

Time credit as of 26.04.2001

New Year 99/00	15 CD
Summer holidays 2000	10 CD
New Year 00/01	15 CD
Time credits debited to client as of 16.10.00	30 CD
Total	70 CD

ACTUAL working days from 15.11.1999 until 26.04.2001

November 99	16 CD	
December 99	31 CD	
January 00	31 CD	
... CD		
... CD		
April 01	26 CD	⇒ **529 CD**

Target construction time based on excavation support types effectively headed

Preparation					19.0CD
Tm	–5.00	to Tm	25.00			43.0CD
Tm	25.00	to Tm	65.50	Type 3 A IV	$\Rightarrow \frac{40.50m}{7.94m} =$	5.1CD
Tm	65.50	to Tm	616.00	Type 1-2 A III	$\Rightarrow \frac{550.50m}{10.75m} =$	51.2CD
Tm	616.00	to Tm	635.00	Type 3 A IV	$\Rightarrow \frac{19.00m}{7.94m} =$	2.4CD
Tm	635.00	to Tm	963.00	Type 1-2 A III	$\Rightarrow \frac{328.00m}{10.75m} =$	30.5CD
Tm	963.00	to Tm	992.50	Type 5 A IV	$\Rightarrow \frac{29.50m}{3.77m} =$	7.8CD
Tm	992.50	to Tm	1084.00	Type 4 A	$\Rightarrow \frac{91.50m}{6.36m} =$	14.4CD
Tm	1084.00	to Tm	1560.00	Type 3 A IV	$\Rightarrow \frac{476.00m}{7.94m} =$	59.9CD
etc.						
Total						293.5CD

Period behind scheduled construction time as of 26.04.2001

ACTUAL working days	529.0 CU
TARGET working days	- 293.5 CD
Time credit	- 70.0 CD
Period behind scheduled construction time	**165.5 CD**

⇒ **approx. 23.6 weeks**

3.3.6 Determining the extensions to set periods (time credits)

The extensions to set periods are determined on the basis given in the work contract and the assumed performance rates in the contractor's bid.

The set periods (and hence the contractual construction time) are usually adjusted due to:

- shutdown times for the driving equipment
- outage of installations provided by the client or others
- delays to construction work due to force majeure (risks assigned to the contractor are not deemed to be force majeure)
- extension or reduction of the construction lots.

As a general rule, delayed deadlines must be recorded immediately by reciprocal agreement.

All the positive and negative time credits, in working days, are totalled. The payment for longer availability is made on the basis of specified principles and unit prices.

3.3.7 Reporting

Reporting uses the normal organisational tools of construction management organisation.

Meetings

Meetings of the senior site supervision management and site supervision management teams take place at regular intervals. The scheduling situation is a standard criterion at all meetings.

Reporting system

The site supervision management compiles daily reports, construction journals, weekly and monthly reports. The scheduling situation is described in all the documents.

Flow of information

The information flow charts regulate the passage of information. In the example of the Gotthard Base Tunnel, two different procedures exist.

The plan illustrated below shows the flow of information during regular construction operations. In case of exceptional events (emergencies, etc.) a special alarm organisation was developed by the ATG.

The regular information flow is as follows:

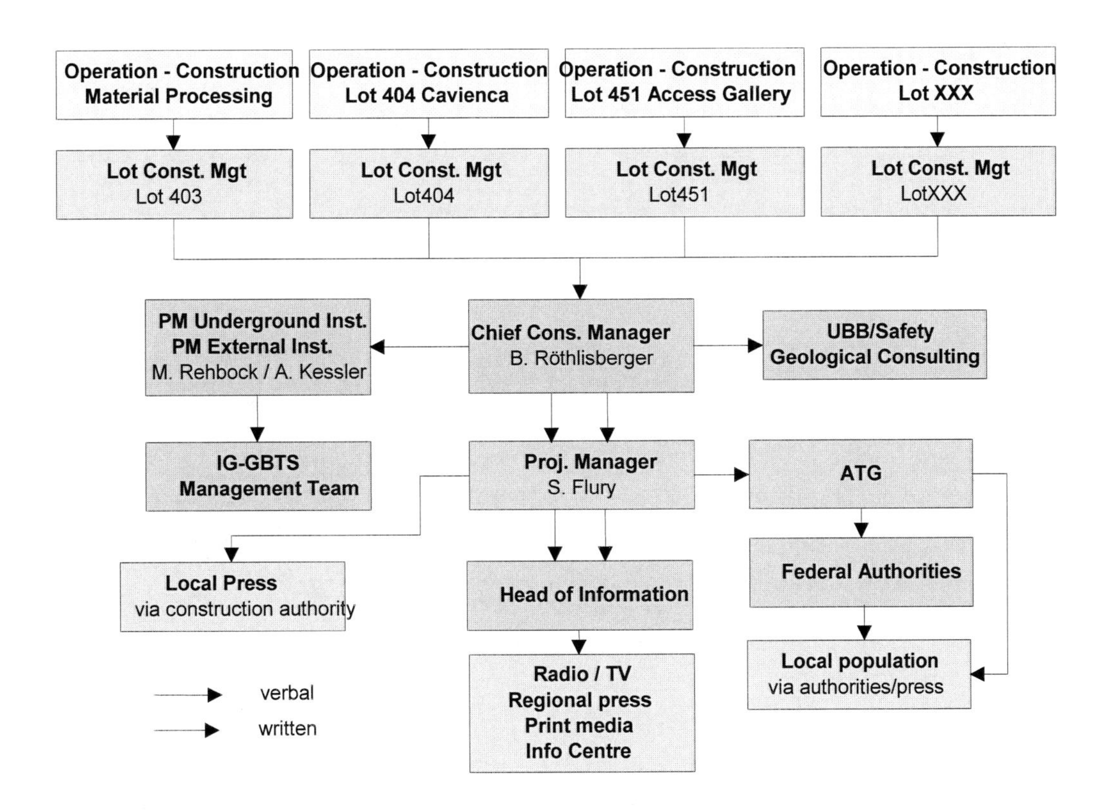

3.3.8 Longer availability

Availability time

All the contractor's installations are to be kept available for the duration of the relevant work. The availability time is defined by the contractual construction programme, and the expenses were calculated into the relevant performance items by the contractor. The availability time is only adjusted under certain specified conditions and according to stipulated rules.

If construction delays occur, the contractor has the right to request separate payment for making the installations available beyond the contractual availability time. This is known as longer availability.

The following rules normally apply to compensation for the items entitled "Longer availability of construction site equipment":

- the conditions for an adjustment of the set periods must be fulfilled;
- if the construction time is shortened, no reduction of the global installation fees is applied (no "longer availability" with negative unit prices);
- the first two months of the extension to the construction period do not qualify for compensation (no payment for "longer availability");
- longer availability is only compensated for the construction site equipment which is needed on the construction site during the relevant phase.

No compensation is made for longer availability in the case of installations which are not allocated an item for extended construction time in the performance schedule.

3.4 Dependencies between deadlines and costs

As the constructor (client) , the ATG has upgraded both aspects to key Q-points. The following extract from the Q-control plan illustrates the definition of the project requirements:

	Project requirement	Description
2.	**Costs**	
	2.1 Investment costs	Minimisation of investment costs (land costs, fees, construction costs, installations, remunerations, ATG's own expenditure, financing costs).
3.	**Deadlines**	
	3.3 Construction time	Ensuring contracts are awarded promptly (selection of contractors) and all execution services (mobilisation, installation, execution, demobilisation) are performed as per the overall scheduling programme, also compliance with milestones and acceptance/commissioning deadlines.

There are virtually no measures aimed at optimising deadlines which do not have direct effects on costs as well. In many cases, the two objectives run counter to each other. Accelerated deadlines usually entail increased costs.

One extreme example is certainly the TBM driving operation in the Vereina Tunnel. The installation for fitting the TH profiles in L1 of the TBM proved to be so expedient that built-in steel was selected as the means of support for extensive sections. This led to the high excavation class V. On the other hand, it was possible to achieve driving rates of more than 30 m/WD with this procedure.

4. Controlling and information management tools used for the Gotthard base tunnel sites

4.1 The context

As already pointed out several times the control of costs, of deadlines and of quality; data processing and the storage and sharing of information are all essential requirements in guaranteeing the control and management of any project, but especially in large underground projects with a long construction time. These activities require a large investment of time and money from all parties involved in a project.

The complexity of an underground project depends on the number of component parts, on the number of companies and parties involved, on the financial planning, on the project management and on the project realisation. The success of the project depends directly on the co-ordination and close collaboration which exists between all parties. The risk of changes increases with an increase in the number of parties. In a project like the Gotthard Base Tunnel the number of parties involved is considerable high.

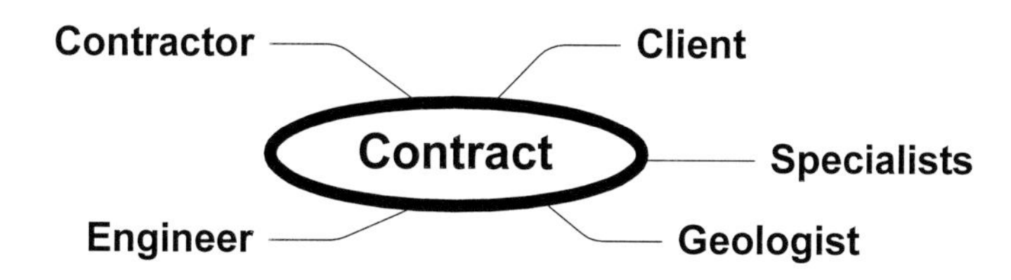

The main partners in a tunnel project

The co-ordination and collaboration of this large number of parties involved can only be guaranteed with the help of a computerised database which, during the entire construction phase, gathers and keeps track of all the project information, being the information exchange hub for all parties. The program also has to be an analytical centre for statistical requirements, for comparisons and correlations and especially has to provide a tool for decision making, for filing, and for the evaluation of cost forecasts and deadlines.

The client of the Gotthard Base Tunnel, ATG, decided to install such a specifically for that type of needs developed software to support and enable the simultaneous treatment and sharing of information, (including graphical databases), which in turn allows complete real-time control of costs, deadlines and quality.

4.2 The software concept

Along with quality, the mastery of deadlines and budgets is the major purpose of project control. The main objectives are above all the management of actual and future costs in order to establish these budgets and deadlines.

To meet these objectives and to obtain an overview of the whole contract, all data and information produced during the construction (costs, geological surveys, progress, site diaries, material testing and placement) must be unified and integrated. The same applies to photos and the associated documents. The chronological recording of the placement of individual elements or events of the project in an INTRANET/ EXTRANET configured database, can guarantee these objectives.

To be coherent, the concept must also evolve with the project and guarantee that, from planning through to production, all data is integrated. To do so, it is essential to use an evolutionary structure which caters for all the changes and adaptations inherent in such complex projects. This structure must be substantial and independent of the type of project to be handled (tunnel construction generates a multitude of ancillary works). This is the key to the longevity of the database and its evolution.

The project management has to evolve from sequential co-ordination to parallel co-ordination, where everything is centralised in order to be handled, analysed, filed and made available simultaneously to all the parties.

Next, management has to guarantee properly structured data and offer the user a clearly defined environment. This data structure is organised around "construction stages" and activities. The construction elements or object types (expressed here as "SISO positions") are the smallest entity of the hierarchical structure. The entire project management (quality, deadlines, costs) follows from this as presented below.

To simplify matters, these "SISO positions" will be regrouped in resources to help project control as described later on. The control of "SISO positions" can be handled spatially (along an axis) as well chronologically and produces a clear graphic representation of the construction elements.

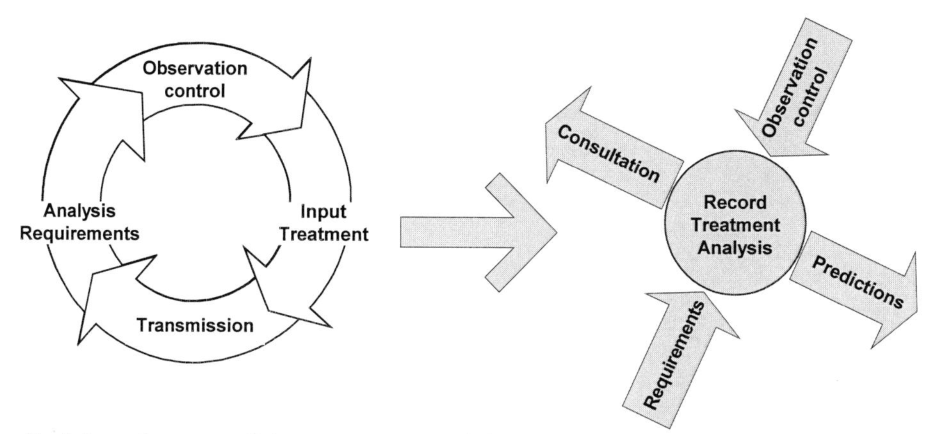

Evolution of a sequential management model towards parallel dynamic management

This method provides not only display, location or interpretation of elements but also the possibility of a geographical scan, i.e. the ability to find, at a given mileage, all the events and information which concern that particular mileage.

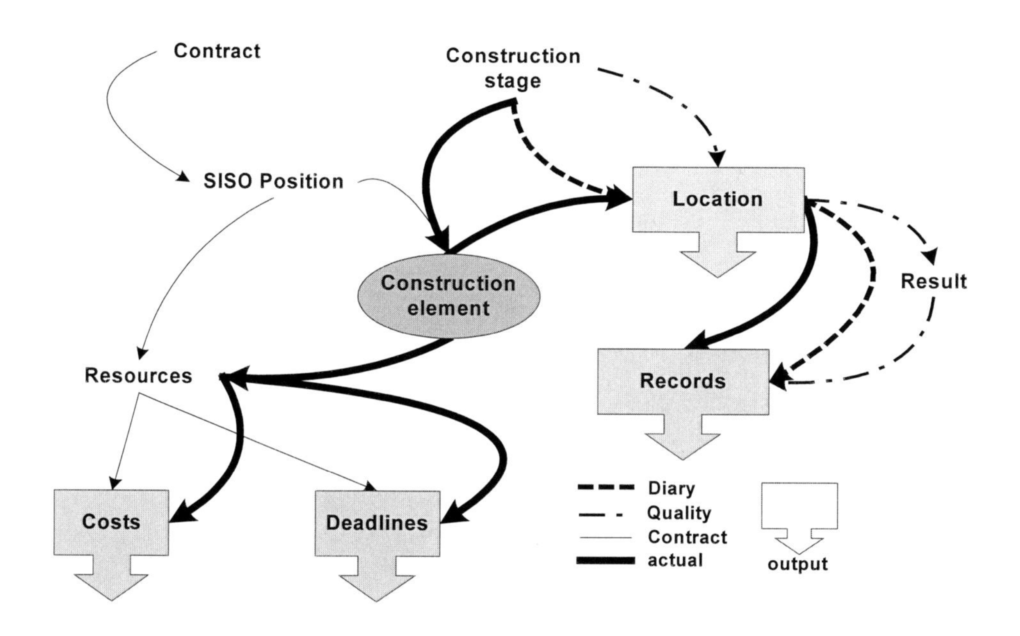

An overview of the software concept

4.3 The database

4.3.1 Concept

The database can integrate all the data which characterises the general activity of a construction site (e.g. matters pending, site diary) as well as physical characteristics (e.g. geology, hydrology, excavation type) or placement characteristics (e.g. support, consolidation). Some of this data can be qualitative or estimated on the basis of an in situ inspection (e.g. geology) or quantitative as in "SISO positions".

The software is structured into various hierarchical levels and includes for the first level a qualitative codification of every piece of data (construction, interruptions, control, etc.). The second level (see figure below) is made up of 9 elements: geology – hydrology – monitoring – progress – support – structure – organisation – preparation of materials – energy – installations – transport. The third level is made up of 85 sub-elements (construction stages), which can be increased or decreased according to the type project.

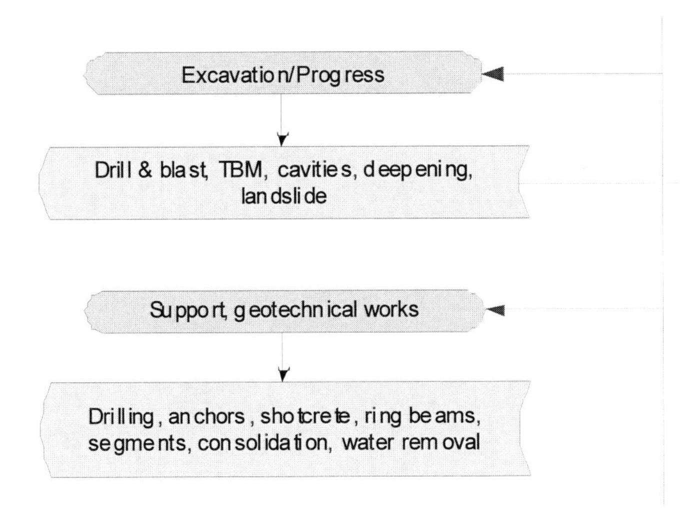

Example of partial structure

The database consists of several completely integrated modules –basic module, graphics module, geological module, cost and time controls etc.

4.3.2 Features

The Software has many functions which allow the integration of all the data for management as well as for the site works. The following list shows some of these functions:

- The simultaneous use of non-numerical data (e.g. dates), measured data (e.g. number and position of anchors) and of imported data from different file sources (e.g. deformation measurement files)
- The geographic location of data linked to actual construction dates
- Structured data allows generation of spacing/element statistics for given situations (e.g. spacing and size of supporting structures with regard to a type of rock or an excavation type)
- Report lists by tunnel section type or by dates
- Graphical display of elements with linked non-graphical data
- Data searches using multiple criteria within the database structure
- Common interface, with instantaneous Web transmission and an integration of all the project information
- Multi-lingual database structure

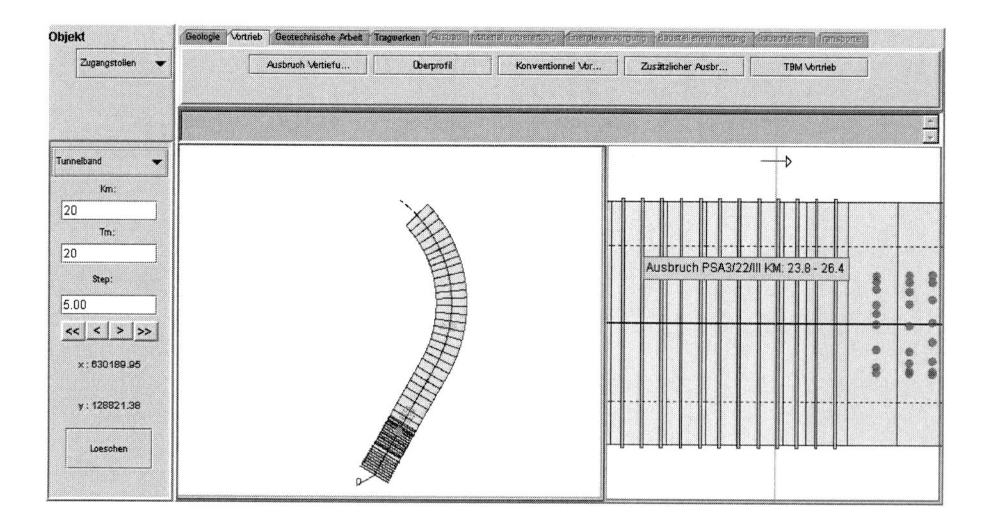

Screenshot of current position with tunnel and chosen elements.

4.4 Project control

4.4.1 Structure

The current contractual conditions at the Gotthard Base Tunnel sites do not facilitate cost management – a large number of items, a complex structure – not to mention final budget figures. A way out consists in simplifying the contractual data while guaranteeing complete control of the follow-up on site.

To confirm this simplification, every object is summarised in RESOURCES (cost and time) as shown in the Figures below. These resources represent the main operations of the site works (e.g. excavation, lining, internal layout). They can be linear for linear objects (tunnels) or fixed for other parts of works or time dependant for certain site installation costs. In this way, single objects, as well as the contract as a whole, can be administered with a limited number of elements.

The resources belong to the hierarchical (pyramidal) works structure, which offers many possible cost combinations, as shown in the Figure below. The work totals are constantly generated.

On one side we find the project organisation (technical structure) and on the other side we find the physical structure (geographical and financial).

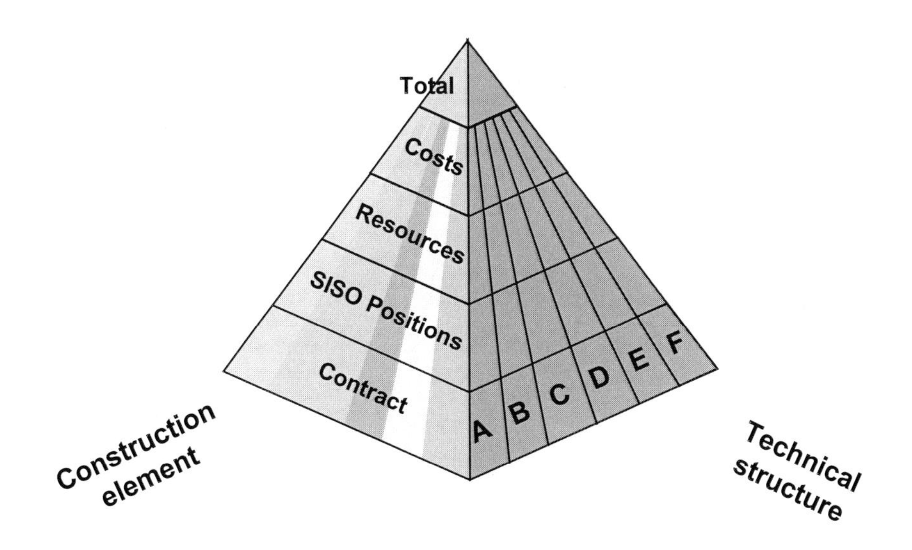

The concept showing hierarchical structure combined with the cost of the works

Results can be obtained by combining vertically or horizontally.

The cost of an underground project can be summarised into 4 categories: fixed costs, time-based costs, linear costs, variable costs. These categories illustrate, for variable costs and time based costs, the risk factor in such projects.

Fixed costs represent the costs of non-linear works or the fixed cost part of the construction site installations.

Time-based costs are directly linked to the duration of works and represent exclusively the construction site installations, which can be either calculated globally (with extra-overs or reductions depending on the duration of works), or with a monthly rental.

Linear costs remain more or less constant throughout the duration of the works and are normally linked to linear works (tunnels). Costs must be applicable to a specific section of tunnel (e.g. waterproofing applied only where it is necessary) or to the entire length of the works or the object.

Variable costs are for the mostly directly linked to the geology encountered which affects the excavations, supporting structures, consolidations, water removal, and work stoppages. It can also include items under transport and materials testing.

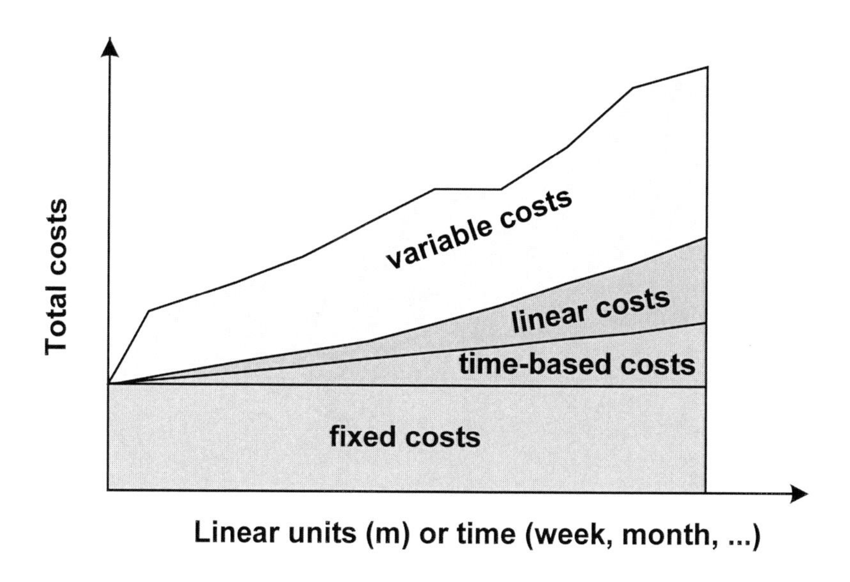

Cost categories in underground works

4.4.2 Management

It is inevitable that changes to the contract will take place throughout the project. These changes are mostly linked to particular events (geological or natural), to corrections of contract (changes to requirements) or to unforeseen factors.

To guarantee complete control of costs and time, these changes must be constantly input.

The works management covers costs and deadlines and takes place on 4 levels:

- the basic contract
- the revised contract (including risks due to e.g. geology, work stoppages) and extra items requested by the client.
- actual performances (as per the monthly reports)
- projected figures

In order to implement day to day control of the works, every object (linear or fixed) will be firstly configured (costs and deadlines) and then linked to other items according to the overall works program.

For linear objects (tunnels), resources are distributed according to the projected geological sections and excavation profiles (excavation types). In the management concept presented in the Figure below, the basic contract is fixed for the whole of the works and the revised contract becomes the controlling instrument as soon as works began.

Difference Δ1 represents the actual delay at a given reference position. Difference Δ2 shows the projected delay at the end of the contract based on the basic contract at this given reference position and then projected to the end of the contract.

Difference Δ2', on the other hand, takes into account the actual performance (time resources) up until the given reference position to further adapt the final forecasted deadline.

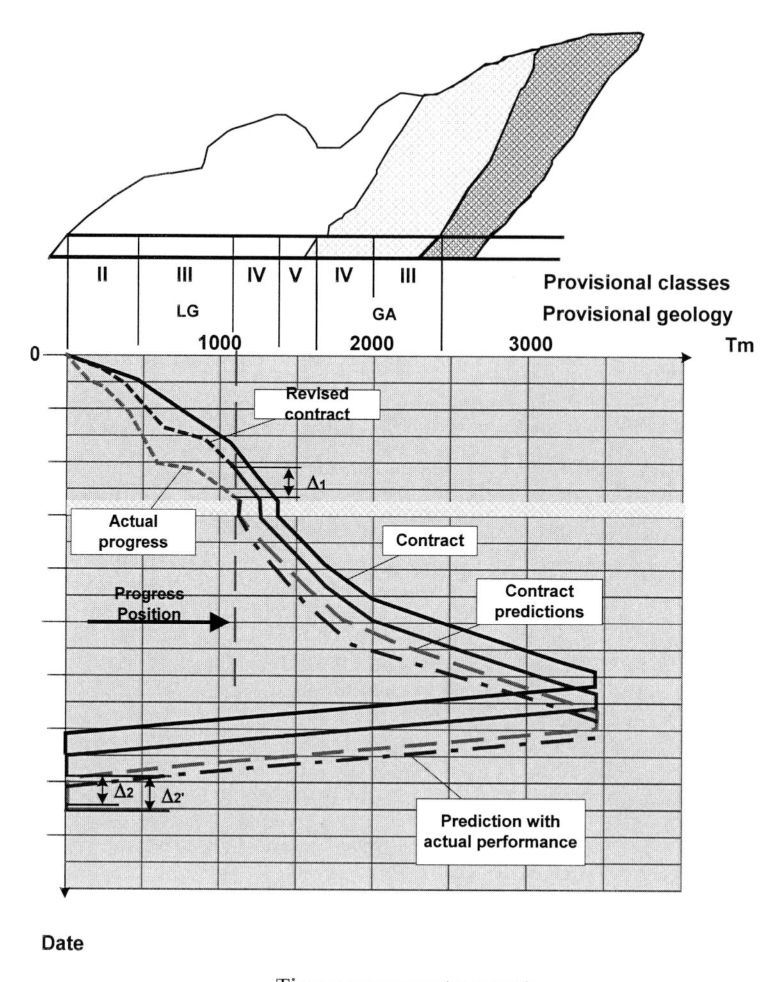

Time management concepts

4.4.3 Cost and deadline controls

Cost and deadline controls are driven by resources. In the case of fixed and time-based resources this would normally be monthly, while for others (linear resources) this would be done continuously (excavation, concrete work, internal layout).

Linear resources are calculated continuously at any level of completion and used as the basis for new cost and deadline forecasts.

By summarising the contract into a limited number of resources, the accuracy of control procedures lays between 95 and 97 %.

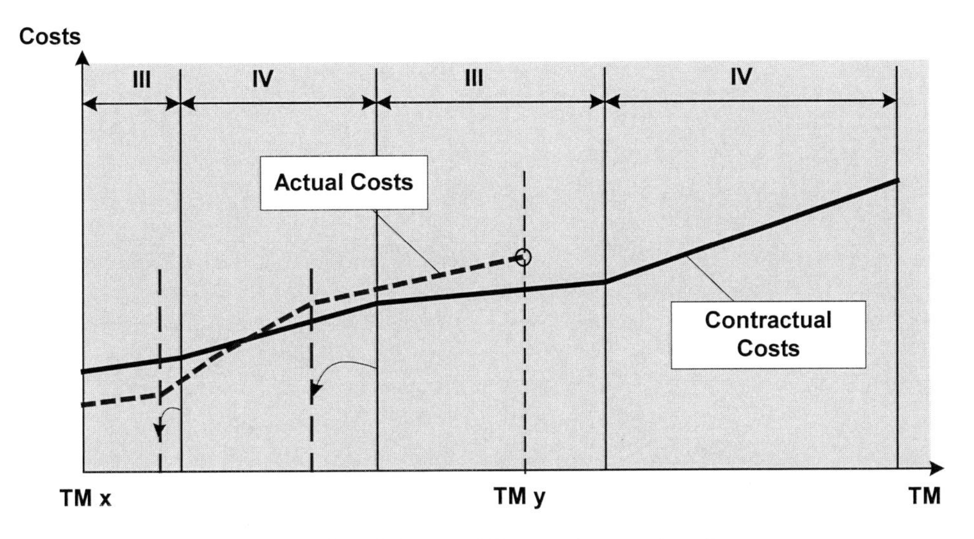

Presentation of the cost differences by tunnel section

The above figure illustrates a control example for a section of tunnel, but such a chart could be constructed for a given class of rock, for a given rock type, or for the entire project, etc. All these statistics are quoted relative to the tunnel mileage.

Another reporting possibility (costs / deadlines) is shown in the figure below. This report could show costs and budgets for the entire contract or for a particular section of work.

For deadlines, the manager has to anticipate any delay or advances as the work progresses with resources which affect deadlines being updated, which further updates performance indicators. The contract stipulates clearly and precisely how deadlines will be handled and calculated.

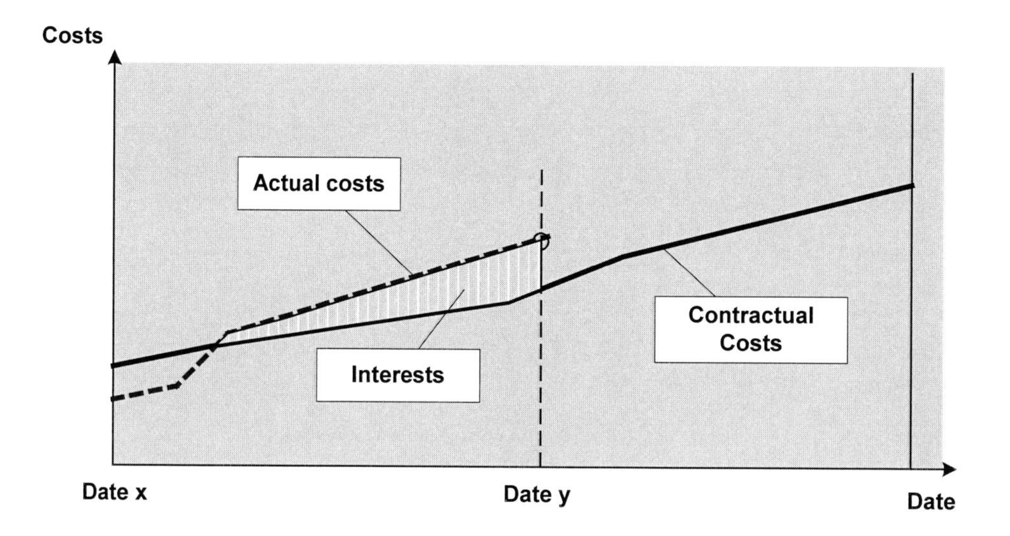

Representation of costs over time

Amberg Engineering AG

Technical features for quality procedures in tunnelling

Claudio Oggeri[1]

[1] Politechnic of Turin, Tunnelling and Underground Space Center, Corso Duca degli Abruzzi 24, 10129 Torino, Italy
e-mail: claudio.oggeri@polito.it

Abstract

The particular and variable context in which tunnel constructions are built in claims for the detailed knowledge of the natural environment, the suitable design approach, the rigorous construction technique. Quality procedures in tunnelling should help to drive along these steps, also by means of assessment and monitoring methods, taking into account that the aims are the respect of expectations in terms of time, costs and overall use of the underground construction.

The various phases, from planning to operation, should follow guidelines for operating staff, in a coordinated project organization and with the due exchange of knowledge: experience, good contracts and responsibility can be the keys for success, thus hopefully limiting the failure events and the controvercies.

1. Introduction

Quality in tunnelling is an approach for the coordination of the design, construction and operation phases, following the purpose of respecting the predictions of the projects, in terms of time duration of the work, construction costs, technical solutions and environmental constraints (Table 1).

All these results can be achieved by a comprehensive and coordinated activity, in which the basic elements are represented by the experience and the self responsibility of the involved parties. In the meantine, in order to limitate the claims and the legal controvercies, the best choice of the contract form is surely recommended. Finally, reference points for permanent collection of case histories and exchange of knowledge should be identified both in the universities and in the national/international scientific and technical organizations, as the experience for the large variety of conditions can only be available with continuous contributions.

The practical tools for operating the quality are both administrative and technical. The first category will not be considered in this note, as it appears mainly as a managerial and financial item: the Quality Management is the organisation aimed to coordinate parties and procedures during a project. The second category presents both routine activities (burocracy) and technical activities (procedures): in this note the last aspect is focused and it is also suggested to intend Quality as the mix of actions oriented to verify the fullfilment of initial requirements (Quality Assurance).

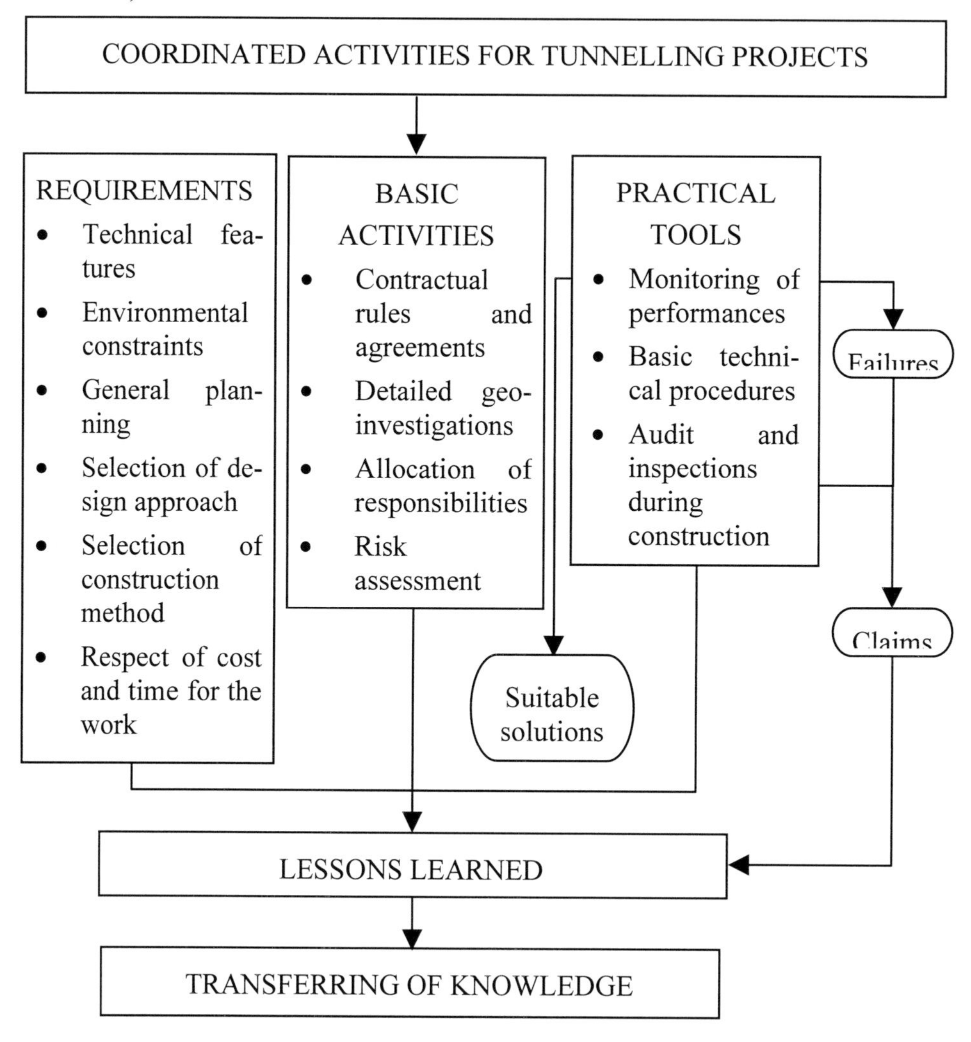

Table 1. Schematic relationships between the various keypoints of the Quality approach

The meaning of "Quality" was defined in ISO 8402-1994 (replaced by ISO 9000:2000) as "The totality of features and characteristics of a product or service that bear upon its ability to satisfy stated or implied needs". As it is difficult to know, and particularly in advance, what the specific needs are, it is of great importance that these are expressed for a tunnelling project: the “stated needs” of the Committent could mean the construction of a tunnel in accordance with the technical quality (i.e. technical specifications), within the estimated budget and predefined time schedule or programme; the “implied needs” are multifold and could be respect for laws and regulations, safety aspects, environmental protection, energy savings and sustainability of the works.

With regard to the ISO 9000 regulations for accreditation and certification, the certification itself may not be a practical quality requirement. It is acknowledged that many employer/client bodies, engineers/consultants and contractors consider that the ISO 9000 certification may go beyond their limits of affordability. It is nevertheless necessary to point out that only some parts of the tunnelling constructions can be clearly considered with the quality procedures of an industrial activity, and just for that reason the control of materials and techniques should be emphasized wherever possible.

Quality is a concept that is constantly evolving. There are unique features in tunnelling due to the great number of involved parties, both in minor and in major projects and due to the non- repeatable nature of tunnels. However, the trend of the concept of quality, in this field, is towards both a good planning / management and suitable technical results.

In December 2000 the ISO committee introduced a revised version of ISO 9000:2000 series for quality management systems based on management principles. These principles are defined in ISO 9000:2000 “Quality management systems - Fundamentals and vocabulary”, and in ISO 9004:2000 “Quality management systems – Guidelines for performance improvements”.

It would be useful to correlate these points to the different aspects of tunnel construction and especially to the operation and maintenance phases in order to create a proper standard management approach to these operations.

As example, operation and maintenance of the tunnel can easily be connected to some of the mentioned points:

1- Customer focus

   It is not possible to do anything without considering the customer aspect; both the client who pays for the construction of the tunnel and the general public who at the end use the tunnel should be considered as "customers".

   It is of primary importance to consider the customer’s needs and satisfaction to plan the operation mode of the tunnel and to decide how and when to carry out the maintenance works, i.e.: not to plan the maintenance of road tunnels during holyday periods

2- Leadership and 3- Involvement of people

It can be useful to define in advance the various responsibilities for the different areas or aspects of the work to ease the solution of possible problems during operation phase or during emergency repairs.

4- Process approach

Definition of procedures for the various activities and co-ordination.

5- System approach to management

Scheduling the maintenance in advance, from the building phase, to be able to operate before real damages occurs to the structure and installations, to be able to allow a continuous functioning of the structure.

6- Continual improvement

The maintenance phase in itself is based on the idea of a continuous improvement and should be approached with such a goal in mind.

7- Factual approach to decision making

All decisions should be converted into interventions on the basis of correction and prevention criteria, following the experience obtained from data control and statistics processing.

8- Mutually beneficial supplier relationship

New materials, new construction technologies and the special requirements for underground works should always be compared, in order to improve safety and usability standards.

Energy savings can also be obtained moving by a simple and rational planning towards a correct organization of transportation, the development of railroads and roads systems etc.

2. Basic technical requirements for quality

All along the project a coordination of activities is necessary in order to reach significative results for: technical requirements, economical results, contractual agreements, environmental effects, safety standards.

All these goals can only be obtained by means of high professionality, self responsibility, clear requirements.

The rational development of scheduled activities is one of the main instrument for success. Other suggestions for good results show the importance of knowledge of state of the art and of case histories which have given problems. In fact, the bad conditions are the most appropriate to understand the reasons for failures, or "non quality": this appears when:

1. time for providing the work are exceeded
2. new costs alter the budget
3. claims within the involved parties occur without resolution
4. controvercies between the project parties and external arise
5. accidents and failures determine loss of lifes or goods

In general the quality procedures should coordinate each action towards:

1. a good knowledge of the geologic formations
2. the evaluation of the interaction between the tunnel structure and the surrounding environment
3. the evaluation of any development of the structure on a long term.

Uncertainties in tunnelling projects arise from the inability to study the complete structure and behaviour of the rock and soil formations in such detail that all possible scenarios can be contemplated. It should be emphasized that tunnel failures have been the result of various reasons such as insufficient site investigation, inadequate evaluation at the planning stage, mistakes during construction and operation phases.

An inventory of inherent risks for the project phases should be undertaken for all tunnelling and underground projects. Risk assessment should be made for:

- Rock and soil conditions versus the stability, physical properties and seepage
- Construction methods versus progress, rock/soil conditions, environment and safety
- Equipment versus suitability, capacity, energy consumption, environmental and safety aspects

The consequences on the total project cost and financial matters should always be included in risk evaluations.

Site investigations and geological exploration and their use in design optimisation, coupled with well matched construction techniques, can prevent most kinds of collapses. The design cannot only be based on a deterministic approach for the evaluation of a risk, a probabilistic approach that takes into account the uncertainties in the geological, geotechnical and construction techniques and external constraints will enable the project management to optimize the final design and define the technical, administrative and financial management of the risk. Detailed paragraphs dealing with risk allocation should be included in all contracts or agreements, and risk should be distributed by taking into consideration both technical and financial aspects.

Contractual procedures should be simple and the role of the various (parties should be made extremely clear, both from a technical and financial point of view. Excessive legal and bureaucratic complexity should be avoided, but specific attention should be given to those contractual aspects that deal with the information obtained on the natural ground conditions.

All project stages are of fundamental importance; the stage of investigations of the geological and geomechanical characteristics, the planning stage, the stage of choosing the equipment and method of excavation, the installation of lining and rock support, and the organisational and financial coordination stage. Quality Systems should reflect the various stages and assess the influence on the final result, mainly focusing those aspects where the interpretation becomes decisive in the technical choices.

Tunnelling and underground projects are usually of significant financial value and are generally considered high risk in terms of final cost and time for construction/delivery. In addition, tunnel construction often requires multinational input for both funding and technical capacity. This consideration arises from the natural circumstances in which the tunnel is built, and the necessary consequence is that great emphasy should be put in geo-investigations and monitoring.

When it comes to the quality issues, it is not considered necessary to maintain a uniform approach for the duration of a project. Some issues might turn out to be more important than others and experience shows that, during the course of construction, some issues may well be abandoned without harm to the project.

Research into the appropriate local legislation and regulations for such matters as health and safety, the protection of the environment and socio-economic consequences, as well as technical standards to be achieved should be carried out. These can then be compared to best international practice, which can be adopted to suit the particular project and then the quality goals can be set accordingly.

All phases of a project and all organisations/groups of organisations involved should have a set of quality goals for the work to be performed. Quality goals should be aimed at giving the project added value with regard to cost, time, improved performance/physical properties and could be set relative to "soft parameters" like improved competence or educational standards for involved parties, environmental aspects, safety standards or benefits for the public in general. When quality goals are determined, these should be related to the factors of success for the project and a strategy on how to achieve this should be included in the Quality Plan. Compliance and the degree of success in achieving the set goals should be communicated to all project members/companies.

Requirements and detailed instructions for quality control for the design and the construction work should be specified in the contract. It is important that procedures for the registration and communication of items such as test results are specified. A project procedure including owner/employer, engineer/designer and contractor for lines of communication and the responsibilities of each party for handling such information should be drawn up. The continuous process of evaluating, for instance, test results and any subsequent changes to the design and/or construction methods should be adequately defined and described.

It is recommended that implementing bodies and contractors make available technical data and statistics from the construction phase and communicate such data to international bodies and universities. By doing so, objective views on the state-of-the-art of methods used and case histories could be obtained, all in line with a generally recommended quality goal regarding the transfer of knowledge.

Methods for procuring, constructing, maintaining and operating private/public infrastructure are numerous. They range from lump-sum contracts for the construction of a designed facility, to complex transactions involving ownership transfer, equity financing and novel revenue features. Understanding and adapting available contract options is crucial for the success of a project. The owner should be aware of how his precise requirements are best served and of the practical considerations associated with different contract strategies.

The details of appropriate methods of financing and remuneration, risk allocation, the nature of the different parties' obligations/responsibilities and the ownership structure are matters specific to the individual projects.

It is possible to resume what have been described in terms of benefits of the application of Quality as follows:

- Good organization and relationship
- Definition of needs and constraints of the work
- Design adapted to needs and constraints
- Control on the evolution of design stages and selection of appropriate contracts
- Evaluation of risks and allocation of responsibilities
- Ability to perform modifications to details
- Historical records of the design and of the performed works

- Reciprocity for communications and problem-solving attitude

3. Practical relationships in the tunnelling phases

For each project it is possible to identify a sequence of phases, which can be summarized in the following table 2.

It is also necessary to outline that different form of contracts can be applied to underground construction, and different roles and responsibilities have the involved parties. Among the others the following contracts can be listed:

BOT: The build-operate-transfer is a contract under which the concessionaire designs and provides the woks ready for operation; after the concession period the works are taken over by the project owner.

BOOT: The build-own-operate-transfer is a contract under which the concessionaire designs and provides the works ready for operation, and then owns, operates and maintain the works during the concession period. At the end the works are revert to the project owner.

DB: The design-build contract is intended for the contractor who design and provides the work.

DBFO: The design-build-finance-operate is a contract under which the concessionaire obtains finance, designs and provide the works for operation, and then operate and maintain the works during the concession period, receiving a payment in respect of the outcome of the operation. At the end the works are taken over by the project owner.

EPC: The engineer-procurement and construct contract is a contract under which the contractor provides the works ready for operation by the employer and may also finance part of the cost of providing the works.

ROT: The repair-operate-transfer is a contract under which the concessionaire rapairs an existing work and then operate and maintain it during the concession period. The work is the taken over by the owner.

For each type of contract specific procedures should be adopted and adapted.

LAYOUT OF THE ADAPTED QUALITY SYSTEM FOR TUNNELLING

STEPS FOR THE LIFETIME OF A TUNNEL

⇓

<table>
<tr><th>QUALITY ITEMS ⇓</th><th>General planning</th><th>Design</th><th>Financing</th><th>Procurement</th><th>Characterization</th><th>Monitoring</th><th>Construction</th><th>Approval</th><th>Claims and controvercies</th><th>Operation</th><th>Maintenance</th><th>Installations</th><th>Statistics</th></tr>
<tr><td>Management responsibility</td><td colspan="13">Authorities and mandates for involved people</td></tr>
<tr><td>Quality planning</td><td colspan="13">Definition of procedures for the various activities and co-ordinator</td></tr>
<tr><td>Contracting</td><td colspan="13">Administrative, legal and technical rules</td></tr>
<tr><td>Control and inspections</td><td colspan="13">All the tests, approval, inspections, verifications</td></tr>
<tr><td>Document control</td><td colspan="13">Issues, control and approval of documents</td></tr>
<tr><td>Purchasing And Contractors</td><td colspan="13">Control of materials and contractual works</td></tr>
<tr><td>Non conformities</td><td colspan="13">Problem definition and notifications</td></tr>
<tr><td>Corrections and preventions</td><td colspan="13">Accurate definition of problems and plannong for preventive and remedial actions</td></tr>
<tr><td>Audits</td><td colspan="13">Indipendent ceck of the organization and methods</td></tr>
<tr><td>Training and information</td><td colspan="13">Internal and external education</td></tr>
<tr><td>Data control and elaborations</td><td colspan="13">Data processing of all phases; economical, statistical, technical, measurements</td></tr>
</table>

This is a general layout of all the possible stages of an underground work (1st row) linked to all the specific actions of quality plans (1st column). Depending on the specific case, an interacting matrix can be implemented following the arguments listed in the various rows.

Table 2. Sequence of interacting stages in tunnelling and quality items.

In table 3 the layout of the matrix for considering the relationship between the various interacting phases is shown. It is then possible to into the desired detail for each item.

TUNNELLING STAGE	Step of the work	Concerned Elements	Negative Effects	Actions	Notes
Work Planning					
Design & Charac-terization					
Financing, Pro-curement & Con-tractual phase					
Construction & Monitoring					
Operation and Maintenance					

Table 3. General layout of matrix aimed to focus the main interacting actions influencing directly or indirectly the success of the work.

The stages of the design can be divided conventionally into:

- Geological and topografical surveys;
- Hydrogeology and environment;
- Design and construction concepts;
- Rock support and reinforcements;
- Sealing and linings;
- Excavation and construction equipment;
- Installations and safety;
- Monitoring, maintenance and statistics.

DESIGN AND CONSTRUCTION CONCEPTS	Step of the work	Concerned Elements	Negative Effects	Actions	Notes
Work Planning	Selection of the geometry of the work and of installations	Type of the work, size and shape, profile. Use of reamer TBM	Inadequacy of the methods or type of work	Experience and lessons learned from similar cases. Adaptability of the project parameters	Discussions and agreements with external parties.
Design & Characterization	Selection of parameters and design methods	Overburden, water, faults, settlements, accessibility. Evaluation of landslides and rock fall.	Accidents, damages, instability, delays and additional cost	Revision of the design and risk assessement	Fundamental step of the whole process
Financing, Procurement & Contractual phase	Detailed description of costs and supplies	Complete knowledge of alternatives; Selection of the appropriate type of contract	Difficulties for the relationship with the involved parties; Resolution of contract	Prepare detailed contracts and establish severe economic constraints	In this step economics, regulations and engineering must agree completely
Construction & Monitoring	Selection of construction method (traditional or mechanized) and site measurements; landfilling or rock waste recovery.	Selection of equipment, personnel, instrumentation, organization of the personnel and phases. Particular conditions for microtunnelling and subsea crossing. Control of blasting cross sections.	Surface occupation, vibrations, dust, settlements, stability of landfills	Detailed and daily control by the field engineers	Fundamental step of the whole process. Daily discussions and claims.
Operation and Maintenance	Provisions for future works.	Provisions for future work. Performance of the installations.	Difficulties for maintenance and inspections	Provisions of the state of the tunnel after long periods of use	

Table 4. Example of the main topics influencing a good tunnelling in form of simplified matrix.

4. Quality and tunnel collapses

There are different levels of importance in the problems which can occurr in tunnelling: for example local irregularities of the section due to poor blasting, damages of the linings due to inadequate preparation of the concrete, delays in construction due to problems in organization. But also some severe accidents can occur: excess of vibrations due to inadequate control, subsidence due to seepage or poor reinforcing. And finally the major accidents, that can compromise the regular constrcution of the tunnel: overall instability due to inadequate design of the support, sudden chimneying to due unpredicted geological conditions, low capacity of advancing of the TBM due to excess of rock stresses, instability at the face due to unbalanced pressures for the shield TBM.

The above mentioned are only schematic conditions which are considered in the Quality procedures among the so called non conformities. The simple cases are solved with agreements or technical claims. The severe cases fall into technical and administrative controvercies and the worst cases, the collapses, could involve also important legal actions and resolution of the contracts.

It is clear that the application of Quality procedures cannot avoid the problems or the collapses, because there are always uncertainties and difficulties to control the construction phases. Nevertheless Quality plans should emphasize the demand for continuity and interaction of planning, investigation, conceptual design, detailed design and construction: each is dependent to a degree on the others.

A site investigation, for example, needs to be directed to obtaining information of particular relevance to a specific form of tunnelling; where unexpected features are revealed, the tunnelling strategy may need to be reconsidered and the site investigation appropriately varied. Conceptual design and construction are particularly interdependent since the former may depend upon quite specific features of the latter for success, with the need to ensure that these are rigorously implemented.

In a similar way, in some cases the due attention is not paid in selecting the most appropriate geotechnical model for the soil or for the rock-support-excavation technique interaction, thus transferring into the tender documents the separate items.

Finally, both in relevant works but also in simple tunnelling, monitoring of the behaviour of the rock formation and of the support structures is fundamental, because it allows one to detect immediately irregular actions during construction and also to prevent undesired negative evolution in the stability conditions.

The collection of data from cases in which collapses have occurred is not easy, but from the technical point of view it is essential to look for the causes. Quality

Figure 1. Collapse of the slope just during the construction of the portal of the tunnel.

Figure 2. Collapse of the first phase support, due to inadequate estimation of rock parameters in design.

procedures cannot be a substitute for understanding in depth the nature of the engineering uncertainties and the best solutions which can give advance consideration of real problems.

4. Conclusions

Technical features of the Quality in tunnelling are identified in the sum of several requirements and in the monitoring of the performance of the structure.

In tunnelling there are two levels of quality as far as performance is concerned. One is related to the use and efficiency of the underground structure, and this is taken into account during the planning and operation phase. The other is related to the characteristics of the works, and this is taken into account during the design and construction phase.

Quality goals as far as expectations are concerned are linked to the adherence to the construction programme, keeping within foreseen costs and limiting the number of claims.

These desired features can be fulfilled through well planned and controlled stages:

In general if the scheduled actions, the design requirements, the indications of the investigations and the contractual rules are observed, a good result for the work is possible.

Each involved party should estimate and describe the uncertainties and risks that are connected to their responsibilities such as the owner for the planning and for the tender and the project manager for the coordination and financing.

The final objective for quality in tunnelling is to limit the number of uncertainties, whether these are related to communication, allocation of risk, technical or financial issues.

Procedures dealing with planning, control and how to manage changes to set procedures and design can ensure that the end product is such that client satisfaction can be achieved, time and cost overruns limited and resulting claims minimised. At the same time proper procedures can ensure that knowledge is transferred not only between parties during the project phases but to parties after completion of a project, including for instance universities and other independent organisations.

Tunnelling practice draws upon experience from similar or comparable projects; the success of new construction is based on these experiences. Similarly success in management largely depends on the ability to draw upon and adapt this experience, leraning the lessons of failures as well as success, and bearing in mind that Quality plans help to organize the works, but they don't offer the key for the practical facing to technical or design problems.

In all case, a list of procedures will not be able to avoid by itself failures: the permanent updating of technical knowledge and the detailed study of each project are probably the only way to fulfill quality requirements, intended as an overall tool to measure the true validity of a tunnel project. In particular experience and both theoretical and practical knowledge of tunnelling can help in finding quick answers in case of unexpected problems, which clearly cannot be solved only referring to heavy procedures.

Bibliography

[1] AFTES G.T. 14 : Les methodes de diagnostic pour les tunnels revetus. Tunnels et Ouvr. Sou. , 131, 1995.

[2] AFTES G.T. : La demarche qualité en travaux souterrains. Tunnels et Ouvr. Sou., 174, 2002.

[3] CEN/TC 250/SC 7 "Eurocode 7 - Geotechnical Design", draft EN 1997, Delft.

[4] FIDIC Contracts, International Federation of Consulting Engineers, Lausanne 1999-2000.

[5] ITA - AITES Report of Working Group 6 "Maintenace and Repair: Study of methods for repair of tunnel linings".

[6] ITA - AITES Report of Working Group 16 "Quality in Tunnelling", Lausanne, 2003.

[7] ISO International Organization for Standardization: the ISO 9000:2000 Standards on Quality, Geneve, Switzerland.

[8] Anderson J.M. "World wide research points to the need for new approaches to control tunnelling risks" Proc. Tunnelling Under Difficult Ground Conditions", Basel, 1997.

[9] Harrison F.L. "Advanced Project Management", Gower, Aldershot, 1981.

[10] Muir Wood A. "Tunnelling: Management by Design", E&FN SPON, 2000.

[11] Pelizza S., Grasso P. "Increasing Confidence in the Expected Safety Factor Using Information from Tunnel Collapses and Closures" Proc. Int. Symp on Ground Challenges & Expectations in Tunnelling Projects, Cairo 1999.

[12] Pelizza S., Oggeri C., Peila D., Brino L. (2000) - "Key points for static approval test of underground works" - Proc. World Tunnel Congr. ITA 2000, Durban, 13-18 maggio, pp 395-398, SAIMM.

[13] Rawlings C.G., Lance G.A., Anderson J.M. "Pre construction assessment strategy of significant engineering risks in tunnelling" Underground Construction in Modern Infrastructure, Stockolm, 1998.

[14] Vlasov S.N., Makovski L.V., Merkin V.E. "Accidents in Transportation and Subway Tunnels" Elex KM. Moscow 2001.

Aknowledgements: The author wish to thank all the member of the ITA-AITES Working Group 16 for the contribution in the development of the Report on which this paper is based.

Part V

Additional aspects

Hard rock TBM advance rates

Michael Alber[1]

[1] Ruhr-University Bochum, Engineering Geology, Universitätsstr. 150, D-44780 Bochum, Germany,
e-mail: michael.alber@rub.de

Abstract:

This paper deals with a practical approach for estimating hard rock TBM advance rates. This approach involves the estimation of rock mass properties, their influence on a TBM`s penetration and utilization rates as well as the design of support systems for hard rock TBM tunnels. As many of the necessary geotechnical parameters are derived from engineering rock mass classification systems, emphasis is also placed on their benefits and pitfalls when classifying anisotropic rock masses.

1. Introduction

The two basic tasks in TBM tunneling, as for any underground excavation, are (i) to excavate and remove the rock and (ii) to maintain tunnel stability until the final lining, if necessary, is installed. Application of rock engineering for TBM drives should be directed towards solving these tasks which define the TBM performance and finally the project delivery time. Performance prediction of TBM drives requires the estimation of both TBM penetration and utilization. The rate of penetration (m/h), defined as the distance mined divided by TBM operating time, depends on the TBM design (thrust, torque, disc size and spacing) as well as the rock mass properties that withstand penetration. TBM utilization is defined as the percentage of shift time during which mining occurs and depends on equipment and non-equipment downtimes [1]. Maintenance of equipment and tunnel are typically scheduled on a regular basis and are mainly fixed quantities. Two major factors in estimating utilization rates are (i) the stability of the rock mass and (ii) the proper support systems as incorporated in the TBM. Figure 1 shows schematically two pairs of interactions between TBM and rock mass that influence performance data. Rock mass strength and disc thrust determine the TBM penetration while rock mass behavior often requires support installation and influences so TBM utilization.

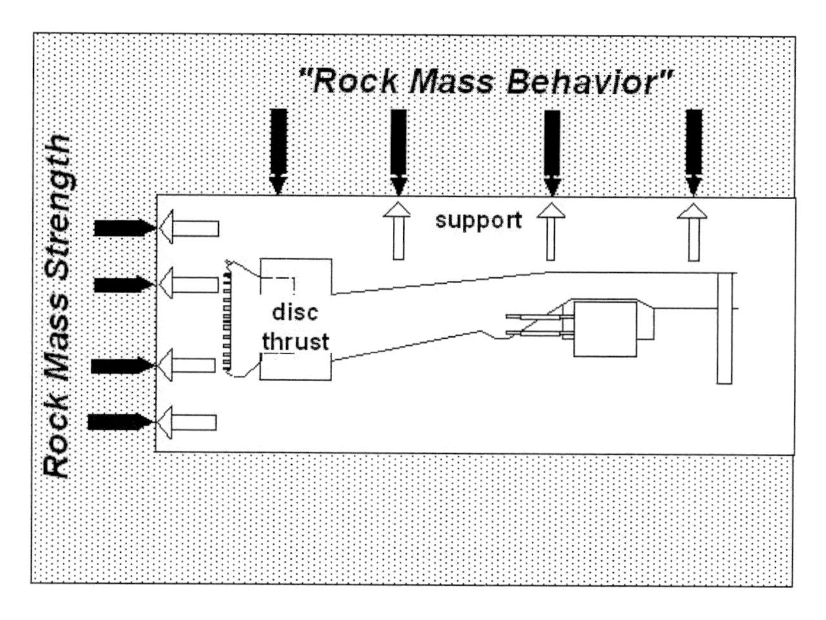

Figure 1: Interactions between rock mass and TBM-system.

The advance rate may be estimated by

$$\text{Advance rate (m/h)} = \text{Penetration rate (m/h)} \times \text{TBM-utilization (\%)} \qquad (1)$$

Rock mass strength, here the uniaxial compressive rock mass strength σ_{cm} is mainly derived from strength criteria based on engineering rock mass classifications. From the author's experience, rock mass classifications are often used and even more often abused and misused, so there is a need for clarification of benefits and flaws of those classifications systems.

2. On the use of rock mass classifications for engineering purposes

There are two basically different classification philosophies available. First there are classification systems that observe the rock mass behavior around an underground excavation and classify this behavior into rock mass classes. Examples of this approach are [2, 3, 4, 5] Their approaches base on observations during tunneling and appreciate the fact that any mechanical calculations in rock engineering are nearly vain efforts. This approach is favored in countries in which the owner takes the risk for the geology.

The second classification philosophy, and on this is focused in this chapter, entertains the idea that a few rock mass parameters may be rated according to their importance to the rock mass behavior and are added or multiplied up to a number reflecting the rock mass quality. Examples of this approach are classification by Deere [6], Bieniawski [7], Barton [8, 9], Palmstrøm [10, 11], and Hoek [12, 13], Some important rock mass properties may be estimated from those rock mass qualities. This approach reflects the need of engineers for input values for design, i.e. numerical or analytical analysis for estimating stability and assessing support systems. It may be noted that this "forward-approach" is favored in countries where the risk for the geology is not taken by the owner and fixed-price contracts are awarded. Consequently, the designer/contractor is forced to minimize his risk by extensive design studies for arriving finally at a fixed price for the complete underground structure. Moreover, as more and more future projects will probably awarded on the basis of fixed prices or Build-Operate-Transfer (B-O-T) this approach may even be more in demand than ever.

It is idle to compare the advantages and disadvantages of both classification approaches as both have their respective history of successful and unsuccessful projects. For this paper the use of forward rock mass classifications is accepted but it is attempted to demonstrate the importance of the understanding of geology within the classification process for estimating rock mass parameters for design process. The rock mass features used by classification systems mentioned above as well as their outputs in terms of geomechanical parameters are given in Table 1.

The rock mass modulus is probably the most important parameter when employing numerical modeling and is the most difficult parameter to estimate. The uniaxial rock mass strength, usually estimated via the Hoek-Brown failure criterion, is the most important parameter for classifying the rock mass behavior in basic contractual classes stable, friable and squeezing [14] and is paramount for estimating advance rates for hard rock TBMs.

A very well investigated regularly jointed limestone will be used for deriving above mentioned geotechnical parameters from rock mass classifications for demonstrating benefits and pitfalls.

2.1 Geological considerations

Rock mass classifications attempt to describe a given rock mass by a numerical value. In this process, the rock mass features are homogenized ignoring thus the anisotropic nature of a rock mass. Any rock engineering literature fosters the fact that rock masses are anisotropic by nature [15, 16] but very seldom this fact is followed up in rock mass classifications process and even less in numerical analysis. It may be shown that application of rock mass classification must be executed carefully for incorporation of the anisotropy of rock masses.

CLASSIFICATION / INPUT ROCK MASS FEATURE	Deere	Bieniawski	Barton	Palmstrøm	Hoek
Intact rock strength σ_C		X	(X)	X	X
Discontinuity Spacing		X	(X)	X	X
No. of Discontinuity sets		(X)	X	X	(X)
RQD	X	X	X	(X)	X
Discontinuity condition					
Persistence		X		X	X
Aperture		X	X	X	X
Roughness		X	X	X	X
Infilling		X	X	X	X
Weathering		X	X	X	X
Groundwater		X	X		X
Applied stresses			X		
Index	RQD	RMR	Q	RMi	GSi
Output					
E_M: Rock Mass Modulus	X	X	X	X	X
σ_{CM}: Uniaxial Rock Mass Strength		X	X	X	X
Stand-up time/max. unsupported span		X	X		

Table 1: Input and output parameters of some popular classification systems.

The anisotropy of a rock mass is mainly due to different discontinuity spacing in different directions. It is proposed that the directional uniaxial compressive rock mass strength as well as the directional rock mass modulus are necessary input parameters for rational tunnel design.

This approach may be demonstrated by classifying a regularly jointed limestone rock mass. The Triassic limestone rock mass features 4 discontinuity sets (1 bedding and 3 joint sets). 1200 data from 30 scanlines yielded the data summarized in Table 2. Additionally, a relation between bedding thickness and joint spacing for each set could be observed (Fig. 2). Based on the field measurements a 3-dimensional reference cube of edge length 10 m was modeled (Fig. 3) and compared with reality. This rather realistic model of volume 10000 m^3 contains roughly 460000 rock blocks with an average block volume of 0.00217 m^3. The limestone may be described with ISRM terminology [17] as made of "small blocks" and is of "blocky shape".

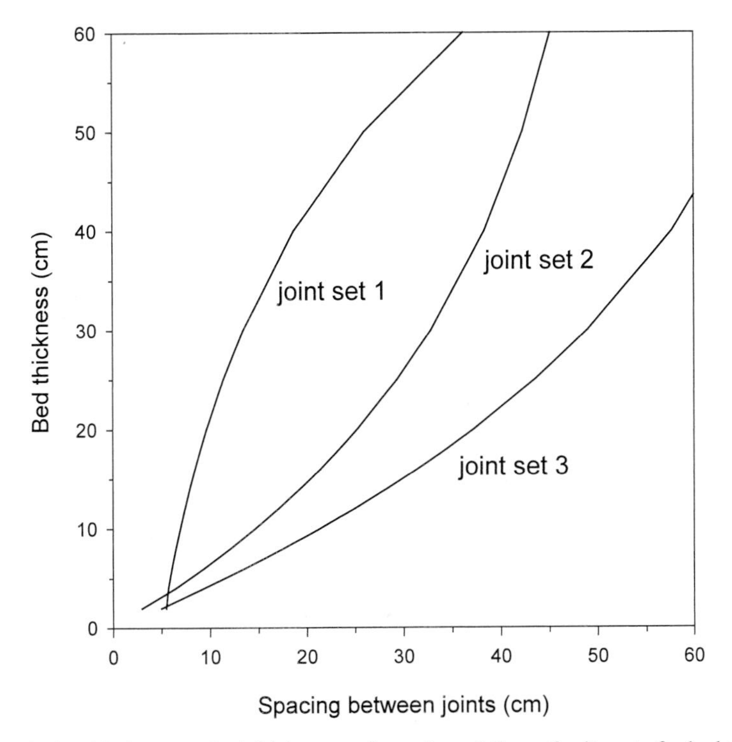

Figure 2: Relationship between bed thickness and spacing of discontinuity sets for bed thickness < 0.6 m.

A way to geomechanically deal with such a rock mass made out of well defined blocks is the use of a distinct element code. This approach is in practice seldom used. Alternatively to dealing with the mechanics of single blocks and their interactions, the limestone rock mass was homogenized by using the average discontinuity spacing as shown in Table 2.

	BEDDING	JOINT SET 1	JOINT SET 2	JOINT SET 3
dip direction/ dip (°)	320/18	173/82	81/86	123/86
spacing distribution	n/a	negative exponential		
mean $\overline{x}$	0.22 m	0.106 m	0.194 m	0.37 m
variance	n/a	0.0112 m	0.0376 m	0.137 m

Table 2: Orientation and spacing of discontinuity sets

Following the procedure as described by [18] and [19], the total discontinuity frequency was calculated for any direction 1° apart in trend and plunge. The resulting anisotropy of the linear discontinuity spacing S is shown in Figure 4. Largest average linear discontinuity spacing of 0.24 m may be observed when drilling at trend/plunge 0°/85°. When drilling vertically in that rock mass the average linear discontinuity spacing is about 0.18 m while an underground structure trending horizontally say NE-SW will encounter a linear discontinuity spacing of about only 0.05 m along the tunnel axis.

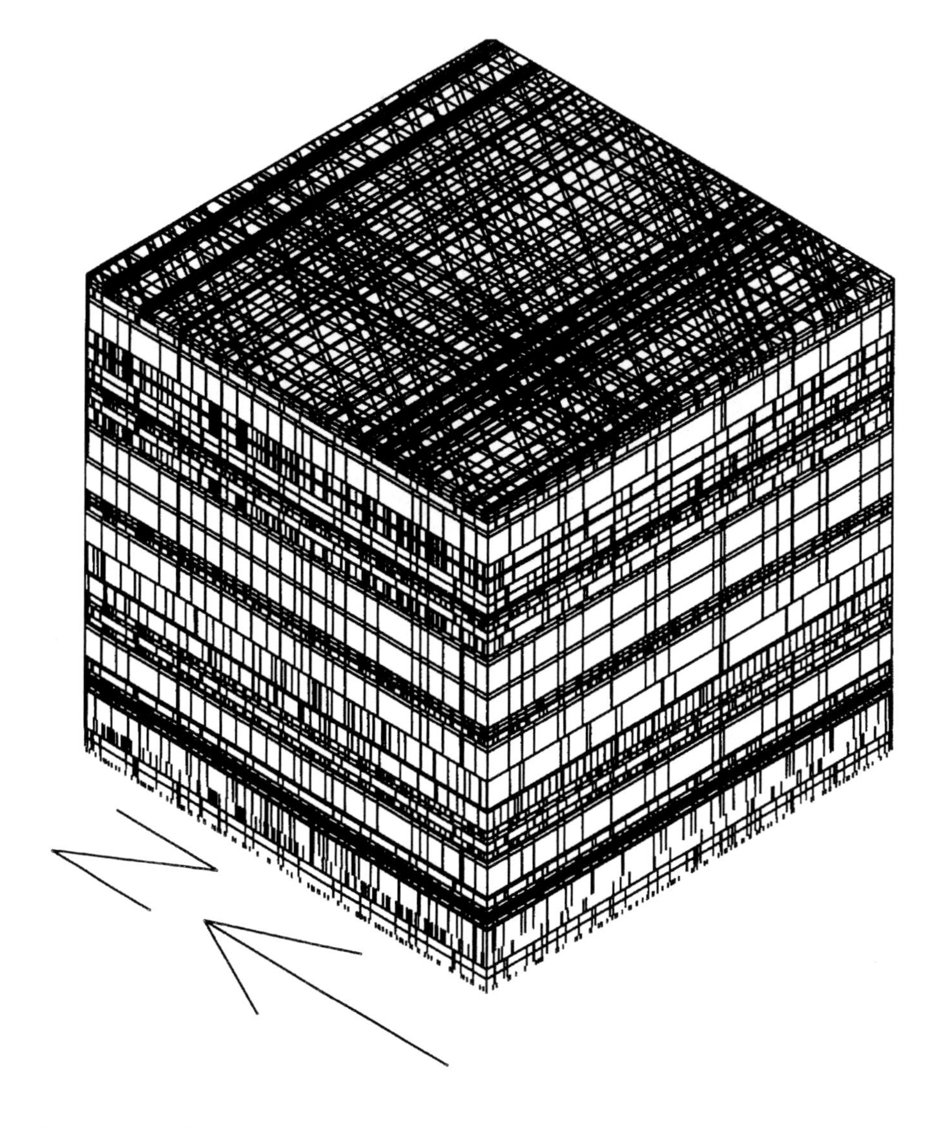

Figure 3: 3-D reference cube of edge length 10 m of the limestone rock mass. For convenience, model was rotated so that bedding planes are horizontal.

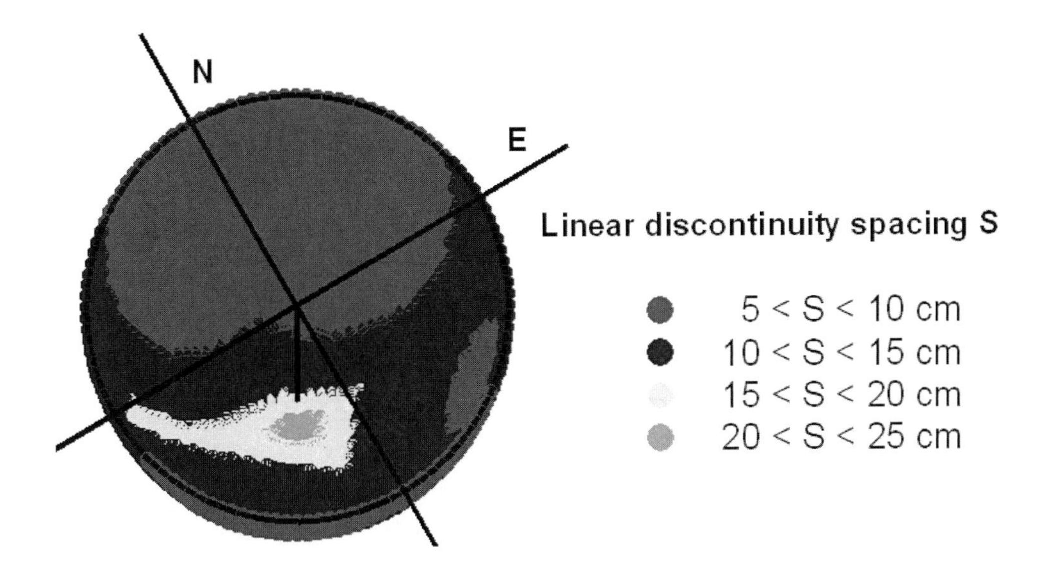

Figure 4: Lower hemisphere projection of linear discontinuity spacing S.

2.1.1 Results from classification systems

Several rock mass classifications, Q [7], RMR [8], GSI [12] and RMi [10, 11] were executed to investigate their sensitivity to the observed anisotropy. The results are shown in tables 3 to 5. It may be noted that RMR and GSI are assessed together as they significantly differ in their ratings only for very low RMR values, which is here not the case. As mandated by [12], GSI may be estimated from RMR by subtracting 5 ratings. It may be noted that Q, RMR and GSI yield the highest values for high RQD and/or discontinuity spacing, which is achieved when subvertical drilling is executed.

Clearly, discontinuity spacing varies with orientation of a sampling line, be it a borehole or a tunnel alignment. The minimum average discontinuity spacing S_{min} = 0.06 m is in orientation 144°/24° and the maximum S_{max} = 0.24 m in orientation 348°/81°. Accordingly, the RQD-values vary with orientation of the sampling line with the extreme values RQD_{min} = 56% and RQD_{max} = 93%. Already two important input parameters for rock mass classification show strong anisotropy. Those two anisotropic input parameters lead to directional rock mass classification indexes as shown in Figure 5.

It may be seen from those examples that rock mass classifications yield indexes that are sensitive to the orientation of the scanline from which their input data were collected. In turn, classifications from typically vertical boreholes yield indexes that need to be transformed into the orientation of an underground excavation.

PARAMETER	VALUE
RQD	55 –93
Joint number: 4	Jn = 12
Joint roughness	Jr = 1.5 – 3
Joint alteration	Ja = 1
Joint water: dry	Jw = 1
Stress reduction	SRF = 1
Q: fair – good	6.875 – 23.25

Table 3: Q-system classification.

PARAMETER	RATING
σ_C = 40 MPa	5
RQD = 55 – 93%	11 – 19
Discontinuity spacing: 0.06 – 0.24 m	2 – 11
Discontinuity condition	20
Groundwater: dry	15
Orientation adjustment: none	-
RMR/GSI: fair – good	RMR: 53 – 69 , GSI: 48 - 64

Table 4: RMR and GSI classification.

PARAMETER	VALUE
Joint roughness	Jr = 1.5 – 3
Joint alteration	Ja = 1
Joint size and continuity	iL = 3.1875 (average)
Joint condition factor	JC = iL·(Jr/Ja) = 2.4 – 2.8
constant $D = 0.37 \cdot JC^{-0.2}$	D = 0.3106 – 0.27
Jointing parameter (block volume Vb = 0.002 m^3)	$JP = 0.2 \cdot \sqrt{JC} \cdot Vb^D$ = 2.4 – 4.8
RMi (for σ_C = 40 MPa)	RMi = $\sigma_C \cdot JP$ = 1.85 – 3.34

Table 5: RMi classification.

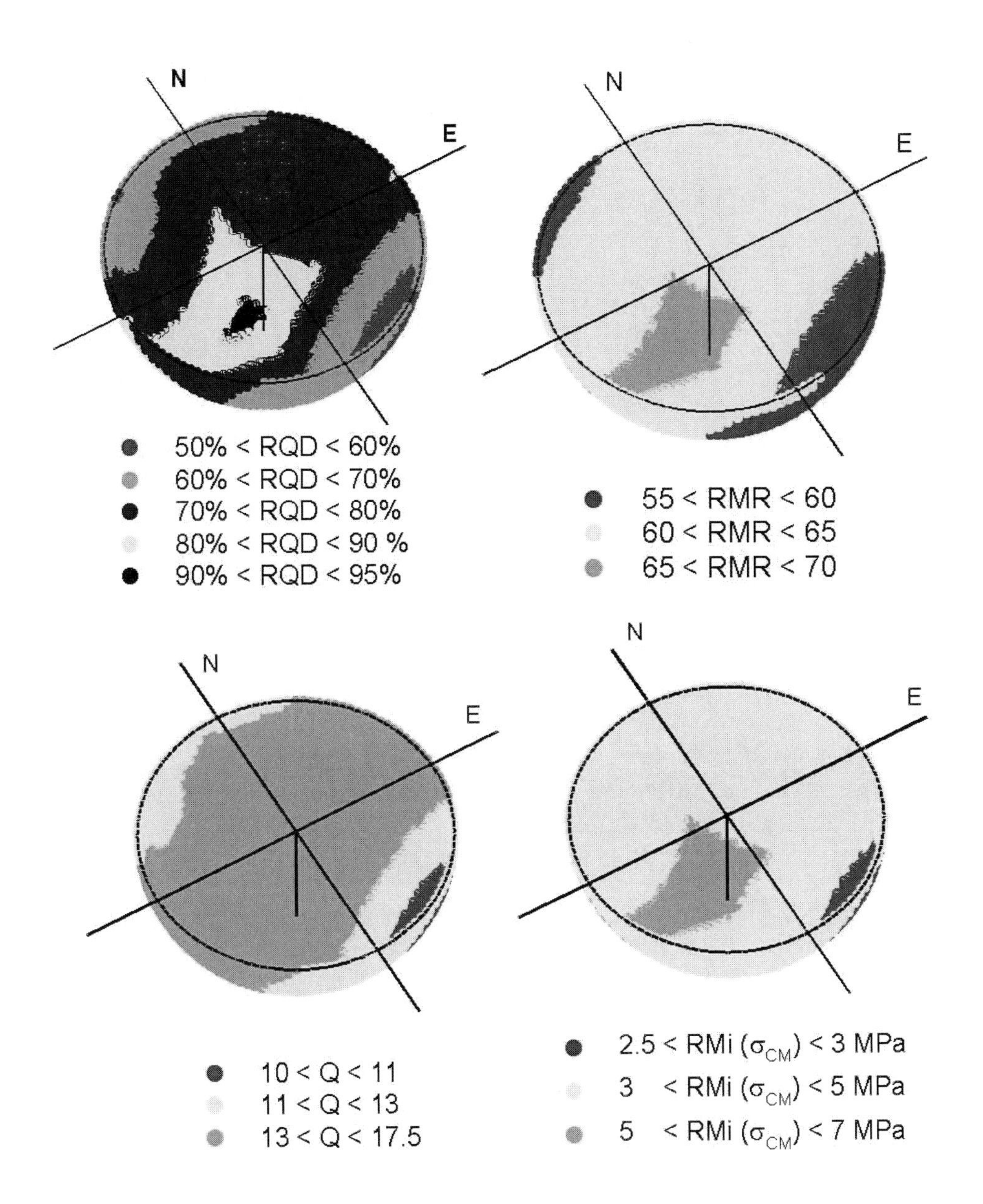

Figure 5: Lower hemispherical projections of directional rock mass classification indexes (RQD, RMR/GSi, Q, RMi) based on discontinuity spacing-induced anisotropy as given in Figure 4.

2.1.2 *Rock mass strength from classification systems.*

The uniaxial compressive rock mass strength σ_{CM} may be derived from the classification systems according to the formulas below. The average intact rock strength σ_C was determined, along with other geotechnical parameters in numerous laboratory tests reported by [20], to be 40 MPa.

For RQD/S [5]:
$$\sigma_{CM} = f(S, \sigma_C) \tag{2}$$

For Q [9]:
$$\sigma_{CM} = 5\gamma(Q \cdot \frac{\sigma_C}{100})^{1/3} \text{ with } \gamma = 2.61 \text{ g/cm}^3 \tag{3}$$

For RMR [7] or GSi [12]:
$$\sigma_{CM} = \sigma_C \sqrt{s} \text{ with } s = e^{\frac{RMR-100}{9}} \tag{4}$$

For RMi [11]:
$$\sigma_{CM} = RMi \tag{5}$$

Those estimates are also plotted in lower hemispherical projection to account for their dependence on the direction of application (Figure 6). For the uniaxial compressive rock mass strength σ_{CM} all but Barton's estimates are of the same magnitude between 2.5 and 7 MPa and all show a factor of 2.5 between the lowest and highest rock mass strength. Barton's estimates of σ_{CM} demonstrate the well known lack of the intact rock strength in the classification parameters.

The typical approach when using rock mass classifications is to report one value or a certain range for the respective index. The engineering geologist or the design engineer should however be well aware of the fact that one index value is misleading by not accounting for the discontinuity spacing induced anisotropy of the rock mass. The same holds true for the estimated uniaxial rock mass strength.

2.1.3 *Rock mass modulus from classification systems.*

The rock mass modulus E_M is probably the most important parameter when employing numerical methods for design of underground excavations in rock. The stiffness of the rock mass determines the deformation around an underground excavation on which, in turn, support measures are based upon. After all, control of deformation is the main task in tunneling. Deformation of a discontinuous rock subjected to stress comprises the deformation of the intact material and of the closure of discontinuities. Accordingly the frequency of discontinuity spacing in a certain direction should be appreciated when estimating the rock mass modulus from classification systems with the formulas shown below.

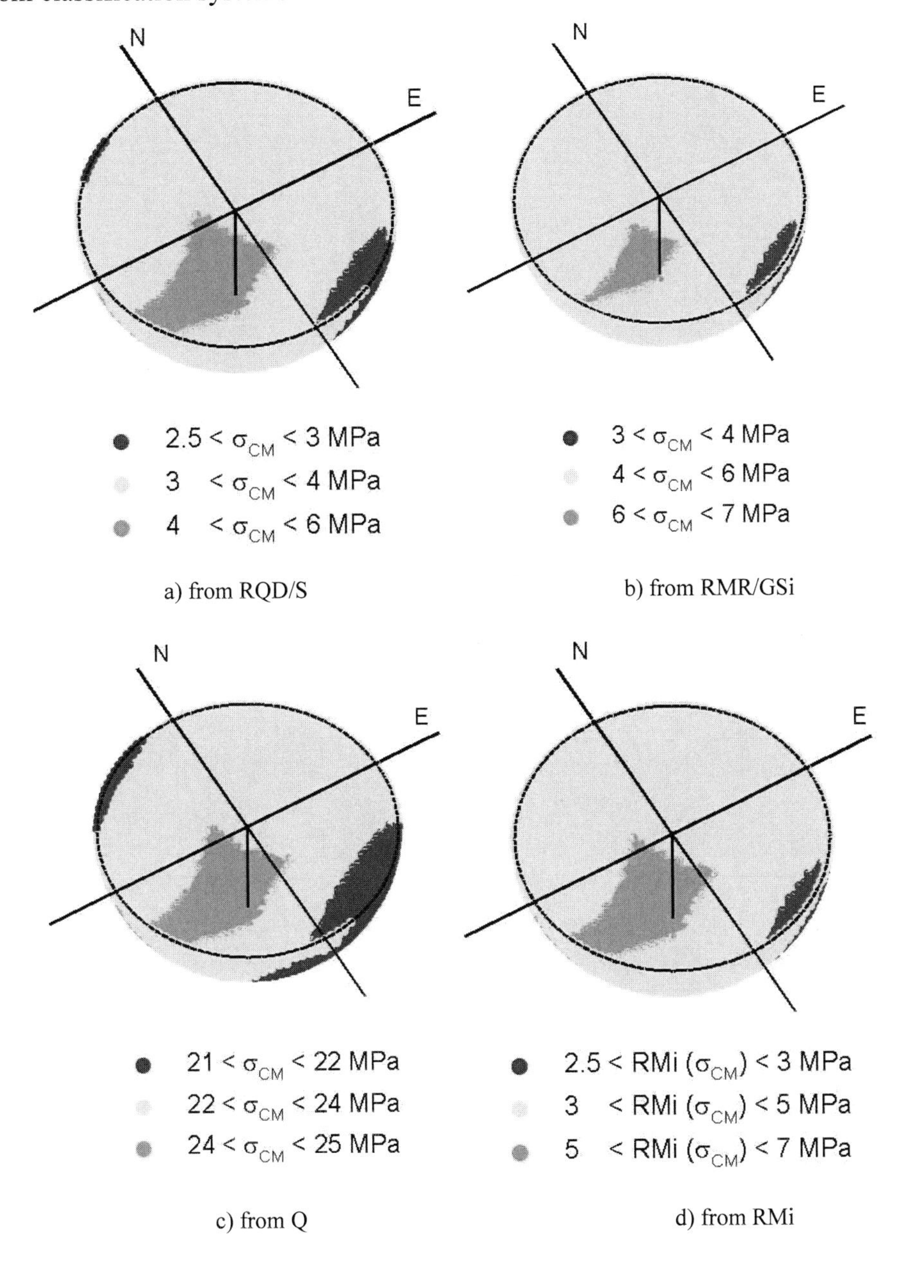

Figure 6: Lower hemispherical projections of directional uniaxial rock mass strength estimated from RQD/S, RMR/GSi, Q and RMi) based on discontinuity spacing-induced anisotropy in Figure 4.

For RQD [21]:
$$E_M = E \cdot (0.0225 \cdot RQD - 1.25) \quad (6)$$

For Q [9]:
$$E_M = 25 \cdot \log Q \quad (7)$$

For RMR[7] or GSI [12]:
$$E_M = 10^{\left(\frac{RMR-10}{40}\right)} \text{ or } E_M = \sqrt{\frac{\sigma_C}{100}} \cdot 10^{\left(\frac{GSI-10}{40}\right)} \quad (8 \text{ and } 9)$$

For RMi [11]:
$$E_M = 5.6 \cdot RMi^{0.375} \quad (10)$$

Those estimates are also plotted in lower hemispherical projection to account for their dependence on the direction of application (Figure 7). In contrast to estimates of the rock mass strength from classifications the estimates of the rock mass modulus show a very large scatter between the different classifications. It may be noted that again estimates from RMR and GSI are plotted in one figure, though the GSI estimates of E_M are slightly lower. The values of E_M also reflect anisotropy, but the values range from E_M = 0.5 GPa (Coon's approach) in the most unfavorable direction to a value of E_M = 31 GPa (Barton's approach) in the most favorable direction. Recalling that the intact rock Young's Modulus E = 25.7 GPa, a rock engineer should be careful to employ RMR, GSi, and Q estimates of E_M as well as some of the higher values of RQD estimates. A more rock mechanical approach by the author [22] involving stress levels, fracture stiffnesses and linear discontinuity spacings suggest upper limits for E_M as given by RMi.

Rock masses consist of intact rock and discontinuities which are controlling strength and deformability of the rock mass [23]. If the intact rock is isotropic then the discontinuity orientation and spacing is the main cause for anisotropic rock mass behavior and should accordingly be incorporated in rock mass classifications.

In practice, however, classifications are applied by homogenizing the rock mass to average conditions and arriving thus at one number for the rock mass quality. It is hardly a surprise for any engineering geologist that such a number is far from reflecting geological parameters which are inherently different in various directions.

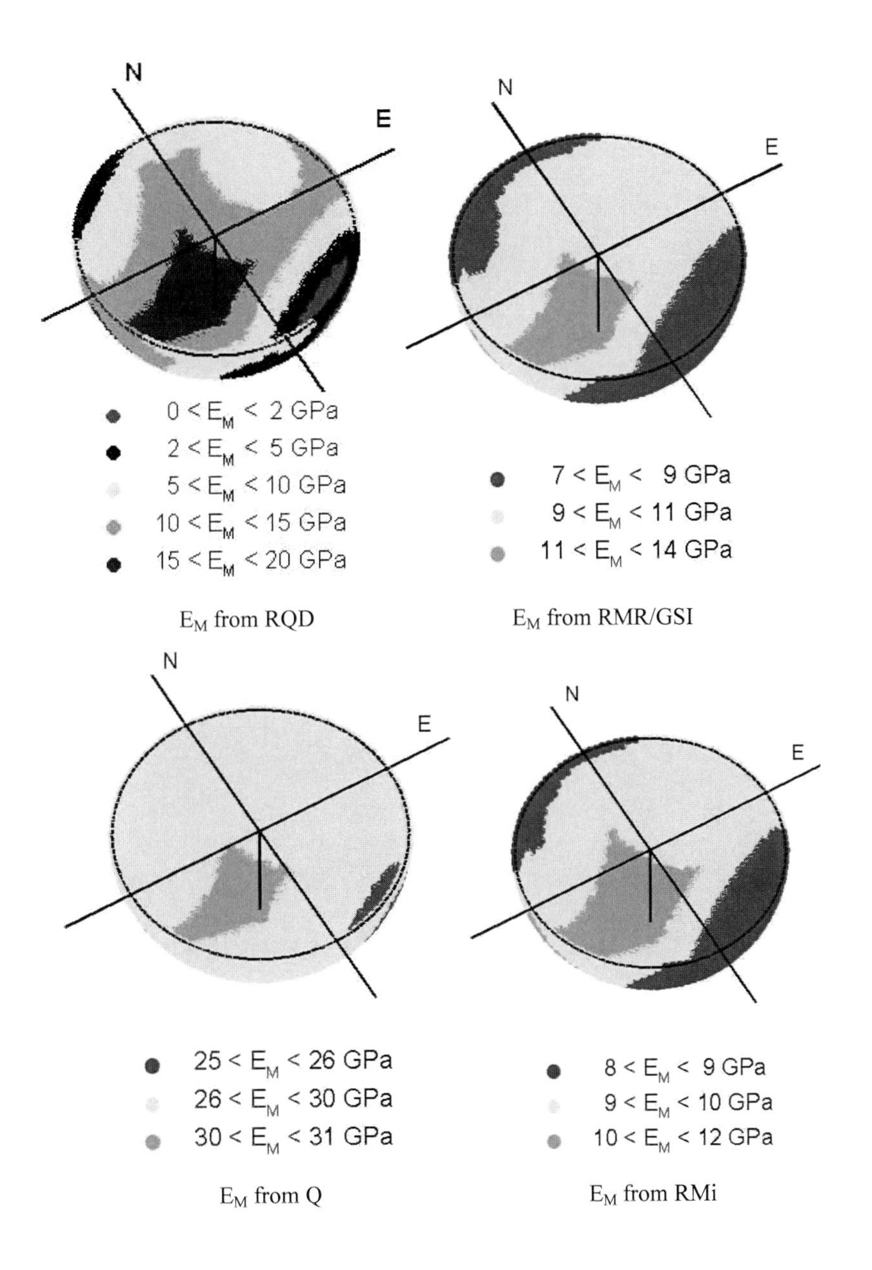

Figure 7: Lower hemispherical projections of directional rock mass modulus E_M estimated from RQD, RMR/GSI, Q and RMi) based on discontinuity spacing-induced anisotropy in Figure 4.

It is believed that rock mass classification of designated strong anisotropic rock masses [12, 23] will benefit from incorporation of directional rock mass features into the classification process.

This example of application of classification may show that the consideration of directional discontinuity spacing in sedimentary rock leads to directional rock mass qualities. Consequently, the derived geotechnical parameters show distinct anisotropy which in turn may lead to grave consequences for numerical analysis of underground structures in rock.

This application of rock mass classifications to an anisotropic rock mass may show that:

- Rock mass classifications may also reflect discontinuity-induced anisotropy.
- Geotechnical parameters derived from so developed rock mass numbers are accordingly anisotropic.
- Knowledge about the local and regional geology is essential when applying rock mass classifications.

Finally, any user of rock mass classification is advised not to depend on one classification system and should carefully cross-check the derived geotechnical parameters.

3. Estimation of hard rock TBM advance rates

Advance rates of hard rock TBMs comprise the penetration rate and the utilization rate. The penetration rate is the distance mined divided by the time during a continuous excavation phase, i.e. the time while the cutterhead rotates and is pushed towards the face. The utilization rate is percentage of the time of a production day while the actual mining takes place. The utilization rate is typically below 66% as the graveyard shift is used for TBM and tunnel maintenance. This maximum utilization rate is further reduced by delay due to support installation, rock jams, power outages, mucking problems and so on.

It has been attempted to connect hard rock TBM advance rates directly with rock mass classification indexes, however, there exits little correlation between advance rate and indexes [24] [25].

3.1 Estimation of penetration rates

The penetration rate of a TBM is typically estimated from some properties of intact rock [26, 27, 28, 29], often the uniaxial compressive intact rock strength σ_C. However, the influence of discontinuity spacing and conditions on the penetration rates is well known [30, 31]. Only the method of the Norwegian Institute of Technology [32, 33] takes rigorously the influence of discontinuity spacing on the penetration rate via the fracturing factor k_s into account.

Alternatively, a prediction method based on the uniaxial compressive rock mass strength σ_{CM} is here used. This rock mass parameter includes both σ_C and discontinuity features and covers thus the main rock mass characteristics that influence the penetration rate of a TBM. The uniaxial rock mass strength σ_{CM} may be estimated by characterizing a rock mass with the RMR-System [7] and the use of the Hoek-Brown failure criterion. The well known formula is

$$\sigma_{CM} = \sigma_C \times \sqrt{s} \tag{11}$$

where

$$s = \exp^{\frac{RMR-100}{9}} \tag{12}$$

and RMR is the rock mass rating of a rock mass with full rating for parameter groundwater and without reduction for discontinuity orientation. It should be noted that the RMR for TBM-excavated drives may be increased [34] by

$$RMR_{TBM} = 0.84\ RMR + 21 \tag{13}$$

to account for the better preservation of the intrinsic rock mass quality compared to drill-and-blast advance, for which the RMR system was developed.

It was found from the analysis of 55 km TBM tunneling involving 5 different TBMs (17" discs size) that there exists a relationship between the penetration rate of a TBM and the uniaxial compressive rock mass strength. To compare the different TBMs with their different cutterhead revolutions per time, different thrusts and number of discs the penetration is expressed as the specific penetration SP:

$$SP = \frac{\text{penetration in cm}}{\text{rev./min} \times \text{MN thrust per disc}} \quad (14)$$

This relationship is depicted in Figure 8. As expected, the specific penetration of a TBM increases as the uniaxial rock mass strength decreases. This holds true down to a rock mass strength of about 15 MPa. Rock masses in that low range of strength are often dominated by closely spaced discontinuities. The formation of rock chips between two discs may be hindered in those very blocky rock masses. Single blocks may be ripped out of the face and reground. In summary, the penetration rate may slow considerably down in those weak rock masses. A similar curve was observed for 50 km TBM drives with 15.5" discs. The approach presented along with the database is discussed in detail elsewhere [34]. Recently, [35] confirmed those findings by extending the Q-System from 1974 by a term for rock mass strength to finally arrive at a Q_{TBM} value, from which a penetration rate may be extracted.

It may be noted that there are many similarities between TBM penetration rates and roadheader instantaneous cutting rates when geology is expressed in terms of rock mass strength, either with the approach for TBMs as presented here or by the rock mass cuttability index RMCI for roadheaders as shown by [36].

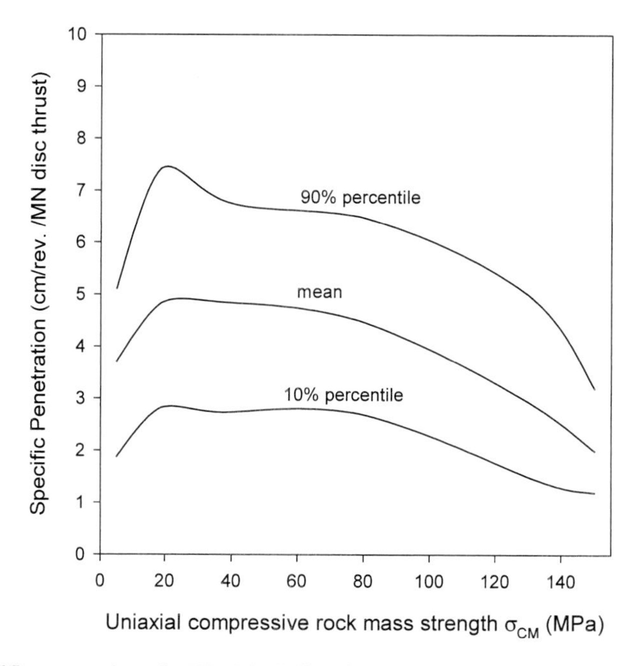

Figure 8: Specific penetration of a TBM (17" disc size) as a function of the uniaxial compressive rock mass strength.

3.1.1 Effects of mixed face conditions on penetration rates

Mixed face conditions, particularly an interbedding of hard and soft rock masses lead, according to Figure 8, to a differential penetration of the discs and consequently to high vibrations at the cutterhead. Besides destroyed discs, the vibrations may finally cause failure of the main bearing. Typically the thrust has to be lowered to a level that leads to bearable vibrations [37]. Figure 9 shows the reduction of the thrust level while a TBM bores in a mixed face situation. The difference in rock mass strengths of the two layers is about 20 MPa and caused differential penetration of the discs with 5 cm/rev /MN disc-load in the dolostone and 4 cm/rev /MN disc-load in the shale. It appears that the reduction of the thrust level to minimize bumping of the discs follows the decrease of the average rock mass strength of the face rock mass, computed according to the percentage of the outcrop of the layers at the face. Table 6 shows that the reduction of the thrust level from 145 to 110 kN/disc lowers the differential penetration from 1.5 mm/rev to 1.1 mm/rev which was apparently enough to avoid the vibrations.

These considerations may help to anticipate the interbedding-induced vibrations. Recently, [38] proposed a mixed-face correction factor for the NIT method [32, 33] based on the proportions of hard and soft layer at the face, respectively.

In this context of TBM problems with rock mass at the face there is another issue to be thought of, namely face instability. There are two failure mechanisms: (i) face instability due to adverse structural geology, which might lead to a jammed cutterhead by wedges, and (ii) face instability due to high rock stresses, which might lead to a closely fracture face might thus impeding high penetration rates. One possible way to lessen the impact of such adverse geological conditions may be to employ a cutterhead with domed shape for evenly distributing the stresses at the face [38].

Thrust level	Dolostone $\sigma_{CM} \approx 35$ MPa	Shale $\sigma_{CM} \approx 15$ MPa
145 kN/disc	5.8 mm/rev	7.3 mm/rev
	$\Delta P = 1.5$ mm/rev	
110 kN/disc	4.4 mm/rev	5.5 mm/rev
	$\Delta P = 1.1$ mm/rev	

Table 6: Penetration P of a TBM in a mixed face situation.

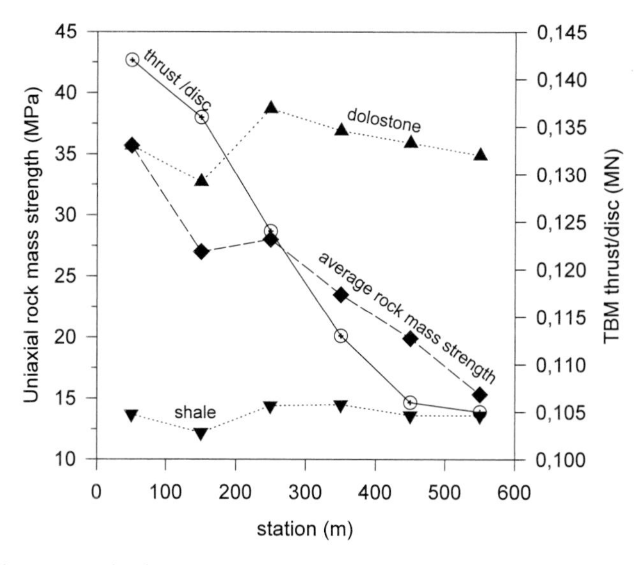

Figure 9: Reduction of thrust level of a TBM in mixed face conditions.

3.2 Estimation of TBM utilization rates

The utilization of a TBM depends mainly on the delay due to ground support. Those delays typically form the basis for contract classifications [39].The prognosis of the utilization rate, defined as the actual boring time as percentage of the available shift time in a given rock mass is barely described in the rock engineering literature. [31] showed for a Swiss project the decrease of the utilization rate with increasing discontinuity frequency, and [40, 41] gave a tentative correlation between utilization and rock classes for rock masses in Hongkong. The estimation of the TBM utilization rate is difficult since it depends on many factors such as maintenance, cutter change, and rock support. [42] proposed a rating system with RQD, water inflow, clay content, maintenance facilities and contractor experience being the main factors for estimating utilization rates for double-shield TBMs. Their rating lead to utilization rates as low as 3 % up to 45%.

In the course of this research utilization rates were mainly seen as a function of support installations. The necessary ground support is a function of the rock mass behavior upon excavation. The respective behavior depends on the ratio of rock mass strength to induced stresses at the tunnel wall. This ratio may be expressed by the Factor of Safety FS:

$$FS = \frac{\text{Uniaxial compressive rock mass strength } \sigma_{CM}}{\text{Tangential stress at wall } \sigma_{\theta}} \qquad (15)$$

A relation between the Factor of Safety at roof centerline and the TBM-utilization was found from the analysis of more than 100 km TBM tunnels [37]. The relation is shown in Figure 10. It may be seen that in stable rock mass behavior the TBM-utilization ranges from 25 to 50% and averages at 40%. For FS between 1.25 and 2 the average utilization reduces to 35%. For FS< 1.25 the TBM's utilization is well below 35% and may even drop to zero in tunnel sections in which the ratio of strength to stress approaches zero.

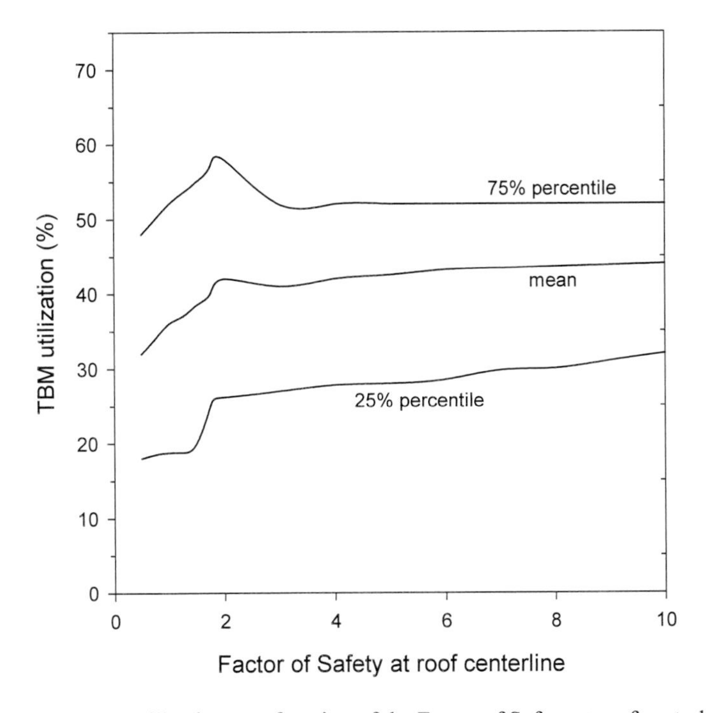

Figure 10: TBM-utilization as a function of the Factor of Safety at roof centerline.

3.3 Design for high speed TBM advance rates

Figure 10 demonstrates the rather wide range of utilization rates as a function of the local Factor of Safety at roof centerline. The main goals for providing support for a TBM-excavated tunnel are (i) to design an efficient support system and (ii) to incorporate this into the TBM production cycle. For example, if the FS at roof centerline is calculated with FS = 0.33, the support works would certainly delay the boring cycle and a utilization rate ranging from 10% to 50%, depending on the degree of integration of the selected support measures into the overall TBM-design.

The next step is to select a support procedure that is able to provide sufficient support. The rock support interaction analysis [42, 43] is, despite its limitations to a hydrostatic stress regime, an excellent tool for this purpose. Common support measures for open hard rock TBM-drives are mechanical or grouted bolts, steel sets and shotcrete. To appropriately apply the rock support interaction analyses for TBM-drives the initial displacement before support installation, reflecting the gap between excavation at the face and the location of support installation, must be assessed. The closest point for support installation is behind the cutterhead i.e. about 2 m from the face. The initial displacement may be estimated with the procedure described by [44] and was estimated for the example shown in Figure 11 to be 0.042 m and the support reaction curves start from this point. The deformation curve calls for an early support installation within the TBM-body. All support measures shown provide a higher pressure than needed and may be employed in this example. Details of the support measures are given in Table 7.

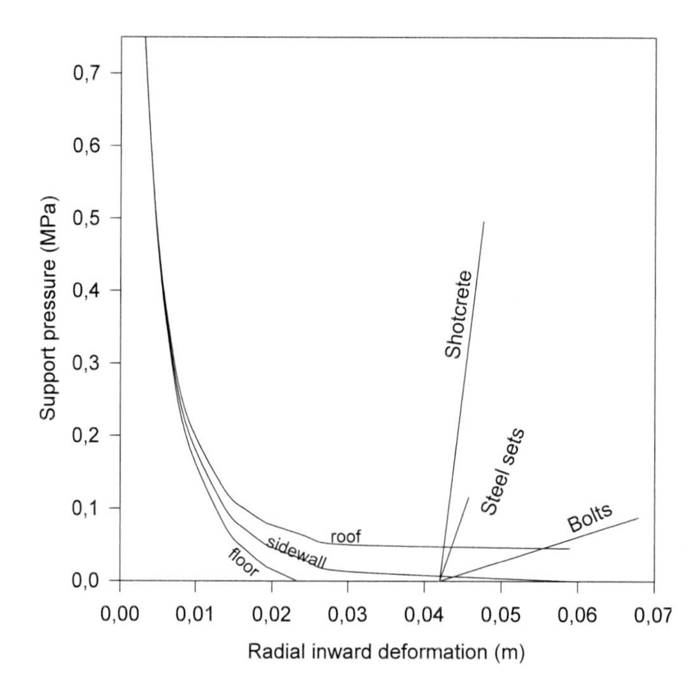

Figure 11: Rock support interaction curve for selection of support systems.

RMR = 45, σ_C = 100 MPa, s = 0.002, m_i =19,σ_v =σ_h = 7.5 MPa, r = 3.5 m	Steel sets: medium steels sets, flange width 0.11 m, spacing between sets 1.5 m
Bolts: mechanically anchored, d = 0.02 m, l = 2 m, pattern 1.5 × 1.5 m.	Shotcrete: σ_C = 35 MPa, thickness 0.05 m, 25 Vol.-% rebound

Table 7: Main input data for rock support interaction analyses.

Support measure [unit]	Weighting factor for support installation within TBM	Support measures needed per lineal m of tunnel	Cumulative support number for example
Swellex bolt [m]	3	19.5 m	58.5
Steel set [m]	4	14.7 m	58.8
Shotcrete [m^3]	50	1.4 m^3	70

Table 8: Selection of support systems based on weighting factors indicating time consumption.

The last step in designing a TBM system for high utilization rate is the selection of the one support system that delays the production cycle the least. The application of shotcrete within the TBM-body, particularly with TBM diameters less than 5 m, is often difficult since the distance from the nozzle to the wall should be about 1.5 m. Moreover, shotcrete is typically used with another support measure which leads to more complex working procedures in a very confined space.

Some quantification concerning the time needed for installation of different support systems for TBM drives may be derived from the Austrian Standard for Underground Works [45] and Table 8 shows the weighting factors as well as cumulative support numbers for the different support measures.

It may be concluded that bolts or steel sets are the least time-consuming support measures for the rock mass under consideration. A well-designed support logistic and a trained mining crew may produce utilization rates of about 40% on average in this overstressed rock mass.

3.4 Project-oriented hard rock TBM advance rates

The findings about the relations between strength of a rock mass and the specific penetration of a TBM as well as between rock mass behavior and TBM-utilization allow to estimate TBM advance rates. The actual penetration rate may be estimated for a TBM with a defined thrust per disc and cutterhead rotational speed, which is penetrating a rock mass with a certain rock mass strength. The average advance rate is computed by reducing the actual penetration rate by the TBM's utilization which, in turn, may be estimated from the Factor of Safety at roof centerline.

The variability in rock mass strength suggests to classify rock masses in three broad groups of low, medium and high rock mass strength, respectively. By the same token it appears necessary to classify the rock mass behavior in three groups, namely stable, friable and squeezing behavior. The Factor of Safety may be used to describe those rock mass behaviors as outlined in Figure 12. A rock mass to be tunneled by a TBM may thus be described by three strength and three behavioral classes (Table 9), which gives nine possible combinations of tunneling conditions or advance classes. It is for practical tunneling purposes sufficient to classify a rock mass in those nine classes. No information from Figures 8 and 12 are lost as probability density functions of specific penetration and TBM-utilization for those classes are provided in Fig. 13. The nine possible advance classes are given by probability density functions (PDF) in Table 10.

Each advance class reflects a defined effort of TBM and crew to mine and support an underground excavation. This approach in classifying rock masses for TBM tunneling avoids ambiguous meanings of advance rates with respect to economic considerations. For example, an advance rate of 1 m/h may mean tunneling in a stable (high utilization) and high strength rock mass (low penetration rate) with efforts only associated with the TBM penetration. It may also mean high TBM penetration (e.g. 3 m/h) in a medium or low strength rock mass delayed by high efforts of the crew to control a friable/squeezing rock mass behavior by installing heavy and time-consuming support.

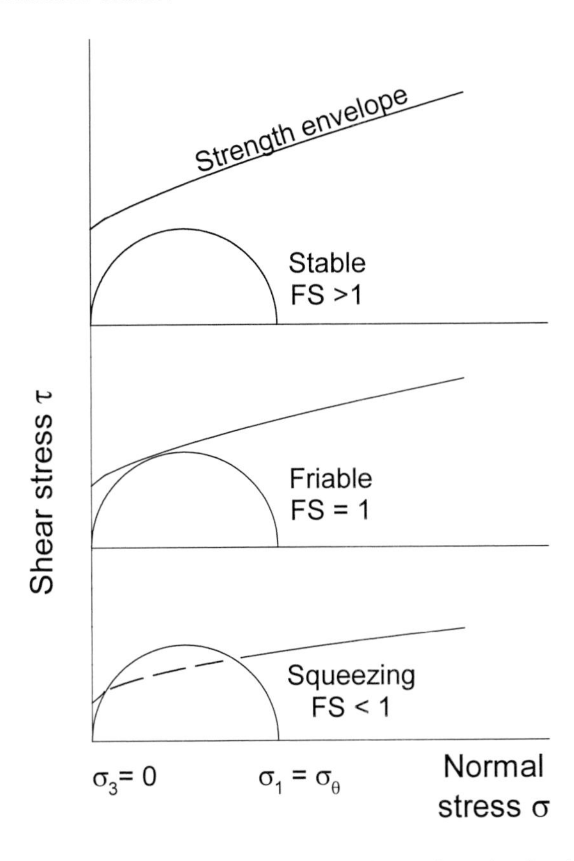

Figure 12: Classification of the rock mass behavior according to the ratio of rock mass strength to tangential stresses at the tunnel wall.

TASK: PENETRATION		TASK: SUPPORT	
Rock mass strength class	characteristic values	Stability class	characteristic values
I low rock mass strength	$\sigma_{CM} < 20$ MPa	A squeezing	FS < 1.25
II medium rock mass strength	20 MPa < $\sigma_{CM} < 80$ MPa	B friable	1.25 < FS < 2
III high rock mass strength	80 MPa < $\sigma_{CM} < 140$ MPa	C stable	FS > 2

Table 9: Classification of the rock mass factors influencing the main tasks in hard rock TBM tunneling.

A TBM advance rate reflects thus a specific tunneling condition. This approach is essential for cost estimates based on the duration of TBM project. It also helps to unambiguously classify rock masses to be tunneled for contractual purposes and reduces the risks of encountering unexpected ground conditions.

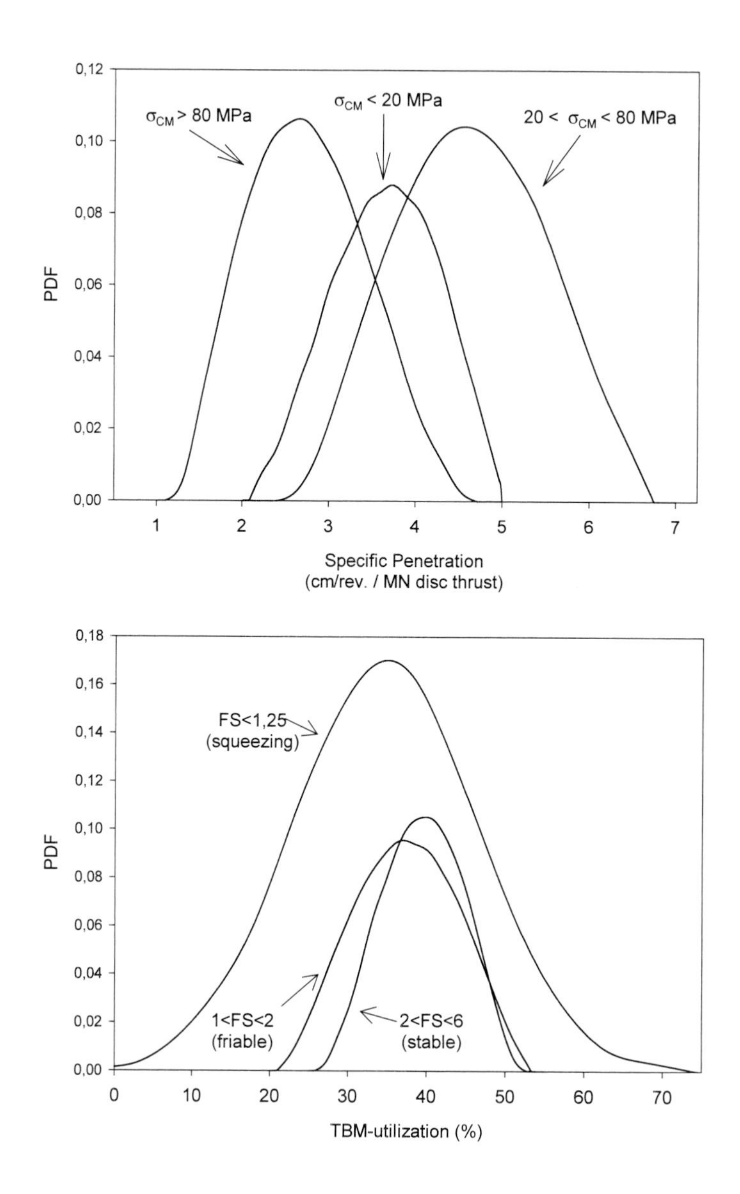

Figure 13: Probability density functions of specific penetration (top) and TBM-utilization (bottom) in the rock mass strength and rock mass behavior classes as defined in Table 4.

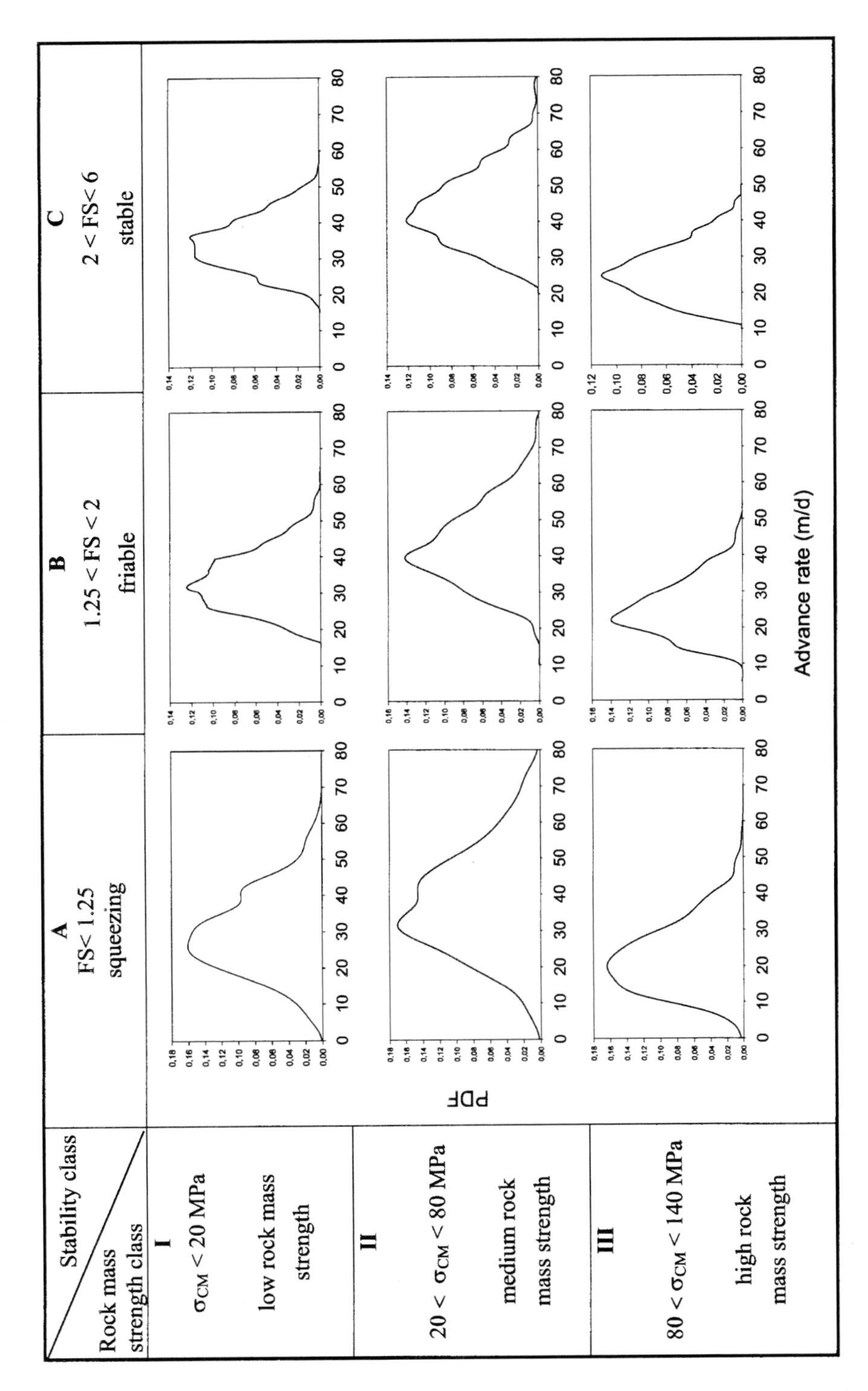

Table 10: Probability density functions of advance classes in the nine possible combinations of stability and strength classes.

3.4.1 Example of application

A practical example may show the application of the presented findings for estimating advance rates. A tunnel with diameter 5 m shall be excavated by a TBM. The geological model is schematically shown in Figure 14 (top). Several regions of similar stability and rock mass strength are defined and are classified along the proposed tunnel in the respective classes (Fig. 14 middle). Each section of the tunnel is now described by combining one rock mass strength class and one stability class to a distinct advance class (Fig. 14 bottom).

A contractor intends to use a TBM equipped with 42 discs (17" size), a cutterhead rotational speed of 8.5 rpm and an average thrust per disc of 0.195 MN. One may estimate then the penetration rates for the three different rock mass strength classes as defined in Table 9. For example, in a tunnel section of medium rock mass strength, class II with 20 MPa < σ_{CM} < 80 MPa, the average penetration rate may found by consulting Figures 8 or 13. The average specific penetration in that class may be 4.6 cm/rev. /MN disc thrust. For the TBM described above, this value of SP translates to an actual penetration rate P:

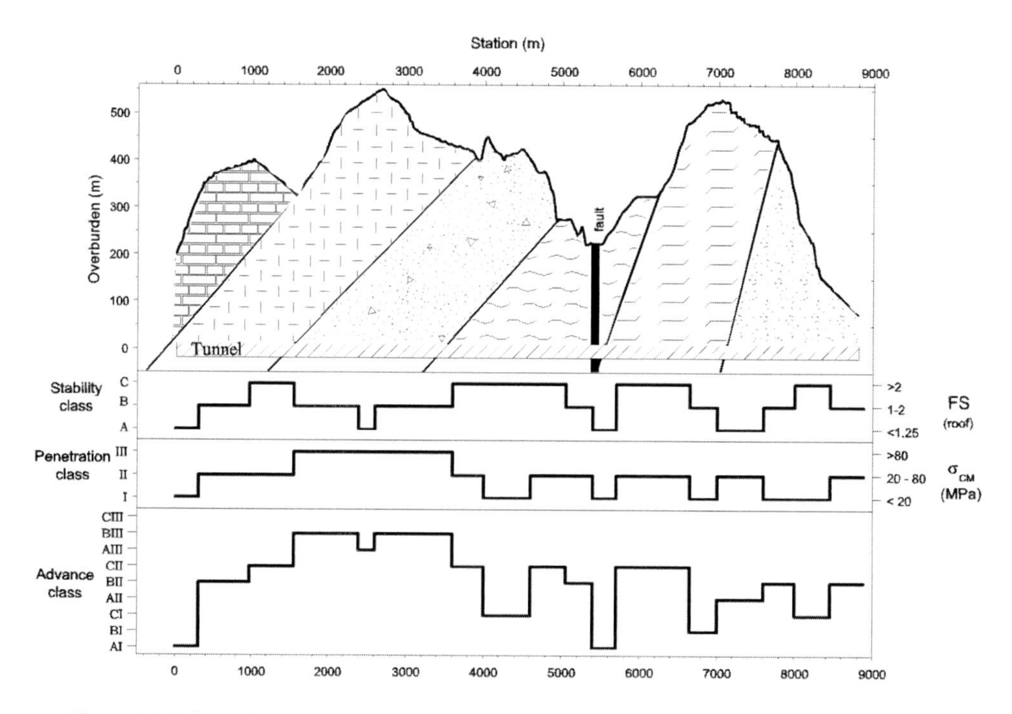

Figure 14: Schematic geological profile, associated stability, penetration and advance classes.

$$P = 4.6 \frac{\text{cm}}{\text{rev. MN disc thrust}} \times 8.5 \frac{\text{rev.}}{\text{min.}} \times 60 \frac{\text{min.}}{\text{h}} \times 0.195 \text{ MN disc thrust} \tag{16}$$

$$= 4.57 \text{ m/h}$$

This maximum penetration is then reduced by the TBM-utilization to the average daily advance rate in a certain rock mass behavior. For example in a friable rock mass the average TBM-utilization U may be estimated from Figures 10 or 13 being about 35%. The advance rate A is then

$$A \text{ (m/h)} = P \text{ (m/h)} \times U \text{ (\%)} = 4.57 \text{ m/h} \times 35\% = 1.6 \text{ m/h} \tag{17}$$

and the daily progress may be v = 38.4 m/d.

Those calculations may be executed for the nine possible combinations of rock mass strengths and stability classes. By taking the variability of penetration and utilization (Fig. 13) into account one may arrive at probability density functions of advance rates for all 9 advance classes.

3.4.2 Estimating the time of completion

Each section i of length s_i of the proposed tunnel is defined by a distinct advance class with an associated distribution of advance rates v_i. The time t_i for mining and supporting one section may be estimated with

$$t_i = \frac{s_i}{v_i} \text{ (d)} \tag{18}$$

The time T for mining and supporting the complete tunnel is then

$$T = \sum_i t_i \text{ (d)} \tag{19}$$

The statistical description of the advance rates v_i allows a probabilistic approach for estimating the time of completion as shown in Figure 15. In this example the expected value for completion T is 294 working days, but the probability that the tunnel may be completely mined and supported within 294 d ± 2 months is only 55%. A highly risk-oriented estimate would be an expected time of completion based on the 10% percentile with T = 198 d. A "low risk" estimator might choose the expected time of completion by following the 90% percentile with T = 403 d. A contractor may now evaluate the risk involved in this project and may select a contractually fixed completion time based on risk analysis.

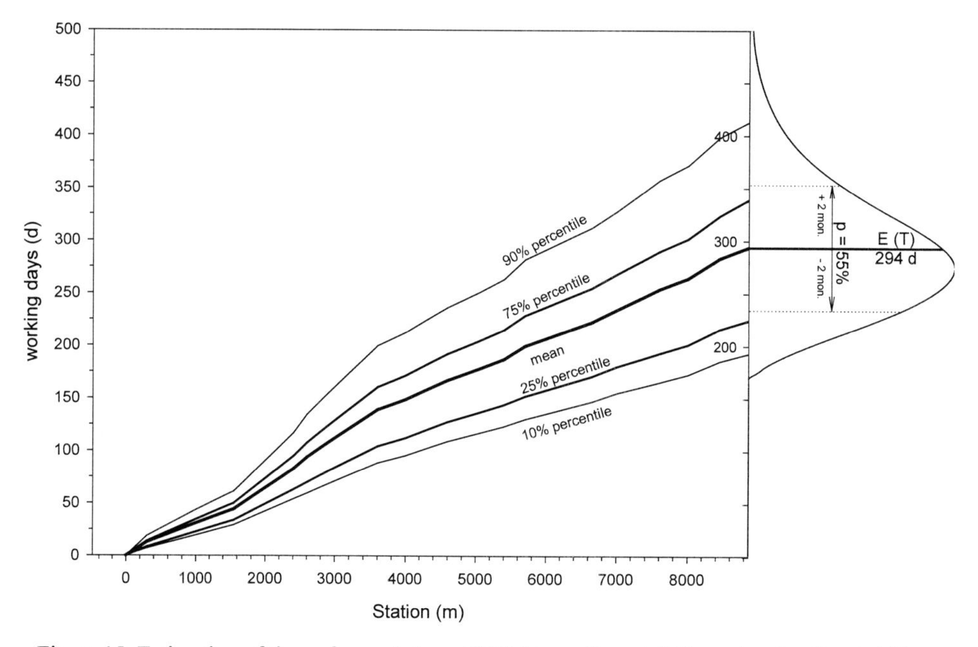

Figure 15: Estimation of time of completion of TBM tunneling at different levels of probability.

4. Conclusions and outlook

It has been shown that TBM penetration rates and TBM-utilization may be estimated with the knowledge about rock mass strength and rock mass behavior along a tunnel. The variability of the parameters involved mandates to classify rock mass strength and rock mass behavior in appropriate groups and assign them probability density functions of penetration and utilization. An important issue in hard rock TBM tunneling is not the achievement of high penetration rates rather than to keep the utilization rates high by designing robust support systems that are fast to install with the overall goal not to delay the penetration process.

An important tool for arriving at the input parameters for estimating advance rates is the use of engineering rock mass classifications. It has been shown that the inherent anisotropy of a given rock mass (i.e. directional linear discontinuity spacing) should be incorporated in the classification process to arrive at meaningful estimates. It may be noted however that all reflections assume that the geology encountered is as expected. The fulfillment of this assumption is still one of the great challenges in tunneling.

Finally, hard rock TBM tunneling may get more rational by conducting applied research. There is an abundance of data from TBM tunneling since most machine data (thrust, cutterhead revolutions, gripper loads, change of discs etc.) are digitally stored on board. Geological and geotechnical documentation of a TBM drive is often also available. However, owners and contractors are for legal grounds reluctant to disclose this information. All participants, geologists, engineers, designers, owners and contractors in such project are encouraged to advance the art of hard rock TBM tunneling by sharing information with the scientific community.

Bibliography:

[1] McFeat-Smith, I. & P. J. Tarkoy. Assessment of tunnel boring machine performance, Tunnels & Tunnelling 12, pp. 23-25. 1979.

[2] Lauffer, H. Gebirgsklassifizierung für den Stollenbau. Geologie und Bauwesen, Vol. 24, No. 1 pp. 46-51, 1958.

[3] Pacher, F.; Rabcewicz, L.v. and J. Golser (1974). Zum derzeitigen Stand der Gebirgsklassifizierung im Stollen- und Tunnelbau, Schriftenreihe Straßenforschung, Bundesministerium für Bauten und Technik, Wien., 1974.

[4] ÖNORM B 2203 - Österreichisches Normeninstitut. Untertagebauarbeiten, Österreichisches Normeninstitut, Wien,A, 25.p, 1994.

[5] Spaun, G. Geologische Gesichtspunkte bei der Klassifizierung des Gebirges, Lehrgang Tunnelbau, Technische Akademie Esslingen, 1984.

[6] Deere, D.U., A.J. Hendron, F.D. Patton and E.J. Cording. Design of Surface and Near Surface Construction in Rock, Proc. 8th US Symp. Rock Mech., AIME, New York, pp. 237-302, 1967.

[7] Bieniawski, Z.T. Engineering rock mass classifications, Wiley, New York, 251p., 1989.

[8] Barton, N. TBM Tunnelling in Jointed and Faulted Rock, Balkema, Rotterdam, 173p, 2000.

[9] Barton, N, and Grimstad, E. The Q-system following 20 years of application in NMT support selection, Felsbau 6, pp.428-436, 1994.

[10] Palmstrøm, A. Characterizing rock masses by the RMi for Use in Practical Rock Engineering - Part 1, Tunnelling and Underground Space Technology, 11, pp. 175-188, 1996.

[11] Palmstrøm, A. Characterizing rock masses by the RMi for Use in Practical Rock Engineering - Part 2, Tunnelling and Underground Space Technology, 11, pp. 287-303, 1996.

[12] Hoek, E.; P.K. Kaiser and W.F. Bawden. Support of Underground Excavation in Hard Rock, Balkema, Rotterdam, 215p, 1995.

[13] Hoek, E. Putting Numbers to Geology – An Engineer's Viewpoint, Felsbau 17, pp. 139–151, 1999.

[14] Seeber, G. Problematik der Gebirgsklassifikation in druckhaftem Gebirge, XXII. Geomechanik-Kolloquium, Salzburg, 1973

[15] Riedmüller, G. and W. Schubert. Critical comments on Quantitative Rock Mass Classifications, Felsbau 17, pp. 164-167, 1999.

[16] Hudson, J.A. & Harrison, J.P. Engineering rock mechanics, Pergamon, Oxford, 444p., 1997.

[17] Brown, E.T. Rock Characterization Testing and Monitoring, Pergamon, Oxford, 211p., 1981.

[18] Priest, S.D. Discontinuity Analysis for Rock Engineering. Chapman & Hall, London, 473p., 1993.

[19] Hudson, J.A. and S.D. Priest. Discontinuity frequency in Rock Masses, Int. J. Rock Mech. & Min. Sci., 20, pp.73-89, 1983.

[20] Alber, M. and Heiland, J. Investigation of a limestone pillar failure – Part 1: Geology, laboratory testing and numerical modeling, Rock Mech. & Rock Eng. pp. 167-186, 2001.

[21] Coon, R.F. Correlation of Engineering Behavior with the Classification of In-Situ Rock, PhD Thesis, University of Illinois at Urbana, USA, 1968.

[22] Alber, M. Anisotropy of a regularly jointed limestone rock mass and its effects on geomechanical parameters, Proc. EUROCK 2001, Balkema, Rotterdam, pp. 199-204, 2001.

[23] Riedmüller, G.; W. Schubert; A. Goricki and P. Pölsler. Investigation Strategies for the Design of the Semmering Base Tunnel. Felsbau 18, pp. 28-36, 2000.

[24] Sapigni, M.; M. Berti; E. Bethaz; A. Busillo and C. Cardone. TBM performance estimation using rock mass classification. Int. J. Rock Mech. Min. Sci. (39), pp. 771-788, 2002.

[25] Alber, M. Geotechnical Aspects of Hard Rock TBM Contract Classification (in German). Mensch & Buch, Berlin, 116p., 1997.

[26] Tarkoy, P.J. Comparing TBMs with drill+blast excavation, Tunnels & Tunnelling, October, pp. 41-44. 1995.

[27] Nelson, P.P. TBM Performance Analysis with Reference to Rock Properties, in Comprehensive Rock Engineering (J.A. Hudson, Ed.), Vol. 4, pp. 261-292, 1993.

[28] Bilgin, N:, Balci; C.; Acaroğlu, Ö.; Tunçdemir, H.; Eskikaya, Ş.; Akgül, M. and M. Algan. The performance prediction of a TBM in Tuzla-Dragos sewerage tunnel, Proc. World Tunnel Congress `99, Balkema, Rotterdam, pp. 817-822, 1999.

[30] Wanner, H. and U. Aeberli. Tunnelling Machine Performance in Jointed Rock, Proc. 4th Symp. Rock Mech., ISRM, Montreux, pp. 573-580, 1979.

[31] Büchi, E. Influence of geological parameters on TBM advance rates (in German), PhD thesis, University Bern, 136p., 1984.

[32] NIT. Hard Rock Tunnel Boring, Norwegian Institute of Technology, Trondheim, 183p., 1988.

[33] NIT. Drillability, Norwegian Institute of Technology, Trondheim, 176p., 1990.

[34] Alber, M. Prediction of penetration and utilization for hard rock TBMs, Proc. ISRM Int. Symp. Eurock'96, Balkema, Rotterdam, pp. 721-725, 1996.

[35] Barton, N. TBM tunnelling in jointed and faulted rock, Balkema, Rotterdam, 173p., 2000.

[36] Bilgin, N; Yazice, SW. and Ş. Eskikaya. A model to predict the performance of roadheaders and impact hammers in tunnel drivages, Proc. ISRM Int. Symp. Eurock'96, Balkema, Rotterdam, pp. 715-720., 1996.

[37] Alber, M. Design of high speed TBM-drives, Proc. 8th Int. IAEG Congr., Balkema, Rotterdam, pp. 3537-3543, 1998.

[38] Steingrimsson, J.H.; Grøv, E. and B. Nilsen. The significance of mixed-face conditions for TBM performance, *World Tunnelling* 15 (9), pp. 435-441, (2002).

[39] Alber, M. Classifying TBM Contracts, Tunnels & Tunnelling, Dezember, pp. 41-43, 1996.

[40] McFeat-Smith, I. Considerations for Mechanised Excavation of Rock Tunnels, Proc. VI Australian Tunnelling Conf., Melbourne, pp. 149-157, 1987.

[41] Ichikawa, H. Planning, design and construction of a utility tunnel using tunnel boring machine and pre-cast concrete method, Infrastructures souterraines de transport. Balkema, Rotterdam, pp. 39-46, 1993.

[42] Bilgin, N.; Balci; C.; Tunçdemir, H.; Eskikaya, Ş.; Akgül, M. and M. Algan. The performance prediction of a TBM in difficult ground condition, Proc. AFTES-Journées d'Etudes Intern. de Paris, pp. 115-121, 1999.

[42] Brown, E.T., Bray, J.W., Ladanyi, B. & E. Hoek. Characteristic line calculations for rock tunnels, . Geotech. Engng. Div. ASCE, 109, pp. 15-39, 1983.

[43] Brady, B.G.H. & Brown, E.T. Rock Mechanics for Underground Mining, Chapman & Hall, London, 571p., 1993.

[45] Panet, M. Understanding Deformation in Tunnels. in Comprehensive Rock Engineering, (J. Hudson, Ed.), Vol. 1, pp. 663-690, 1993.

[47] ÖNORM B 2203 - Österreichisches Normeninstitut. Underground Works – Works contract (in German), Österreichisches Normeninstitut, Vienna, Austria, 25.p., 1994.

Tunnel Refurbishment

Anton W. Ackermann [1], Christopher Hunt [2]

[1] Amberg Engineering Ltd., Tunnel Refurbishment Division, Trockenloostr. 21, Regensdorf-Watt, Switzerland
e-mail: aackermann@amberg.ch
[2] Amberg Engineering Ltd., Tunnel Refurbishment Division, Trockenloostr. 21, Regensdorf-Watt, Switzerland
e-mail: chunt@amberg.ch

Abstract:

There are numerous reasons why a tunnel structure must be refurbished, of which structural deficiencies, increasing space requirements, deficient waterproofing, modernization, changes in service and new loading conditions are most common. This paper provides an overview of selected aspects of the tunnel refurbishment process. Attention is given to the topics of refurbishment tasks and objectives, inspection, common damage, and important considerations when developing refurbishment concepts. Case studies from two current refurbishment projects in Switzerland are presented including the enlargement of the single-track tunnels of the TRAVYS railway line and the refurbishment of the two-lane San Bernardino national highway tunnel.

1. Road and railway tunnels in Switzerland

The economic, cultural and environmental importance of road and railway tunnels is evident by the sheer number of existing structures and the current trend to construct ever more facilities underground.

In Switzerland, more than 188 national motorway tunnels of over 170 km length are currently in service. By 2015, 265 national motorway tunnels of over 280 km total length will be in operation [1]. Road tunnels are on average 30 to 35 years old.

Over 700 railway tunnels of over 395 km length are currently in operation in Switzerland [2]. The oldest railway tunnel in Switzerland, the 148 year old Nord-de-Mormont, began operations in 1855. The average age of all Swiss railway tunnels is approximately 80 years.

Assuming a new construction investment value of 40'000 CHF/m' (27'000 €/m'), the new investment value of the above mentioned Swiss road and rail tunnels is approx. 22.6 billion CHF (approx. 15.4 billion €). A value preservation percentage of 2.0 – 2.5% of the new investment value yields a necessary yearly investment of approx. 450-565 million CHF (305 – 385 million €) for structural upkeep of tunnels. From this perspective, the structural maintenance and refurbishment of tunnel infrastructures can be clearly recognized to be of notable economic relevance.

2. Tunnel refurbishment

2.1 Definition

Although the term "Tunnel Refurbishment" contains a degree of ambiguity, it can be defined for the purposes of this paper to encompass the wide range of tasks associated with structural improvements, modifications and renewal of existing underground structures. From the engineering perspective, tunnel refurbishment tasks include tunnel inspections, preliminary and feasibility studies, engineering design and planning, project management, tendering, construction supervision, project reviews and the development of comprehensive maintenance strategies.

2.2 Reasons for tunnel refurbishment

There are numerous reasons why a tunnel structure must be refurbished, of which structural deficiencies, increasing space requirements, deficient waterproofing, modernization, changes in service and new loading conditions are most common. Two or more problems can often be solved within the same refurbishment project, e.g. renewal of a damaged lining and the enlargement of a tunnel profile.

Tunnel refurbishment tasks must ensure existing structures fulfill all project boundary requirements with regard to structural safety, serviceability, durability, human safety and the environment. Because every tunnel structure is unique, refurbishment objectives and requirements are determined for each structure together with the client on a case-by-case basis.

3. Tunnel inspection and investigations

In order to enable the development of a tunnel refurbishment concept, the current state of the tunnel must be known. Tunnel inspections and investigations provide information which gives insight into the current state of an entire tunnel or of particular elements or features. The inspection provides a basis upon which the determinations can be made as to which measures are necessary, when and where.

Inspections can also provide invaluable information for global planning tasks for larger tunnel systems or tunnel networks such as the prioritization of repairs and rehabilitation works, financial planning and scheduling of works. As investigations of the current condition increase in detail and accuracy, the evaluation of construction measures can be carried out more precisely. The corresponding quantities and construction costs can then be better estimated.

Tunnel inspections carried out by qualified professionals provide owners and operators with reliable information, thus providing a key element towards increasing the success of all later tasks which are dependent upon inspection results. High-quality inspection information enables well founded decisions. In the end, long-term safety and serviceability as well as financial sustainability over the service life of the structure are at stake.

3.1 Inspection preparations

Essential for successful and effective inspections is to make adequate preparations. The first step is commonly to study of existing documentation in the office. The scope, goals and procedures of the inspection must then be considered in detail and agreed upon with the client. They vary widely depending on the circumstances of each project such as the clients needs, project constraints, structural condition, available budget, existing data, established inspection procedures, etc. Important aspects of inspection preparation for all structures include the safety measures for all personnel involved on site, determination of the correct inspection equipment, availability of the necessary equipment and personnel, and competent scheduling.

3.2 Qualification of inspection personnel

The inspection and data collection in the field should be led and/or carried out by a qualified engineer with tunnel inspection experience. The inspection personnel should be trained in inspection methods and be able to correctly recognize, interpret and classify damage. They must also determine where immediate construction measures or more detailed investigations are necessary. Some clients have established minimum qualifications for inspectors.

The inspection results must be correctly analyzed, interpreted and applied, and can be most accurately assessed by an experienced engineer who has been on site. For this reason, the ideal person to interpret the results of a given inspection is an experienced tunnel engineer who has carried out or taken part in the inspection itself.

3.3 Inspection methodology

Traditional methods of tunnel inspection involve visual inspection and data recording by writing observations and field notes directly in notebooks or on prepared inspection forms and checklists. Inspection vehicles or wagons with hydraulic platforms and sufficient lighting are essential to enable inspection of the entire tunnel structure. Typical information recorded during an inspection include the location, types and dimensions of damage, construction materials, and installations as well as localization of unsound or hollow areas of the lining through systematic hammering of the surface etc.

Additional detail investigations of particular parameters concerning tunnel structure and surroundings are often necessary as part of or subsequent to a tunnel inspection. Detail investigations include rock and material tests, water analysis, geotechnical analysis, core borings, potential measurements, visualizations and many others. The correct choice and necessary quantity of investigations varies highly for each project. All investigations should directly support realization of the project goals and requirements.

In railway tunnels, clearance analysis is often carried out. This involves investigations of the tunnel profile conditions in relationship to the desired clearance profile of vehicles operating in the tunnel. These typically involve systematic profile measurements (at intervals or continuous) using one of a number of different profiling systems and the subsequent analysis of the clearance profile.

All results from tunnel inspections and detail investigations including additional sketches, photos, calculations etc. are integrated into the state assessment report. From this basis, suitable construction measures can be developed during the project design process.

3.4 Tunnel inspection using digital technologies

An ever increasing number of digital technologies are being applied during inspections for the collection, processing, and documentation of tunnel infrastructure data. Digital systems enable efficient evaluation of data and simplified data management. Digital technologies such as tunnel maps or scanning technology are often applied to increase efficiency of inspections and their results provide graphical representations of inspection data. It is important to note that digital technologies provide increasingly accurate and efficient methods of collecting and managing data. They do not, however, eliminate the need for on-site inspection of the facility by an experienced engineer.

3.4.1 Tunnel mapping

Tunnel mapping is a methodology used to chart and provide an overview of the condition of the tunnel surface. It involves the observation and input by inspection personnel of structural data into tunnel maps, including the type and location of damage, installations, construction materials and other structural features. The inspection personnel can input data of varying degree of detail depending on the project scope. Tunnels are typically divided into 10 m long zones to enable accurate mapping. Zones are marked with spray paint or by hanging permanent numbered markers before inspections.

Digital equipment and software has been developed to rationalize data input and help manage information in the tunnel maps. One modern method of tunnel mapping involves equipping the inspection engineer with a sturdy and portable PC. The PC hard drive is stored in a backpack, leaving the hands free to digitally enter observations directly onto the PC screen and into the digital "Tunnel Map". Mapping software has a number of preprogrammed tools which can be selected on the screen to simplify the input of data. The tools enable an accurate input of damage, construction details, materials etc. The efficiency and effectiveness of the inspection is thereby increased significantly.

3.4.2 Scanning

Digital scanning systems have been developed that provide continuous, rapid and objective data collection in tunnels. Scanning systems record comprehensive data from the tunnel surface to provide digitized visual images and profile measurements. The data results are typically provided to the customer together with software with which data can be viewed in the office.

Visual image scans can be made in combination with profile scans in a single run. Profile scanning creates a file consisting of continuous cross sections of the object. The standard distance between adjacent profiles is about 1 cm, i.e. profiles are available at any location in the tunnel. The profile data allow evaluations of current clearance conditions as well as simulations for vehicles of all sizes.

4. Common damage in road and railway tunnels

4.1 Common damage in tunnels

The repair of damage in tunnels is one of the main goals of refurbishment projects. Several types of damage which are commonly observed in tunnels include:

- Water Inflow
  (humid areas, wet areas, dripping water and flowing water)
- Joint Damage
  (masonry - empty joints/ mortar decay; concrete - damaged construction joints)
- Cracking
  (from loading, constructive deficiencies, frost, shrinkage, settlements, chemical processes, impact damage etc.)
- Scaling
  (localized flaking or peeling of material near the surface)
- Spalling
  (detached fragments of material from a more massive piece of material)
- Deformations
  (local or large-scale movements in side walls or crown, or heaving invert)
- Weathered surfaces
- Accumulation of water in invert
  (wet and muddy rail substructures, wet road pavements)
- Surface deposits and discoloration
- Exposed reinforcement
  (reinforcement bars or mesh)
- Corroded reinforcement
- Cavities behind lining
- Honeycombing
  (voids in concrete with collection of coarse aggregates & lack of fine particles)

4.2 External and internal causes of structural damage

The deterioration of tunnel materials is a complex process usually caused by one or more external and internal influences. In order to make effective evaluations and successful refurbishment concepts, it is important to have a basic understanding of the mechanisms which lead to aging and damage of existing structures. Several examples of external and internal influences are listed in Table 1 [3].

EXTERNAL AND INTERNAL CAUSES OF STRUCTURAL DAMAGE	
EXTERNAL INFLUENCES	INTERNAL INFLUENCES
principle loading (design loads, self-weight, traffic loads)	incompatible materials
geological and hydrological influences (earth pressure, settlement, water inflow, water pressure, ice pressure)	unstable substances
mechanical influences (impact, mechanical damage)	effects of moisture
climatic influences (temperature changes, frost, humidity, fire)	
chemical influences (road salts, sulfate attack, alkali-carbonate reaction)	
electrochemical influences (corrosion of reinforcement)	
biological influences (plant growth, fungi, microorganisms)	
construction errors	

Table 1 – External and internal causes of structural damage [3].

The damage observed of during inspections should be understood as a result of one or more influences. Initial damage to the structure becomes a cause of further damage, which in turn further complicates the process. It is extremely important that the refurbishment concept addresses the causes of the damage, and not only the resulting damage. The optimal refurbishment solution can only be chosen after the cause has been correctly determined.

5. Considerations for tunnel refurbishment concept

The refurbishment of tunnels is a process defined by the interaction between project boundary conditions and project constraints, the current structural state, refurbishment design, and execution of the works. All these aspects must be given sufficient attention to ensure project success.

Tables 2 - 5 list a select number of important considerations that are generally relevant during refurbishment projects.

5.1 Boundary conditions and project constraints

CONSIDERATION OF:	COMMENTS / APPLICATION / EXAMPLES
Refurbishment goals	e.g. structural goals (structural improvements, profile enlargement, modernization, waterproofing) or operational goals (improvements or changes in service, increase in capacity, closure)
Traffic	Number of lanes, direction of traffic flow, traffic flow and capacity during refurbishment works and after refurbishment, additional safety requirements
Scheduling	Coordination with other works, prioritization in connection with global goals, traffic conditions, time available
Costs	Cost limitations, available funds from responsible authorities, evaluation of cost effective alternatives must be considered
Climatic Conditions	Temperature and climate can limit the possible construction period and be favorable for frost and ice
Current and future use	technical requirements and current standards for future use (incl. consideration of modern vehicles, new equipment etc.)
Human Safety	safety measures during construction and after refurbishment, new standards, equipment
Environment	Measures for environmental protection & sustainability, protection of overlying structures, reduction of noise & dust
Accessibility	accessibility by road, rail, water; space considerations for storage of materials, etc.
Time available for works	equipment selection, number of construction crews, minimization or avoidance of traffic interruptions

Table 2 – Important boundary conditions and project restraints

5.2 Structural state

CONSIDERATION OF:	COMMENTS / APPLICATION / EXAMPLES
Geology	Geology varies in every tunnel. The interaction of the soil or rock with structure is important for evaluation of loading & hazard scenarios.
Hydrology	Water is the source of most damage in tunnels & important for environmental considerations. Source, quantity & quality are important.
Age	Age of construction gives insight into original construction methodology and material quality
Construction methods	The method of initial excavation & tunnel construction gives indicators for predicting conditions behind lining
Current and future use	technical requirements and current standards for future use (incl. consideration of modern vehicles, new equipment etc.)
Lining type	Common are concrete, shotcrete, natural masonry and brickwork, each with distinct characteristics. Many tunnels have 2 or more types.
Invert condition	Stability, wet or dry, design, effectiveness of drainage, existing conduits, etc.
Type and location of damage	evaluation of safety & serviceability, planning purposes, determination of repair measures
Amount of damage	quantification, cost analysis, determination of repair measures
Causes and development of damage	probability of further development, determination of repair measures long-term planning; setting of refurbishment priorities
Severity of damage	evaluation of safety & serviceability, determination of repair measures

Table 3 – Important considerations concerning structural state of existing structures

5.3 Project design

CONSIDERATION OF:	COMMENTS / APPLICATION / EXAMPLES
New clearance profile	In most cases, new clearance profiles determine space requirements. It must be valid for new and future wagon/vehicles and be defined in close cooperation with the responsible authority. The clearance profile varies with speed, track geometry, and vehicle/wagon size and is be determined for straight and curved sections for intended service conditions. Extra space should be planned between the clearance profile and the new inner tunnel profile after refurbishment as a reserved space for later wagon enlargements.
Drainage concept	Drainage concepts may involve further use of the existing system or design of a completely renewed system.
Waterproofing	Amount of allowable water inflow must be determined, as well as appropriate measures to prevent or drain water inflow
Conduits	A determination of which new conduits for running electric wires or various cables must be made
Electromechanical equipment	Space requirements and design changes to enable installation of new electromechanical equipment must be considered. Time of installation must also be coordinated to prevent unnecessary obstructions.
Railway installations	Design changes to enable installation of catenary system, communications equipment, signalisation, cables. etc.
New niches, cross-overs, and galleries	New niches, cross-overs and galleries for safety and maintenance are often built within the scope of refurbishment projects. The corresponding space, design, costs and coordination are often significant.
Profile design	Profile design must take into account the goal of refurbishment and relevant boundary conditions. Of utmost importance is integration of construction process into design.
Construction Equipment	In rail tunnels, availability of construction trains and/or requirements to use such trains. Clarification necessary of whether or not other projects of client are using the same equipment. If equipment not supplied by client, contractor must choose equipment and construction methodology. This can be advantageous, as contractor is able to choose existing machines which are familiar, thus potentially increasing productivity and reducing execution costs.

Energy Source	Source of energy and extent to which energy to be provided by client or by contractor must be clarified
Location of installation sites	Location of installation site (affects access time to site) to be clarified
Temporary support measures	Critical areas which must be identified which must be protected with temporary support measures during refurbishment
Ground stability	The stability of the ground behind the tunnel lining during removal of old linings must be investigated
Methodology of lining removal	Effective equipment, temporary support and final support in limited time frame must be considered
Permanent support measures	Permanent support measures must be designed for varying conditions and enable all refurbishment goals to be met must be (e.g. concrete, shotcrete, steel supports, grouting, rock bolts etc.)

Table 4 – Important considerations during project design of refurbishment projects

5.4 Project execution

CONSIDERATION OF:	COMMENTS / APPLICATION / EXAMPLES
Monitoring during construction	Monitoring during construction, measure deformations, to verify material quantities etc.
Quality control	Materials must be tested in accordance with a project-specific quality control plan
Profile controls	After construction, tunnel profiles must be measured with a sufficient degree of accuracy to ensure minimum profile requirements have been met
Construction supervision	Construction supervision by experienced professional staff is an extremely important aspect of refurbishment project success and ensures proper execution of the design project, increased project quality, and a ensures an economic execution of the works.
Qualified Contractor	The importance of having a qualified contractor with an experienced construction team cannot be underestimated. It is unwise and often very costly for the client to hire an unqualified contractor that has submitted the cheapest bid.

Table 5 – Important considerations during execution of tunnel refurbishment projects

6. Case study 1 - The San Bernardino road tunnel

6.1 Tunnel description

The San Bernardino road tunnel is located on the A13, an important component of the Swiss motorway network between northern Italy and southern Germany. The single-tube tunnel is 6'600 m long with two 3.75 m wide lanes. It climbs from the north with 0.95 % and from the south with 0.40 % to a culmination point at the middle of the tunnel. The tunnel was built over a six year period between 1961 and 1967. In 1999, over 5900 vehicles used the tunnel every day, or 2.2 million vehicles a year.

The standard tunnel cross section is horse-shoe shaped with an excavation area of 85 m^2. The tunnel was constructed of an approx. 35 cm thick single-shell concrete lining. The cross section is subdivided into three sections. The exit air duct is located in the upper profile section. It is separated from the traffic space by a 10-15 cm thick intermediate roof. The input air duct and utility channels are located below the roadway. Running along both side channels are gutters for the groundwater and roadway runoff. In addition, the west channel contains all electrical cables required for operation of the tunnel. The fire fighting water system, which includes hydrants every 50 – 80 m, is housed in the east channel.

The San Bernardino tunnel has four ventilation units. Two are situated next to the portals and the other two are located a third of the way from the portals in underground chambers near the tunnel tube. Two inclined ventilation shafts lead to the surface.

6.2 Tunnel inspection and state assessment

6.2.1 Tunnel inspection and investigations

A systematic visual inspection and various technical investigations indicated that although the tunnel is still operable, structural safety is clearly threatened by progressive damage. The following detailed investigations were carried out to assess the condition of the tunnel structure:

- systematic visual inspections
- determination of the concrete's carbonization depth, density, strength, sulfate and chloride concentrations

- electrochemical investigations employing potential measurements to determine the condition of the reinforcement
- shifting and deformation measurements on the roadway structure

6.2.2 Assessment of inspection and investigation results

Various parts of the tunnel and electromechanical equipment are in urgent need of repair or renewal. The existing concrete surface slab exhibits corrosion damage, some of it very serious. The damage is concentrated mainly at the lateral joints spaced 2.50 m apart, which are provided with shear reinforcement (round bars). The bottom layer of reinforcement is badly damaged at the joints, where conditions are very conducive to corrosion. The worst rust damage coincides with insufficient coverage of the concrete reinforcement. Except for the joint areas, the top layer of reinforcement is practically intact despite high chloride concentrations. This indicates that, even in the case of high-chloride concrete, the reinforcement is corroding only where it has been exposed to oxygen. The cause of the damage is clearly traceable to the spreading of salt. Especially hard hit are areas around the dilatation and construction joints.

The intermediate roof is generally in good condition. Only the reinforced cut-and-cover sections and the first 250 m of intermediate roof at the north and south portals are in need of repair due to the effects of salt spreading.

The unreinforced tunnel vault has suffered local damage from aggressive groundwater seepage. The two gutters have been badly damaged by the groundwater, which often contains high levels of sulfates and carbonic acid.

6.3 Refurbishment concept

6.3.1 Goals of the client

The client has drawn up the following list of primary project goals:

- Operational safety is to be guaranteed at all times
- The durability and quality of tunnel materials should remain sound and secure for 50 years

- The service life of electromechanical equipment is to be at least 25 years
- The work should not burden other Swiss cantons and regions with additional traffic
- The renewal must be carried out during tunnel operation.
- The operation and servicing of installations is to be eased, thus reducing maintenance costs

6.3.2 Construction of concrete surface slab

In the course of execution, the following requirements and conditions have to be observed:

- 1-lane traffic flow must be guaranteed during all work. Continuous 2-lane traffic is to be guaranteed on weekends, holidays, and during 4 weeks in summer (main travel season).
- Defined traffic rules to apply whenever work is being done. To achieve optimal traffic flow, construction traffic must observe strict rules regarding entering the tunnel and exiting the tunnel work sites.
- Working hours are from Monday 04:00 hours to Friday 13:00 hours.
- Maximum length of the work site is 800 m and only 1 site is permitted in the tunnel at a given time. Maximum 20 minutes tunnel closure time are permitted for setting up and clearing the work site.
- Separation of road traffic from site area with sturdy barrier

To satisfy the prescribed conditions, the work will be done over one-third of the tunnel length at a time, and will be broken down into 3 stages:

- preliminary work (demolition of wall lining, shoulder and pavement)
- concrete surface slab renewal (demolition and reconstruction of slab, roadway drainage, hydrant system)
- roadwork (sealing, shoulder, pavement)

To reduce the tunnel work site dependence on road traffic, the contractor's proposed installation of muck train tracks has been adopted. The tracks are to be laid in the middle passageway under the roadway and will be used primarily for removal of debris and supply of concrete mix.

The new surface slab consists of a jointless in situ concrete slab supported laterally by three walls and resting on two side supports. Frost-resistant B40/30 (Swiss Norm) concrete is used.

The slab renewal occurs in two locations, each about 300 m long. One-half of the surface slab will be replaced, as the other half is in use. Including the distance required for changing lanes between the two work sites and the entries into the site, the total single-lane traffic stretch is 800 m long. For the work sites, a fixed weekly cycle was decided upon. Within each week, the following work is carried out to replace 90 m of concrete surface slab:

- secure the existing surface slab for demolition on one side
- sever the slab structure
- demolish and remove the resultant slab elements
- install formwork and reinforcement, pour concrete for the new half-slab
- treat the concrete surface

The work as been planned so that the fresh concrete sections can be driven over by Friday at 13:00 hours. The concrete formulation and formwork must be planned accordingly. Comprehensive concrete controls, specialized curing and frost protection measures ensure the slab quality and serviceability. The concrete is required to be at least 24 hours old.

6.3.3 Invert lowering

The invert lowering and renewal of the gutters in the west and east channels is done underneath the roadway without interfering with the traffic. The invert being lowered by 50 cm, which will eliminate groundwater damage and provide the overhead clearance necessary to permit use of the central channels as escape and life-saving passageways.

The cramped space necessitates the use of special, compact construction equipment. Excavation machines remove concrete and rock. All cables now installed on open cable trays in the west channel are to be transferred to a new conduit under in the new concrete invert. A drainage system (air gap membranes and seepage lines) is also being built. The existing drainage channels are being renewed by installing prefabricated half-pipe gutters.

6.4 Tunnel safety

Within the scope of the San Bernardino refurbishment, the tunnel's ventilation, emergency escape, lighting, signaling and general control systems are all being modernized.

The operational and fire ventilation systems will be adapted to meet current standards. The main modifications and additions include:

- fresh air fed directly into the traffic space at both ventilation machinery rooms and at portals
- high-volume fire exhaust system employing controlled fire dampers every 96 m
- 12 jet-fans mounted in niches along the traffic space for controlling the axial air flow
- fire detection system above the traffic space

The existing wall-lining panels will be replaced with light-colored panels that are designed to improve lighting conditions.

Most electromechanical installations including the lighting, power supply and cabling systems are being brought up to state-of-the-art technical standards. This will substantially increase the complexity of the control and automation systems. The entire control and instrumentation system as well as the visualization equipment at the traffic and operation control stations will therefore be renewed as well.

6.4.1 Construction measures for operational and fire ventilation systems

Installation of the new ventilation systems requires the cutting of 12 niches for the jet-fans and cutouts in the intermediate roof for the fire-ventilation fire dampers. Because this work must be done during traffic operations, it is to be carried out exclusively during the nighttime hours.

The niches, which are broken out with pneumatic picks, are 21 m long, 2.25 m high and 1.1 m deep. The work site is restricted to a length of 200 m and is allowed to obstruct just one traffic lane. The extremely cramped conditions and safety precautions make the logistics very costly. The excavation and concrete work will be carried out in 7 stages and will be continuously monitored by means of vibration and deformation measurements on the existing structure.

For the fire dampers, openings measuring 2.2 x 2.2 m spaced about 96 m apart have to be cut in the intermediate roof. This will be done with a truck-mounted concrete saw. Erection of the jet-fans and fire dampers and laying of the corresponding cables will be done during construction works. Work on the tunnel control system will proceed at the same time.

6.4.2 Construction of escape galleries

17 galleries are to be built on the west side of the tunnel. The galleries will be lined with single-shell shotcrete. Their longitudinal gradient is between 7.2% and 8.7%, their length between 30 and 40 m. This makes the escape galleries suitable for wheelchairs.

The excavation work will be monitored continuously with vibration measurements. A separate monitoring and alarm system is also installed that responds whenever certain limits are exceeded or damage is detected. Shock waves up to 60 mm/sec are admissible.

6.5 Construction schedule and costs

The construction period is currently estimated to last 8 years. Works in the invert and all cable re-laying began in 1998 and were completed in the autumn of 2002. Construction work required to improve tunnel safety will be completed at the beginning of 2003 with the start-up of the new ventilation system. The renewal of the concrete surface slab will be started in the middle of 2003 and is expected to be completed by the end of 2006. The wall panels will be installed in the latter half of 2006. The construction costs are currently estimated at CHF 200 million.

7. Case study 2 - The railway tunnels of TRAVYS

7.1 Project description

Starting in Yverdon, Switzerland, the TRAVYS narrow gauge railway winds its way through mountainous terrain towards Sainte Croix, passing through 5 tunnels with a combined length of 462 m along the way. The railway, opened in 1893, was once the most important transport route for industry based in St. Croix. Today the private railway (TRAVYS) is especially important for tourist access throughout the region. Additionally, the line is heavily used on workdays by local commuters. Over the years, the tunnel structures have become deficient and the existing tunnel profile no longer satisfies modern clearance requirements.

Tunnel	Length (m)	Radius (m)	Incline (%)
Les Murets	146	- 69.41 - 101.89	4.40
Cochâble I	60	- 265.00 straight	4.20
Cochâble II	74	A = 63.36	4.20
Covatannaz	153	+ 157.51 straight - 100.00	4.20
L'Onglettaz	29	A = 47.63 Straight	4.00
TOTAL	**462**		

Table 6 – Overview of the tunnel structures

7.2 Tunnel inspection and state assessment

In October 1997, an inspection of the tunnels were carried out. The structures are for the most part lined with the natural limestone masonry. Rough edged stones were used in the crown and cut stones on the side walls. In most all the tunnels there is an existing 2 and 3 m wide shotcrete layer in the crown which acts as a seal against water infiltration. The foot of the vault lies loosely on the tunnel rock invert only a few cm below the top of the rail, and is poorly secured.

The tunnel profile cross section of the 5 tunnels is 19 m^2. Using a profile measuring system, the actual cross sections were reviewed and the rail axes was measured using a fix point network.

Damage observed during the inspection in all five over100 year old tunnels included:

- Numerous water inflows through the existing tunnel linings
- Damage to masonry joint mortar caused by frost-thaw cycles
- Stability problems resulting from disintegration of the masonry
- Structural cracks attributed to localized vault deformations

The results of the inspection and subsequent state assessment indicated that the existing tunnel linings are in a poor state and that the size of the tunnel profile no longer meets the client's requirements. On the basis of the inspection, the condition of tunnel sections were classified from 1 to 5, whereby "poor" sections (classification 4) were found to be present in all five tunnels. The following refurbishment priorities were established based on the structural classification and economic viability (budget, installation and logistics):

Cochâble I and Cochâble II	Priority 1	Total Length = 134 m
Les Murets	Priority 2	Total Length = 148 m
Covatannaz and L'Onglettaz	Priority 3	Total Length = 184 m

7.3 Refurbishment concept

The state assessment of the tunnels clearly indicated that the substance of the existing tunnel linings are heavily damaged and would not be economic to repair. Due to space restrictions, a "light" refurbishment of the damaged vaults is also not possible. It was consequently decided to completely remove and replace the tunnel lining and simultaneously enlarge the profile.

On the basis of the endoscopy investigations and experience with similar structures, it was known that loosened ground was to be expected behind the lining. Extreme care must therefore be taken during removal of the existing tunnel lining.

The refurbishment includes the following tasks:

- Step by step demolition of the existing natural stone lining
- Removal and excavation of loose backfill behind the lining
- Excavation and enlargement of the rock profile as necessary
- Construction of new single-shell shotcrete lining reinforced with 2 layers of wire mesh
- Replacement of the tunnel drainage

7.4 Construction process

The entire refurbishment is performed during the night. Normal railway operations continue during the day. The five tunnel structures are not easily accessible and all personnel and materials must be transported by rail. During work in the tunnel, a path will be built next to the rails which enable operation of motor vehicles.

At the start of the demolition works, small holes in the crown allow the condition of the lining and the local rock characteristics to be systematically assessed. This permits the site management to arrange for additional support measures as if necessary. Demolition of the natural stone vault is performed in longitudinal stages between 0,5 and 2 m in length at several working locations. Where the quality of the ground is very poor, the site management can plan additional cross section stages with intermediate supports.

The new tunnel lining,a single-shell shotcrete lining approximately 25 cm thick, requires less space than the 45 cm thick natural stone masonry lining. If this difference is not sufficient, the profile is enlarged by excavation of the rock surface. Because small volumes of shotcrete are needed during each refurbishment shift, dry-mix shotcrete is used.

Infiltrating groundwater is collected and diverted with half pipe drains. These drains are connected to the new tunnel drainage system being constructed at the side wall in the invert.

7.5 Construction schedule and costs

The construction phases were fixed as follows:

1998	Construction phase 1	Tunnel: Cochâble II	L = 31 m
2001	Construction phase 2	Tunnel: Cochâble I and II	L = 103 m
2003	Construction phase 3	Tunnel: Les Murets	L = 146 m
2004	Construction phase 4	Covatannaz and L'Onglettaz	L = 184 m

Table 7 – Overview of the construction schedule

A refurbishment performance of 1 to 2 m per night can be achieved in the available night shifts. The construction period totals approximately 75 to 80 weeks. In Switzerland, it is only possible to work in the mountains during the summer because it is too cold to apply shotcrete in the winter. Construction was therefore distributed over several years, including an extended break during the Swiss Expo in 2002. The total cost of the refurbishment is approximately 8 million Swiss Francs.

Bibliography:

[1] ASTRA Tunnel Task Force – Bundesamt für Strassen: Schlussbericht, p.6, 23. Mai, 2000

[2] BAV - Bundesamt für Verkehr, Die Schweizer Bahnen in Zahlen – Mai 2002, from Website: www.bav.admin.ch, 2003

[3] Bundesamt für Konjunkturfragen; IP Bau – Zustandsuntersuchung an bestehenden Bauwerken, Leitfaden für Bauingenieure, p.12, 1992

Fire protection in tunnelling

Volker Wetzig[1]

[1] VersuchsStollen Hagerbach AG, Rheinstrasse 4, 7320 Sargans, Switzerland
e-mail: vwetzig@hagerbach.ch

Abstract:

Since several years VSH Hagerbach Test Gallery Ltd. is involved in investigations concerning fire resistance of building materials like concrete. Next to the research work on materials there is also a demand for a training and test facility for fire fighters, detecting and extinguishing systems. Within the last year a first step of a fire research centre has been realised with a 200 m long two lane motorway tunnel. First training courses and investigations on detecting extinguishing units have confirmed that there is an urgent need for such a facility. The situation of an accident in a tunnel is much more complicated than on an open road. The increase of heat is faster, communication is more difficult, the temperatures are hotter and toxid fumes make the access to the place of accident more difficult.

1. Introduction

The accumulation of serious and catastrophic fire accidents in European road rail tunnels since 1999 has greatly raised awareness concerning aspects of the operational safety of tunnel constructions. In this regard, operational safety (as a comprehensive term) is influenced by a wide variety of parameters such as

- infrastructure (design of the construction)
- operating equipment (vehicles, loading)
- users (vehicle drivers)
- operation and organisation (ventilation, extinguishing system, fire fighters)

Optimisation of one parameter makes a major contribution towards increasing safety in the tunnel, but it is unable to guarantee this result on its own. The following proceedings will focus on the subjects of infrastructure and operation and organisation of tunnel. This comprises also the education of fire fighters under realistically conditions.

2. Fire resistant constructions

2.1 Definitions

The definitions of the terms that denote the fire resistance of concrete contradict one another to some extent. The terms reflect differing objectives and requirements. All the definitions are formulated in relation to site-mixed concrete. However, there are no obvious reasons for not transferring the same nomenclature to shotcrete.

It seems that the definitions as per DIN 1045 are the most appropriate in order to describe the conditions in the construction industry. The classification contains the following groups, according to the ambient temperatures:

Designation as per DIN	**Temperature range [°C]**
Concrete for normal temperature	up to 80
Concrete for use at high temperatures	80 - 250
Refractory concrete	above 250

This classification corresponds to the damaging mechanisms described below, and it is also justified from this viewpoint.

A completely different classification can be found in the Cement Bulletin (CB) from TFB of Wildegg (Switzerland). The temperature ranges listed here lead one to suspect that the classification originates from the blast furnace industry.

Designation as per CB	**Temperature range [°C]**
Heat-resistant materials	< 1,520
Fire-resistant materials	1,520 – 1,830
Highly fire-resistant materials	> 1,830

The values cited in this classification are far in excess of the temperatures which occur in the general construction sector, so the classification is inappropriate for this sector.

As well as the designation of types of concrete based solely on the fire temperature, there is also a classification based on fire classes or fire resistance classes.

As a general rule, the assignment of fire classes specified in DIN 4102 is irrelevant to concrete that is used as a construction material. This classification is based on an assessment of a material's sensitivity to fire. Normal concrete is assigned to class A1 - non-combustible.

A re-assessment of the fire class is required for concretes that are modified by the addition of lightweight aggregates in the form of polystyrene or artificial resins as bonding agents.

As well as an assessment of the building materials, there is also an assessment of the components which involves various fire resistance classes. An assessment is made of the duration for which a component or group of components withstands a fire without impairing its ability to function; this assessment is based on the standard temperature curve, which is of little relevance to the tunnel construction sector. The usual classes are F30 to F90, which denote a resistance period of 30 to 90 minutes respectively.

In principle, the various fire load curves used to assess the fire resistance of tunnel structures are based on the idea of a division into fire resistance classes. In each case, a fire load is defined which the concrete must withstand for a defined period. However, no provision is made for a graduated rating of the fire resistance; the assessment is either "requirement met" or "requirement not met".

2.2 Fire load curves

Throughout Europe, a wide variety of fire load curves have already been defined and new ones are constantly being added.

Time-temperature progressions are used for the dimensioning of protective precautions against fire loads; some of these progressions differ significantly from one another.

Fires involving hydrocarbons (which are frequently present in a vehicle fire) are inadequately represented by the progression of the standard temperature curve (STC) to ISO 834. The warming phase which is responsible for the slow rise in temperature is reduced to a few minutes. In all vehicle fires, the maximum temperature is already reached after a few minutes. The other curves which are usual in Europe, and which are shown in Figure 1, take account of this rapid increase.

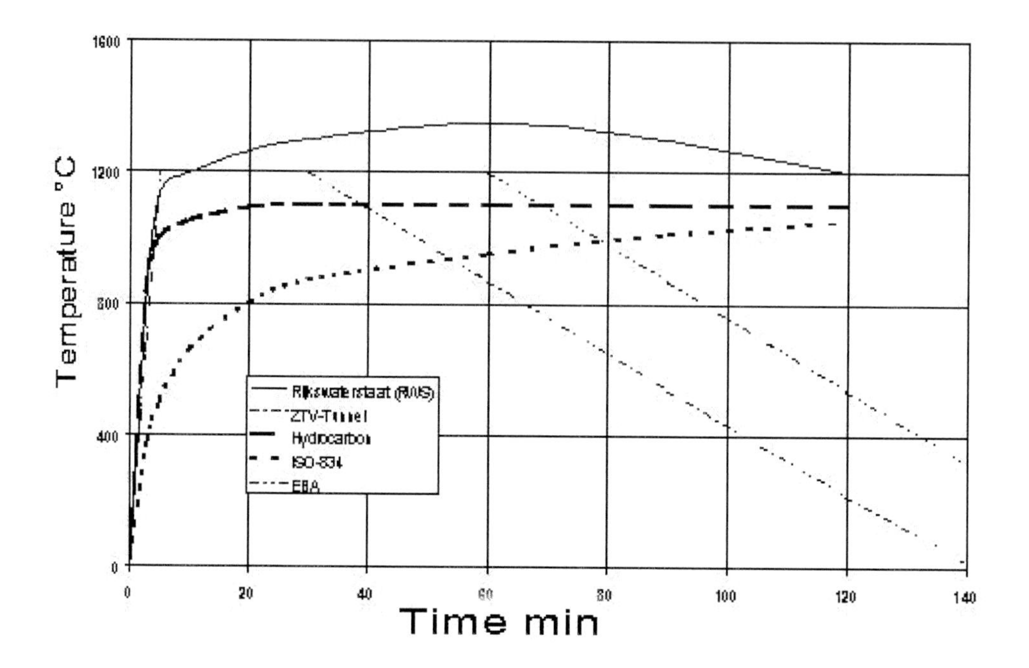

Figure 1: Fire load curve

The highest requirements are imposed in the Netherlands, by the Rijkswaterstaat (RWS). This requirement curve is based on a fire in a fuel truck in a tunnel. The exposed position of the Netherlands, with tunnels in the groundwater and a large proportion of the land area below sea level, results in maximum safety requirements to prevent the possible destruction of structures in cases where damage could have catastrophic consequences for the entire country.

In Germany, fire loads as per ZTV-Tunnel or RABT and EBA are used. The only difference between them is the duration of the fire, which is 30 or 60 minutes respectively. In contradistinction to all other fire load curves, these requirements also take account of the behaviour after the fire, during the cooling-off phase. Depending on the particular material and the requirements imposed for the specific application case, the behaviour during and after the cooling-off phase can also be an important factor in the assessment.

For consistency's sake, it would also be necessary for the test to include the action of extinguishing agents which lead to abrupt cooling.

Even after the significantly longer fire durations in recent tunnel accidents, the fire load curves presented here do not necessarily have to be adapted. The longer fire durations - sometimes in excess of 50 hours - are attributable to flashovers, leading to burn-out of various vehicles at different intervals in time and space.

Assessment of fire load curves

When specifying and dimensioning components, and therefore also when assessing test requirements, it is necessary to estimate the hazard potential and the resultant risk.

In the first phase of a fire accident, it must be ensured that people present in the fire zone are able to leave it without danger. However, even after the fire has developed for just a few minutes, there is no longer any chance of survival for any living creature at temperatures of several hundred °C. As the heat cannot escape from the closed areas, it accumulates and produces a dramatic rise in temperature.

In order to guarantee the stability of the components during the phase when people are escaping, the fire loads as per RWS, ZTV or EBA must therefore be assessed as equivalent.

For the subsequent phase of the fire when only the protection of property has to be assessed, the risk should be considered according to the scope of protection required for the maximum credible fire accident. If it is only the concrete structure which is damaged, cost-effectiveness calculations should be used to assess the expenditure which can be justified in order to achieve greater fire resistance.

If personal protection and rescue is the only relevant assessment criterion, it is sufficient to design the lining according to the EBA fire load. The structure is proven to have adequate strength during the escape phase.

The following aspects should be given particular consideration here:

- whether the structure crosses below the course of a river, and/or whether it is situated in the groundwater
- whether the structure runs through unstable rock mass, a short distance below existing constructions
- whether escape bays or chambers would be made inaccessible if the tunnel construction were to be destroyed

In the cases mentioned above, the subsequent damage arising from the fire could result in immeasurable losses which might have been prevented by the selection of suitable protection systems.

Regardless of the security measures that are selected, which include not only the constructional aspects but also the other factors mentioned at the outset, a certain residual risk will always remain and its consequences cannot be prevented. As with other types of catastrophe, where one speaks of events that happen once in every 100 or 1000 years; this implies that there are even more disastrous events which can occur, infrequently. In the same way, we must also expect incidents to occur in the fire protection sector for which the protective precautions that have been taken are inadequate. Nevertheless, the protective goal to be achieved must be defined and accepted in political and social terms. It is then the purpose of all fire protection precautions to achieve this goal in return for the lowest possible expenditure. At the same time, everyone involved must be aware that in any case, there is the possibility of an accident scenario which is not covered. 100 % security against accidents will never be possible.

2.3 Damaging mechanisms

The extent of the destruction of concrete in the event of a fire can be ascribed to several damaging mechanisms in the case of conventional concrete. As a general rule, the individual damage patterns overlap one another, and they may be broken down into the following categories:

- formation of water vapour
- chemical conversions
- failure of reinforcements
- thermal length changes

2.3.1 Formation of water vapour

Water in concrete has a variety of effects, but the damaging influences are prevalent. The transition between the liquid and gaseous phases of water (which occurs at 100 °C, as is well known) entails an energy absorption of 2,257 kJ/kg. In view of the quantities of energy which are released in a fire, this absorption of energy is insignificant, and it only causes a cooling effect for a few moments.

As well as the absorption of energy during the transition between the liquid and gaseous phases, there is also a massive increase in volume which is responsible for concrete components chipping or flaking off. Under identical pressure conditions, the vapour volume amounts to approx. 1,100 times the same mass of liquid water. Since this volume is not available in the concrete, pressure is built up in it: this rises until the tensile strength of the concrete is attained and is released abruptly when the concrete flakes off. This is a dynamic system, since the formation of water vapour takes place as a function of time, and part of the water vapour can escape via the pore structure that is present. Damage only occurs when more water vapour is created in the component than can escape via the pore structure. This means that

- the fire intensity and
- the porosity of the concrete

influence the damaging process. In the case of a small fire with a low introduction of energy and correspondingly high porosity, less damage may be expected than from a large fire in the vicinity of impermeable concrete. The requirement for concrete with high strength and impermeability, in order to attain adequate resistance, is a justified one on the basis of structural design considerations; as regards resistance to fire, however, it proves to be counter-productive [2].

2.3.2 Chemical conversions

Chemical conversions of minerals from the hardened cement paste and the aggregates occur in relation to the temperature level in the concrete.

Dehydration

From 400 °C upwards, dehydration of the calcium hydroxide in the hardened cement paste occurs in concrete, according to the following reaction mechanism:

$$Ca(OH)_2 \rightarrow CaO + H_2O$$

Decomposition causes the strength of the hardened cement paste to reduce until the concrete collapses completely. The water which is released in the form of vapour additionally accelerates the destruction process, as has already been described.

Transformation of quartz

In the case of quartz, temperatures of 575 °C lead to mineral transformation, which is associated with an increase in volume. As a consequence, this causes bursting of the concrete structure and the aggregates containing quartz.

Decarbonatisation

In the temperature range above 800 °C, decarbonatisation of limestone rocks occurs according to the following reaction equation:

$$CaCO_3 \rightarrow CaO + CO_2$$

Rocks containing limestone collapse. The separated CO_2 escapes from the concrete structure as a gas. If the escape routes are inadequate, disorders in the structure will occur.

2.3.3 Reinforcement

As the temperature increases, the strength characteristics of metal materials reduce. Steels with a low C-content, which generally include reinforcement steels as well, show irregularity (blue brittleness) in the range between 200 and 300 °C. These phenomena may be ascribed to precipitation-related sliding blockages in the grid.

In the temperature range up to 200°C, the stress/expansion behaviour of steel shows no dependency on temperature. Above the temperature limit of 200°C, there is a decline in the load-carrying capacity; at 700 °C, this is only approx. 20% of the value at normal temperature. The load-carrying capacity of a component is massively impaired for this reason, and for this reason, limits are placed on the temperatures to which the reinforcement is exposed. In practice, slight losses of load-carrying capacity are tolerated, so that temperature limit values in the range of 250°C to 300°C are stipulated.

2.3.4 Thermal length change

Like all materials, concrete also shows a temperature-related change in length. Even under climatic conditions by temperate climates, these changes in length have to be taken into account by means of appropriate equalisation sections or expansion joints. Reports about tracks bent by the summer heat can also be attributed to this phenomenon.

In case of fire, significantly higher temperatures occur, and the changes in length are also correspondingly greater. If the space allowed here is insufficient (and several centimetres may be needed depending on the temperature and absolute length of the component), secondary bending may occur, entailing the collapse of the structure. This is a risk to which (for example) an intermediate roof in a tunnel is exposed, and it should not be underestimated.

In addition, the heating of the components which proceeds from the outside inwards can create internal and secondary stresses. The internal stresses can cause the concrete covering to flake off like a shell, so that the armouring loses part of its protective function and the composite is impaired. Moreover, the temperature-related expansion behaviour of steel and concrete is only identical in the range from approx. 0 °C to approx. 400 °C. At higher temperatures, damaging stresses are also induced here.

2.3.5 Evaluation of damage mechanisms

The damage mechanisms described here already attack the concrete structure at low temperatures, from 100 °C upwards. Investigations on conventional concrete have shown that the destruction process takes place at high speed. The degradation rate in these processes attained values of 0.33 mm/min or 20 cm/h. As the structure chips off and breaks open, the damage front eats its way into the component. Reinforcements laid only a few centimetres below the surface are exposed, and they completely lose their ability to function in the fire zone due to the extremely high temperatures.

2.4 Test procedures

Tests on tunnel linings are only carried out at a few locations throughout Europe. In each case, one of the temperature curves mentioned above is tracked. The temperature required for the fire load curve is measured at a distance of 10 cm in front of the surface of the slab. In many cases, a slab size of 1.6 x 1.6 x 0.15 m is used for routine examinations. This slab is placed on a shaft furnace and is flamed from below, using an oil burner. The temperatures are measured with special thermal elements in the slab and on its underside. The measured data are recorded online at a frequency of 1 Hz. A test set-up of this sort was also used to carry out the investigations described below.

The resistance or protective effect of a system is assessed on the basis of the temperatures measured on the armouring which faces the fire side, and of any instances of flaking or chipping on the surface.

Figure 2 shows the temperature progressions in the fire area, and at two different depths below the surface of the component which faces the fire, as typically attained in the air-placed concrete with increased fire resistance which is presented below.

2.5 Fire resistant concrete

For concrete and shotcrete the same protection mechanisms are working. Whenever concrete is mentioned in this chapter also shotcrete is comprised. All the types of concrete were examined according to the RWS requirements.

For this reason, any other assessment of the concrete at lower thermal loads is reserved.

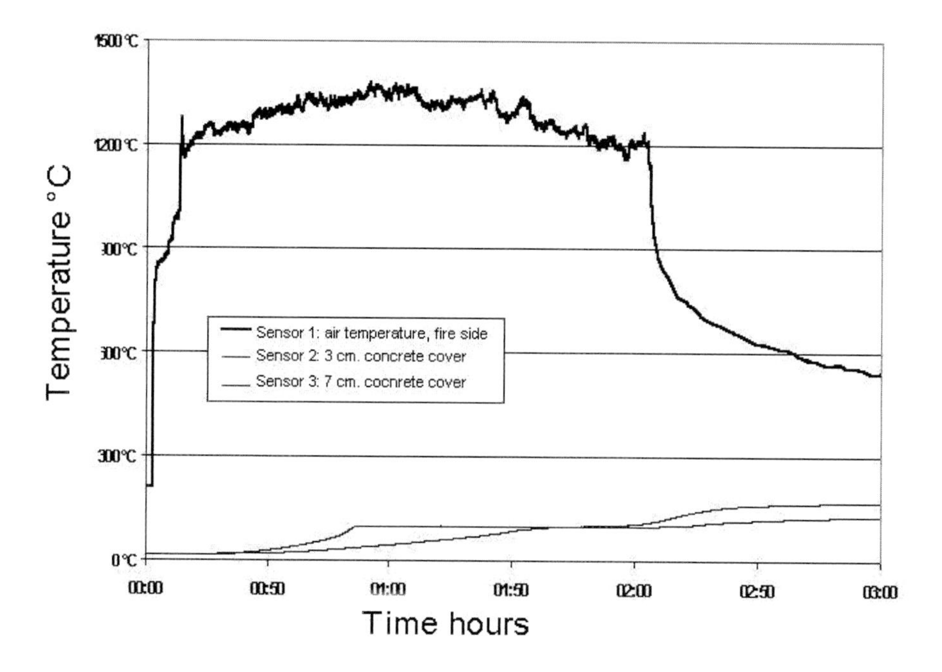

Figure 2: Temperature progression in shotcrete

The following parameters or systems were examined on the basis of aptitude tests to determine their influence on the fire resistance of the concrete:

- standard concrete
- steel fibres
- polypropylene fibres
- aggregates
- cements

For this purpose, slabs with an edge length of 1.6 m and a thickness of 0.15 m were prepared. After storage for 28 days in the gallery, at air humidity of 70 %, the fire tests were carried out on the untreated surfaces. As well as the temperature in the fire area, the temperatures at depths of 3 and 7 cm below the surface were measured.

The tests with elements made analogues standard concrete in the furnace showed massive incidences of spalling on the entire surface, starting within the first 2 minutes of testing. After 12 minutes, the test was halted because the spalling was continuing unabated, with the danger that the test object would be completely destroyed.

In the context of the tests carried out to date in the Hagerbach Test Gallery, the standard formula described above was modified with steel fibres. The fibres were added to the initial mixture at a rate of 30 kg/m³ in concrete and 45 kg/m³ for shotcrete.

During the fire test, a reduction in spalling was observed as compared with standard shotcrete. However, it was impossible to prevent incidences of spalling completely.

After the tests had been completed, the manner in which the steel fibres act in the event of fire could be clearly seen on the surface of the test object. As in the case of standard concrete, this type of concrete also shows instances of spalling. To some extent, however, the steel fibres act as anchors in the subsurface concrete which is not yet disrupted, causing:

- the tensile strength of the concrete to increase;
- the severed concrete sections to remain on the surface like a sheet suspended in front of it.

As the duration of testing increased, either the steel fibres or the concrete structure were completely destroyed by the temperatures, so that the concrete sections fell off. At this time, the concrete surface was again exposed to the test temperatures without protection, which led to renewed instances of spalling.

To prove this point in concrete, fibre contents varying between 2 und 4 kg/m³ were dosed into the basic mixture. Both fibrillated and monofil fibres were used for this purpose.

2.6 Sprayed protection

Sprayed fire protection systems are mainly mortar based. Next to the fire resistance of the system it has also an isolating effect. Thicknesses between 30 an 50 mm almost fulfill the requirements of fire protection. The quality of the systems is mainly influenced by the nozzle man and the quality of spraying. He has to achieve the required thickness and homogenous of the mortar.

The high isolation effect is realized by a high content of air and special aggregates like for example vermiculite in the mortar.

High attention has to be given to the tensile strength at the tunnel lining. Specially on casted concrete or prefabricated elements problems could occure.

2.7 Prefabricated fire protection system

There are several prefabricated protection system available at the market. The quality of the system cannot be influenced the applicator. Special attention has to be given to the junction of protection elements.

3. Fire detecting

In housing there is a quite long experience on fire detection available. But there is one big difference in tunnels compared with houses, that is the ventilation of the tunnels an the wind velocity which can exceed rates of 10 m/s. Under those conditions normal detection systems does not work or they can not detect the actual place of the fire.

Figure 3 gives an indication of the temperature field with high and low wind velocity. With this dates it is easy to understand how difficult it is to detect the actual place of a fire.

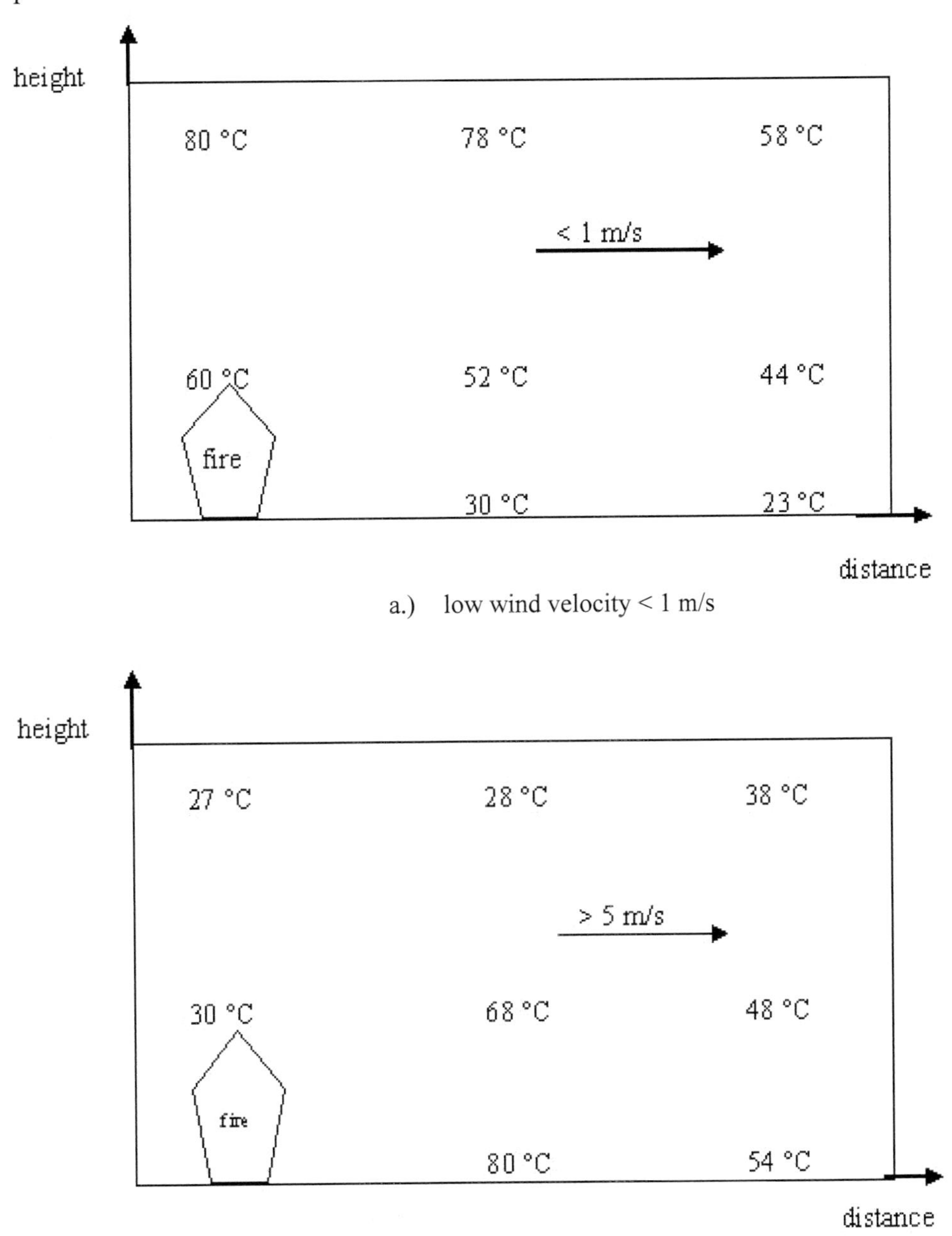

Figure 3: Air temperature in a tunnel as function from the distance from the fire

In a tunnel where these conditions can be built up detection equipment like sensors and video analyses are tested.

Parameters investigated are:

- time for detection
- accuracy of fire place detection
- reproducibility

4. Fire extinguishing

Until the fire brigades arrive at the fire accident in a tunnel the temperatures raised to a level where an attack is impossible. For this reason several systems for fire extinguishing are under construction.

To prove the design of such system investigations are conducted several tests with real wooden fires. Subjects of those tests are:

- nozzle configuration (Type and number)
- water consumption
- influence of wind velocity
- influence on the temperatures in the tunnel
- behaviour of the fire

The last aspect is a very important one, because fires can not be extinguished completely by a water spraying system. Sprinklers only manage to keep the fire on a certain level up to the moment the fire fighters arrive.

4.1 Basic principles of water mist

Water mist systems differ fundamentally from sprinkler systems not only in terms of water consumption and drop size, but also in terms of the principle underlying their operation. Water mist, which is discharged under pressure through specially developed nozzles (Illustration 4), consists of tiny water droplets measuring less than 100 µm in size that form a huge water surface in the fire zone. This dramatically improves the heat transfer onto the water, making it possible to attain highly efficient vaporisation of the water mist, which in turn leads to two decisive effects for successful fire-fighting: the local displacement of oxygen in the flame zone, and the cooling effect on the surrounding area.

When water vaporises, its volume is increased by a factor of 1.675, leading to a reduction of the oxygen concentration in the immediate area of the flames, thereby checking the fire. As well as the expansion of the water, additional energy is also required for the vaporisation and this is taken from the thermal energy of the fire. These two actions are especially effective in the case of water mist, because virtually all the water can vaporise due to the fine droplets. The combination of these two effects accounts for the particular efficiency of water mist for fire-fighting purposes.

4.2 Safety in transport tunnels

In times where traffic is increasing every year, the potential hazard for every road user is becoming significantly greater, especially in confined areas such as tunnels. An accident in a tunnel leads to a traffic standstill, which entails an accumulation of people in the hazard zone. If a fire breaks out during an accident of this sort, a life-threatening situation is ensured for all the road users involved.

The tunnel fires in the Mont Blanc, Tauern and Gotthard tunnels have vividly demonstrated the main problem of fire-fighting in a tunnel: the particularly rapid development of extreme temperatures due to the circumstances in the tunnel prevents the fire brigade from advancing.

Based on these experiences, the target requirements for a water mist system in tunnels were drawn up together with leading experts: rapid fighting of the fire in the developing phase should limit the temperatures to a specified level, in order to protect people and the tunnel structure and to restrict the development of smoke by preventing the spread of the fire. Accordingly, the use of this water mist system also makes it possible for the fire brigade to advance to the seat of the fire and to extinguish the fire.

The only possible way of meeting these target requirements is to install the water mist system directly in the tunnel.

The water mist system offers the following benefits: not only is it possible to achieve a substantial reduction in personal injuries and damage to the tunnel structure as the consequences of a fire; the operational shutdown of the tunnel itself can also be restricted to a minimum.

4.3 The configuration of the water mist system in a tunnel

The Aquasys water mist system is installed in the tunnel and it comprises:

- pump units at each portal
- a main pipe through the entire tunnel
- nozzle pipes under the ceiling, and a
- control unit, which also acts as the interface with the detection system.

4.4 Test results

During 2001, a series of full scale tests were carried out with a fully-loaded truck in the Hagerbach Test Gallery. The extensive experimental experience gained in this way made it possible to subject the water mist system to an approval procedure which was conducted by two accredited testing organisations.

The result of the approval tests was that the Aquasys water mist system successfully met all the required criteria.

These criteria were as follows:

- maximum of 50° C at a distance of 20 m from the burning truck in order to guarantee that the seat of the fire can be reached by the fire brigade.
- maximum of 250° C at a distance of 5 m from the burning truck in order to guarantee that sparkover of the fire can be successfully prevented.
- below 100° C at 10 mm depth in the concrete of the intermediate ceiling, to prevent the concrete from flaking off.
- reproducible test results with different wind speeds in the tunnel.

The specified requirements were satisfied. Repeated tests were able to prove the reproducibility of the results. Even at a distance of 5 m from the burning truck, temperatures in the range of 50° C were maintained. This value is about 20 % of the tolerated variable. The rapid rise in temperatures after the water mist system was switched off proves the existence of the fire potential which was previously cooled by the water mist system.

The water mist system is arranged in sectors of about 30-40 m, of which only a maximum of three sectors are ever activated in the area of the fire location, so the stored supply of water that is required is also reasonable. These three sectors always comprise the sector where the fire was detected and the two adjacent sectors. In principle, the quantity of water required always depends on the circumstances and general conditions in the particular tunnel to be protected, but for guidance purposes, the water quantity for the approval tests can be used: this was approx. 2,000 l/min for all three activated sectors. Also, the stored quantity of water ultimately has to be dimensioned according to the time taken by the fire brigades to reach the location.

Pipes, nozzles and all the components required in the traffic area are manufactured from non-rusting steels which withstand the aggressive atmosphere of the tunnel air.

5. Fire fighter training

A tunnel is a seldom and very special challenge for the fire fighters. Boundary conditions are completely different from a normal fire. Special here are mentioned:

- smoke distribution
- highest temperatures
- difficult communications
- closed rooms

To train these situations special courses have been built up

- for recognising and communication
- for fire fighting and rescuing persons
- for handling harmful goods and fluids in a tunnel

The first courses have shown that there is a great demand for further education's. Fire fighters learned how to behave in a real tunnel and get confidence to there own abilities.

The training course comprises several stages to familiarise the participants with the problems of operations in tunnels, providing in-depth training in the correct behaviour patterns.

Operations in artificially smoke-filled tunnel sections are used to train participants to identify the problems involved in tunnel operations, so that they can learn the correct behaviour patterns. Communication becomes very difficult in the confined conditions of a tunnel system, with cross-connections and secondary areas such as ventilation and transformer stations. Radio contact is interrupted after the second branch of a tunnel, if not before, so communication must be ensured by suitable measures such as relay stations, etc. Provided that the importance of the communication problem in underground facilities has been recognised, it is basically possible to implement improvements by installing the appropriate equipment.

The limited reserves of air for breathing protection equipment impose severe restrictions on the radius of movement for the fire brigade personnel. Even though the Westerschelde tunnel has cross-connections positioned every 250 m for access to the other tunnel tubes, the configuration leads to very difficult operating depths of 250 m in extreme cases. All the participants were made vividly aware of this problem by the personal rescue exercises using breathing protection in sections of real tunnel that are artificially filled with smoke.

Simulated vehicle fires give the course participants a realistic idea of the general conditions prevailing during a tunnel fire. In contrast to open air conditions, most of the heat cannot escape upwards in this case. The tunnel ceiling and walls reflect the heat back into the driving area, which reaches tremendous temperatures. The structure of the Westerschelde tunnel is fitted with insulating fire protection mortar, but if this is not the case, the intervention forces have to expect concrete to flake off, with massive damage to the tunnel construction leading to the collapse of structural components in extreme cases.

The final exercises combine all the training goals, and the intervention team is confronted with a staged incident which they have to bring under control on their own (under the supervision of the instructors).

Briefings after each course offer a chance to discuss the experience gained and to optimise the firemen's operational behaviour.

Impressions gained from the various underground operational exercises have shown that it is essential for the intervention forces to train under the most realistic conditions possible. Operations with breathing protection equipment pose an added difficulty in every case. The smoke and heat conditions can hardly be compared with events in the open air, where operating depths are usually very limited. The secret of success in a real emergency is for firemen to learn during "peacetime" what it means to ensure communication, to discover their own physical and mental limits, and to use their equipment quickly and efficiently even under harsh conditions.

Exercises give participants a demonstration of the possibilities and limitations of fire fighting in tunnels, and the firemen are quite often stretched to the limits of their personal capabilities. A decisively important aspect for fire brigade staff is the personal experience they gain from these realistic exercises: as a result, they can implement the knowledge they have acquired from training when a real incident occurs, so as to bring the emergency rapidly under control.

Bibliography

[1] Kordina/Meyer-Ottens, Betonbrandschutzhandbuch [Manual of Fire Protection for Concrete], VBT, Düsseldorf 1999

[2] Wetzig, Zerstörungsmechanismen beim Werkstoff Beton im Brandfall und-Schutzsysteme [Destruction mechanisms in concrete as a material in the event of fire, and protection systems], Tunnel 7/2000

[3] Wetzig, V., Fire resistant concrete and shotcrete, ITC-Conference 2001 Madrid

[4] Mägerle, R., Branddetektion und Löschung von Tunnelbränden im Test, S+S Report 2/2000

[5] Shani, W., Fire fighting in tunnels – a reality training facility, Tunnel 3/2001, May.